Advanced Optical-Fiber-Related Technologies

Advanced Optical-Fiber-Related Technologies

Guest Editor

Galina Nemova

MDPI

Basel • Beijing • Wuhan • Barcelona • Belgrade • Novi Sad • Cluj • Manchester

Guest Editor
Galina Nemova
Department of Engineering Physics
Polytechnique Montréal
Montréal
Canada

Editorial Office
MDPI AG
Grosspeteranlage 5
4052 Basel, Switzerland

This is a reprint of the Special Issue, published open access by the journal *Applied Sciences* (ISSN 2076-3417), freely accessible at: https://www.mdpi.com/journal/applsci/special_issues/8H28M40X0O.

For citation purposes, cite each article independently as indicated on the article page online and as indicated below:

Lastname, A.A.; Lastname, B.B. Article Title. *Journal Name* **Year**, *Volume Number*, Page Range.

ISBN 978-3-7258-5991-7 (Hbk)
ISBN 978-3-7258-5992-4 (PDF)
https://doi.org/10.3390/books978-3-7258-5992-4

Contents

About the Editor

Galina Nemova

Galina Nemova is a research fellow at Polytechnique Montréal. She received her M.Sc. and Ph.D. degrees from the Moscow Institute of Physics and Technology. Dr. Nemova is a Senior Member of Optica, formerly the Optical Society of America. She has edited two books and authored more than 100 papers. She is the author of *Field Guide to Laser Cooling Methods* (2019), *Field Guide to Light-Matter Interaction* (2022), and *Field Guide to Optical Levitation and Levitodynamics* (2024). Her research interests cover a broad range of photonics topics, including rare-earth-doped materials, nanophotonics, fiber lasers and amplifiers, Raman lasers, nonlinear optics, laser cooling of solids, and quantum optics.

applied sciences

MDPI

Editorial

Editorial for the Special Issue: "Advanced Optical-Fiber-Related Technologies"

Galina Nemova

Department of Engineering Physics, Polytechnique Montréal, 2500, Ch. De Polytechnique, Montreal, QC H3T 1J4, Canada; galina.nemova@videotron.ca

An optical fiber is a flexible glass or plastic fiber that can transmit light from one end to the other. An optical fiber consists of three basic concentric elements: a core surrounded by cladding, which is then surrounded by an outer coating [1,2]. The core is usually made of glass or plastic, although other materials are sometimes used, depending on the transmission spectrum desired. The core and cladding form a waveguide that is essential for transmitting light. The cladding is usually made of the same material as the core, but with a slightly lower index of refraction. This difference in index causes total internal reflection, which is responsible for transmitting light along the fiber without it escaping through the sidewalls. Optical fibers do not need a grounding connection. They are immune to electromagnetic interference. They emit no radiation themselves that might cause interference. Fiber-optic transmission systems are superior to metallic conductor-based systems in many applications. Copper wire is about 13 times heavier than fibers. Fibers are also easier to install and require less duct space. Contrary to metallic conductors, the bandwidth of optical fibers permits them to transmit signals that contain considerably more information. In addition, optical fibers have low power loss. Optical fibers have numerous applications. They have gained wide-spread acceptance in communications including telecommunications, local area networks, industrial control systems, avionic systems, and military command and control systems. The fibers themselves can be used as a distributed sensor to measure a number of environmental effects, such as temperature, strain, and acoustic signals. They can deliver light from a remote source to a detector to obtain pressure, temperature, or spectral information. Optical fibers can deliver high levels of power for tasks such as laser cutting, welding, marking, and drilling. They can illuminate areas that are difficult to reach. Optical fibers are widely used as the active gain medium in a special form of solid-state lasers known as fiber lasers.

This Special Issue presents the latest advanced optical-fiber-related technologies and their applications. It includes 10 contributions that collectively offer insights into the actions and strategies of current fiber-optics. Here, I will briefly summarize the content of the contributions to this Special Issue.

1. Brief Review of Recent Developments in Fiber Lasers [contribution 1].

This paper reviews the recent achievements in high-power rare earth (RE)-doped fiber lasers, Raman fiber lasers, and Brillouin fiber lasers. RE-doped fiber lasers operate in several wavelength ranges including 1050–1120 nm (ytterbium-doped fiber lasers), 1530–1590 nm (erbium- and erbium–ytterbium-doped fiber lasers), and 1900–2100 nm (thulium- and holmium-doped fiber lasers). White spaces in the wavelength spectrum, where no RE-doped fiber lasers are available, can be covered by Raman lasers.

2. Advances in High–Speed, High–Power Photodiodes: From Fundamentals to Applications [contribution 2].

This paper reviews the fundamentals of high-speed, high-power photodiodes, mirror-reflected photodiodes, microstructure photodiodes, photodiode-integrated devices, and related equivalent circuits and design considerations. These characteristics of photodiodes and the related photonic-based devices are analyzed and reviewed in detail, and the devices are compared, providing a new path for the application of these in short-range wireless communications in 6G and beyond.

3. Water Pipeline Leakage Detection Based on Coherent φ-OTDR and Deep Learning Technology [contribution 3].

Leakage in water supply pipelines remains a significant challenge. It leads to resource and economic waste. In this paper, the authors have developed several leak detection methods, including the use of embedded sensors and pressure prediction.

4. Optical Cable Lifespan Prediction Method Based on Autoformer [contribution 4].

In this paper, a novel method for predicting the service life of optical cables based on the Autoformer model combined with the calculation method has been proposed. The method developed in the paper validates the superiority and stability of the Autoformer model in predicting cable lifespan, which can offer a more reliable approach for ensuring cable technology reliability and the management of associated industries.

5. A Multi-Format, Multi-Wavelength Erbium-Doped Fiber Ring Laser Using a Tunable Delay Line Interferometer [contribution 5].

This work demonstrates the use of an erbium-doped fiber amplifier (EDFA), a tunable bandpass filter (TBF), and a tunable delay line interferometer (TDLI) to form a ring laser that produces multi-format, multi-wavelength laser beams. The proposed system enables dual-wavelength spacing ranging from 0.3 nm to 3.35 nm, with a switchable wavelength position at approximately 1527 nm to 1535 nm, providing flexible tunability.

6. Wavelength-Dependent Bragg Grating Sensors Cascade an Interferometer Sensor to Enhance Sensing Capacity and Diversification through the Deep Belief Network [contribution 6].

This paper used machine learning to enhance the number of fiber-optic sensing placement points and promote the cost-effectiveness and diversity of fiber-optic sensing applications. The framework adopted is the FBG cascading an interferometer, and a deep belief network (DBN) is used to demodulate the wavelength of the sampled complex spectrum.

7. A Hybrid GAN-Inception Deep Learning Approach for Enhanced Coordinate-Based Acoustic Emission Source Localization [contribution 7].

In this paper, a novel approach to coordinate-based acoustic emission (AE) source localization has been proposed to address the challenges of limited and imbalanced data from fiber-optic AE sensors used for structural health monitoring (SHM). A hybrid deep learning model has been developed, combining four generative adversarial network (GAN) variants for data augmentation with an adapted inception neural network for regression-based prediction.

8. Simultaneous Vibration and Temperature Real-Time Monitoring Using Single Fiber Bragg Grating and Free Space Optics [contribution 8].

This work showcases simultaneous temperature and vibration measurement using a single-fiber Bragg grating (FBG) sensor. The concurrent interrogation of vibration and temperature by the FBG sensing system can be integrated with free space optics (FSO), which reduces the costs associated with fiber-optic cables and overcomes terrain barriers.

9. Side-Illuminating Optical Fiber for High-Power-Density-Mediated Intraluminal Photoacoustic Imaging [contribution 9].

In this work, the design, method of fabrication, and characterization of a new compact, side-fire optical fiber that can deliver high-energy laser pulses for PA imaging have been described. Side-fire illuminators were fabricated using UV laser ablation to create windows on the side of a 1.5 mm diameter single-core, multi-mode optical fiber with a reflective silver coating and a beveled end. Devices with 10 mm, 20 mm, and 30 mm window lengths were fabricated and their beam profiles were characterized.

10. Novel Fiber Bragg Grating Sensing Structure for Simultaneous Measurement of Inclination and Water Level [contribution 10].

This paper presents a pair of fiber Bragg grating (FBG) subsidence sensor systems designed to simultaneously measure tilt and water levels and explore the system's potential to detect temperature variations. The configuration of the FBG subsidence sensor is intentionally skewed to enhance measurement sensitivity. The system is capable of concurrently detecting a 0.5 cm variation in water level and a 0.424° change in tilt, with tilt measurements spanning from −1.696° to 1.696°.

I believe that these works will collectively help to clarify the current state of optical-fiber-related technologies.

Conflicts of Interest: The author declares no conflicts of interest.

List of Contributions:

1. Nemova, G. Brief Review of Recent Developments in Fiber Lasers. *Appl. Sci.* **2024**, *14*, 2323. https://doi.org/10.3390/app14062323.
2. Chen, Q.; Zhang, X.; Sharawi, M.S.; Kashyap, R. Advances in High–Speed, High–Power Photodiodes: From Fundamentals to Applications. *Appl. Sci.* **2024**, *14*, 3410. https://doi.org/10.3390/app14083410.
3. Zhang, S.; Xiong, Z.; Ji, B.; Li, N.; Yu, Z.; Wu, S.; He, S. Water Pipeline Leakage Detection Based on Coherent φ-OTDR and Deep Learning Technology. *Appl. Sci.* **2024**, *14*, 3814. https://doi.org/10.3390/app14093814.
4. Niu, M.; Li, Y.; Zhu, J. Optical Cable Lifespan Prediction Method Based on Autoformer. *Appl. Sci.* **2024**, *14*, 6286. https://doi.org/10.3390/app14146286.
5. Yao, C.-K.; Dehnaw, A.M.; Peng, P.-C. A Multi-Format, Multi-Wavelength Erbium-Doped Fiber Ring Laser Using a Tunable Delay Line Interferometer. *Appl. Sci.* **2024**, *14*, 6933. https://doi.org/10.3390/app14166933.
6. Bogale, S.D.; Yao, C.-K.; Manie, Y.C.; Zhong, Z.-G.; Peng, P.-C. Wavelength-Dependent Bragg Grating Sensors Cascade an Interferometer Sensor to Enhance Sensing Capacity and Diversification through the Deep Belief Network. *Appl. Sci.* **2024**, *14*, 7333. https://doi.org/10.3390/app14167333.
7. Huang, X.; Han, M.; Deng, Y. A Hybrid GAN-Inception Deep Learning Approach for Enhanced Coordinate-Based Acoustic Emission Source Localization. *Appl. Sci.* **2024**, *14*, 8811. https://doi.org/10.3390/app14198811.
8. Yao, C.-K.; Kumar, P.; Liu, B.-X.; Dehnaw, A.M.; Peng, P.-C. Simultaneous Vibration and Temperature Real-Time Monitoring Using Single Fiber Bragg Grating and Free Space Optics. *Appl. Sci.* **2024**, *14*, 11099. https://doi.org/10.3390/app142311099.
9. Singh, N.; Roa, C.-F.; Cherin, E.; Lilge, L.; Demore, C.E.M. Side-Illuminating Optical Fiber for High-Power-Density-Mediated Intraluminal Photoacoustic Imaging. *Appl. Sci.* **2025**, *15*, 3639. https://doi.org/10.3390/app15073639.
10. Yao, C.-K.; Chung, Y.-J.; Xu, Y.-C.; Kumar, P.; Peng, P.-C. Novel Fiber Bragg Grating Sensing Structure for Simultaneous Measurement of Inclination and Water Level. *Appl. Sci.* **2025**, *15*, 4819. https://doi.org/10.3390/app15094819.

References

1. Tsao, C. *Optical Fiber Waveguide Analysis*; Oxford University Press: Oxford, UK, 1992.
2. Buck, J.A. *Fundamentals of Optical Fibers*, 2nd ed.; Wiley-Interscience: Hoboken, NJ, USA, 2004.

applied sciences

MDPI

Review

Brief Review of Recent Developments in Fiber Lasers

Galina Nemova

Department of Engineering Physics, Polytechnique Montréal, P.O. Box 6079, Station Centre-Ville, Montreal, QC H3C 3A7, Canada; galina.nemova@videotron.ca

Abstract: This review covers the recent achievements in high-power rare earth (RE)-doped fiber lasers, Raman fiber lasers, and Brillouin fiber lasers. RE-doped fiber lasers have many applications such as laser cutting, laser welding, laser cleaning, and laser precision processing. They operate in several wavelength ranges including 1050–1120 nm (ytterbium-doped fiber lasers), 1530–1590 nm (erbium- and erbium–ytterbium-doped fiber lasers), and 1900–2100 nm (thulium- and holmium-doped fiber lasers). White spaces in the wavelength spectrum, where no RE-doped fiber lasers are available, can be covered by Raman lasers. The heat power generated inside the laser active medium due to the quantum defect degrades the performance of the laser causing, for example, transverse-mode instability and thermal lensing. It can even cause catastrophic fiber damage. Different approaches permitting the mitigation of the heat generation process are considered in this review. Brillouin fiber lasers, especially multiwavelength Brillouin fiber lasers, have several important applications including optical communication, microwave generation, and temperature sensing. Recent progress in Brillouin fiber lasers is considered in this review.

Keywords: fiber laser; rare earth-doped fiber laser; Raman laser; Brillouin fiber laser; radiation-balanced fiber laser

1. Introduction

The basic principle for light confinement and propagation in optical fibers is total internal reflection. Total internal reflection was first studied by Johannes Kepler in 1611 [1]. In 1621, it was mathematically described by Willebrord Snellius (Snell's law), but remained unpublished during his lifetime. In 1690, Christiaan Huygens in his Treatise on Light showed how Snell's law of sines could be explained by the wave nature of light [2]. Optical fibers are one of the major technological successes of the 20th century.

The first uncladded glass fibers were fabricated in the 1920s [3]. In the 1950s, a cladding layer was added [4–6]. The cladding layer considerably improved the fiber characteristics. The early fibers were very lossy (~1000 dB/km). The situation changed in the 1970s when silica fibers with a loss of ~20 dB/km were developed [7]. In 1979, a loss level of ~0.2 dB/km near the 1550 nm wavelength was achieved [8]. It was limited mainly by Rayleigh scattering. Such low-loss fibers have led to a revolution in optical fiber communications and to the advent of nonlinear fiber optics, fiber lasers and amplifiers, and fiber sensors.

Stimulated Raman and Brillouin scattering in optical single-mode fibers were investigated both theoretically and experimentally in 1972 [9–11]. In 1973, it was suggested that optical fibers can support solitons [12]. Almost at the same time, optically induced birefringence, parametric four-wave mixing, and self-phase modulation were observed in optical fibers.

The use of rare earth (RE)-doped optical fibers for gain in glass fiber lasers and amplifiers was first demonstrated by Eli Snitzer in 1961 [13]. He used a neodymium-doped silicate glass fiber emitting at 1060 nm. In the 1990s, erbium-doped fiber amplifiers and lasers were under intensified investigation [14]. These devises were commercialized in 1992 and employed in undersea fiber-optic communication systems by 1995.

The state-of-the-art design of optical fibers is a careful trade-off between optical losses, optical nonlinearity, group–velocity dispersion, and polarization effects.

In 1978, Pochi Yeh and colleagues proposed the idea of a photonic crystal fiber (PCF) [15]. They proposed to clad a fiber core with Bragg grating, which is similar to 1D photonic crystal. In 1992, Philip Russell invented a photonic crystal fiber made of 2D photonic crystal with an air core. In 1996, the first fabricated single-mode PCF was reported at the Optical Fiber Conference (OFC) [16]. Photonic crystals are periodic dielectric structures with the period being less than the wavelength of the light. They affect the motion of photons in the same way as an ionic lattice affects the electrons in semiconductors. PCFs are among the most specialized optical lightguides. For example, PCFs can range from fibers with low levels of nonlinearities supporting high-power pulses to highly nonlinear PCFs for supercontinuum generation.

From a theoretical point of view, an optical fiber with a circular core has no birefringence, such that the polarization of a beam sent through the fiber may not change. In reality, however, a small amount of birefringence is always present in an optical fiber due to external perturbations or manufacturing imperfection. Such birefringence causes random power coupling between two polarization modes (polarization crosstalk) in an optical fiber, which is undesirable in a number of applications. A polarization-maintaining fiber (PMF) with intentionally induced birefringence along the entire fiber length maintains two polarization modes without polarization crosstalk. Larger birefringence better prohibits the polarization mode coupling. Birefringence can be induced by loss in rotational symmetry in the refractive index profile or by the stress distribution known as stress birefringence. The first commercially available PMF was introduced in 1983.

Heat generation in fiber lasers is briefly discussed in Section 2. Sections 3–5 are devoted to the latest achievements in RE-doped fiber lasers, Raman fiber lasers, and Brillouin fiber lasers, respectively. Fiber laser applications are considered in Section 6.

2. Heat Generation in Fiber Lasers

For most lasers, the laser wavelength, λ_L, is longer than the pump wavelength, λ_P, (Stokes shift). This is unavoidable for all optically pumped schemes where one pump photon can generate one laser photon. The energy difference between the pump and laser photons can be estimated using the so-called quantum defect:

$$\eta_q = 1 - \frac{\lambda_P}{\lambda_L} \tag{1}$$

The quantum defect is an unavoidable source of heat generation in lasers. The process of generating coherent radiation through laser action is never 100% effective. The quantum defect causes a substantial part of the pump energy to be converted into phonons instead of photons. Heating due to the quantum defect manifests itself in many ways including transverse-mode instability, thermal lensing, and catastrophic fiber damage.

Heat management is one of the most critical issues for scaling higher output laser powers. Although the very high surface-to-volume ratio and optical guidance have provided tremendous progress in the power scalability of high-power fiber lasers, limitations on the power handling capacity of fiber lasers are now encountered.

Under the steady-state operation ($\partial/\partial t = 0$), the heat conduction equation in an isotropic medium can be written as follows:

$$\nabla^2 T(r,z) = -\frac{Q(r,z)}{\kappa} \tag{2}$$

where T is the temperature, κ is the thermal conductivity, and Q is the dissipated heat density per unit volume. In the center of the fiber core ($r = 0$), the following equation

$$\left.\frac{\partial T_{co}}{\partial r}\right|_{r=0} = 0 \tag{3}$$

must be satisfied. Here, r is the fiber radius, and T_{co} is the temperature of the fiber core. The core and cladding temperatures and their derivatives must be continuous across the borders.

$$T_{co}(r_{co}) = T_{cl}(r_{co}) \tag{4}$$

$$\kappa_{co} \frac{\partial T_{co}}{\partial r}\bigg|_{r=r_{co}} = \kappa_{cl} \frac{\partial T_{cl}}{\partial r}\bigg|_{r=r_{co}} \tag{5}$$

$$\kappa_{cl} \frac{\partial T_{cl}}{\partial r}\bigg|_{r=r_{cl}} = h[T_c - T_{cl}(r_{cl})] \tag{6}$$

where T_{cl} is the temperature of the fiber cladding. r_{co} and r_{cl} are the core and cladding radii, respectively. κ_{co} and κ_{cl} are the thermal conductivities of the core and cladding, respectively. h is the heat transfer coefficient. T_c is the environmental temperature. Considering $Q(r,z) \cong Q(z)$ and taking into account the boundary conditions (3)–(6), one can estimate the temperature of the fiber [17].

$$T_{co}(r,z) = T_0(z) - \frac{Q(z)r^2}{4\kappa_{co}}, \text{ if } 0 < r < r_{co} \tag{7}$$

$$T_{cl}(r,z) = T_0(z) - \frac{Q(z)r_{co}^2}{4\kappa_{co}} - \frac{Q(z)r^2}{2\kappa_{cl}} ln\left(\frac{r}{r_{co}}\right), \text{ if } r_{co} < r < r_{cl} \tag{8}$$

$$T_0(z) = T_c + \frac{Q(z)r_{co}^2}{2hc} + \frac{Q(z)r_{co}^2}{4\kappa_{co}} + \frac{Q(z)r_{co}^2}{2\kappa_{cl}} ln\left(\frac{r_{cl}}{r_{co}}\right) \tag{9}$$

Here, T_0 is the temperature at the core center. As seen from (7)–(9), the temperature profiles are significantly affected by pump evolution along the fiber length [17]. The temperature distribution is uneven along the fiber. In the forward pump mode, the maximum temperature is attained at the input side of the laser cavity, which corresponds to the intense amplification of the laser signal, accompanied by a tremendous increase in the number of the phonons ("superfluous" phonons), causing heating of the laser medium [17,18]. In the two-end pump mode, the temperature evolution is more even. Under these conditions, the maximum temperature is attained on both sides of the laser cavity [17,19].

The radial distribution of the temperature profile is more even than the axial distribution. It can be ignored while significant heat is traveling through the axial direction of the fiber [17,19].

Apart from the quantum defect, photodarkening in high-power fiber lasers can also lead to increased background absorption, resulting in excess heating. Photodarkening is color-center formation through some type of structural change in the glass which may or may not be reversible [20,21].

3. Rare Earth-Doped Fiber Lasers

The wide range of industrial and defense applications has fueled the development of RE-doped fiber lasers. Fibers used for data transmission are known as passive fibers, since they do not change the power of the propagating signal. RE-doped fibers can amplify the propagating optical signal. They are called active fibers. The successful development of low-loss optical fibers doped with various RE elements, such as ytterbium (Yb), erbium (Er), thulium (Tm), holmium (Ho), praseodymium (Pr), or dysprosium (Dy), opened the door to creating fiber lasers. Each RE-doped fiber laser has a specific emission window defined by its dopant. For example, Yb^{3+}-doped fiber lasers have a window of emission wavelengths ranging from ~1050 nm to ~1120 nm; Er^{3+}-doped fiber lasers have a window of emission wavelengths ranging from ~1530 nm to ~1590 nm.

A laser beam can be characterized by its brightness:

$$B = \frac{P}{A\Omega} = \frac{P}{(M^2)^2\lambda^2} \tag{10}$$

where P is the source power divided by the product of the mode area, $A = \pi w_0^2$, and the solid angle, $\Omega = \pi\theta_0^2$. Here w_0 and θ_0 are the beam waist and half-angle divergence, respectively. λ is the wavelenght of the laser beam.

$$M^2 = \frac{\pi w_0 \theta_0}{\lambda} \tag{11}$$

Equation (11) is used to calculate the beam quality factor.

3.1. Rare Earth Ions for Laser Applications

RE atoms include Sc, Y, and two groups: lanthanides with atomic numbers 57 through 71 (characterized by filling the 4f shell); and actinides with atomic numbers 89 through 103 (characterized by filling the 5f shell). In solids, the trivalent (3+) level of ionization is the most stable for lanthanide ions. RE ions have the sharp spectral lines. The sharp spectral lines in the optical spectra of trivalent RE ions are a consequence of the screening of 4f valence electrons by the 5s and 6d electrons due to the effects of the environment. An ion in a solid can be considered an impurity (a dopant) embedded in a host material. The ion replaces the host ion and forms an optically active center. The host materials play a fundamental role in the spectral nature of doped ions.

Laser glasses can be divided into oxide, halide, oxyhalide, and chalcogenides [22]. Glasses are amorphous over a long range but contain a local order. Glasses are insulators with bandgaps greater than 5 eV (the deep UV). All transitions in ions occur within the host bandgap. The energy levels of ions typically are identified by three principal quantum numbers: L, S, and J. L describes the total orbital angular momentum, S the total spin angular momentum, and J the total angular momentum. There are several possible ways to obtain a given set of values for L, S, and J. The possible states are referred to by the symbol $^{2S+1}L_J$, where $L = 0,\ 1,\ 2,\ 3,\ 4, \ldots$ corresponds to letters $S,\ P,\ D,\ F,\ G\ldots$, respectively. Schematical diagrams of energy levels with transitions for Yb^{3+}, Er^{3+}, Tm^{3+}, and Ho^{3+} ions are presented in Figure 1.

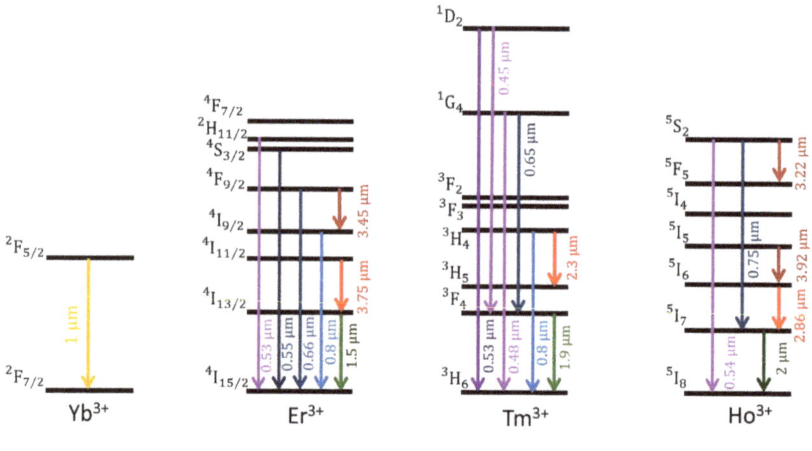

Figure 1. Diagrams of energy levels and transitions for some Yb^{3+}, Er^{3+}, Tm^{3+}, and Ho^{3+} ions.

Several states can correspond to the same atomic energy level (degeneracy). In RE-doped ions, this degeneracy is lifted through interaction with the local environment. Fiber lasers have been realized with a wide variety of RE ions including ytterbium (Yb^{3+}), erbium (Er^{3+}), thulium (Tm^{3+}), holmium (Ho^{3+}), praseodymium (Pr^{3+}), and dysprosium (Dy^{3+}). The most used RE ions for laser applications are Yb^{3+}, Er^{3+}, and Tm^{3+}. As one can see in Figure 1, only Yb^{3+} ions have only one excited state, that is, only Yb^{3+} ions are free from excited state absorption, which can serve as a source of heat generation.

Pump radiation excites electrons from the ground level to higher levels. Excited electrons can undergo spontaneous (radiative or nonradiative) decay or stimulated transition. The spontaneous decay process can be characterized by the lifetime of the excited level τ_{ex} as follows:

$$\frac{1}{\tau_{ex}} = \frac{1}{\tau_r} + \frac{1}{\tau_{nr}} \tag{12}$$

$$\frac{1}{\tau_{nr}} = C\left[\widetilde{n}(T) + 1\right]^p exp(-\alpha\Delta E) \tag{13}$$

where C is an empirical parameter that depends on the host.

$$\widetilde{n}(T) = \left[exp\left(\frac{\hbar\Omega}{\kappa_B T}\right) - 1\right]^{-1}, \quad p = \frac{\Delta E}{\hbar\Omega}, \quad \alpha = -\frac{ln(\varepsilon)}{\hbar\Omega} \tag{14}$$

ΔE is the energy gap, ε is an electron–phonon coupling coefficient, $\hbar\Omega$ is the maximum phonon energy of the host, $\hbar$ is the reduced Planck constant, T is the temperature of the host, κ_B is the Boltzmann constant, and p is the number of phonons required to bridge the ΔE gap. The empirical parameters of relation (13) for some host materials can be found in [23–25]. The spontaneous decay process can be characterized by the mean fluorescence wavelength, λ_F, or the mean fluorescence frequency, ω_F.

$$\lambda_F = \frac{2\pi c}{\omega_F} = \frac{\int \lambda I_F(\lambda) d\lambda}{\int I_F(\lambda) d\lambda} \tag{15}$$

where c is the speed of light in the vacuum, and $I_F(\lambda)$ is the fluorescence intensity at the wavelength λ.

If the upper level is more populated than the lower level, a population inversion takes place in the system. Population inversion is a necessary condition for amplification or lasing. It is impossible to attain population inversion in a two-level system [26]. The system must have at least three levels in order to be suitable for laser amplification. A commonly used four-level laser scheme is presented in Figure 2a for Yb^{3+} ions. In this scheme, two levels of the ground manifold and two levels of the excited manifold are involved in a laser cycle.

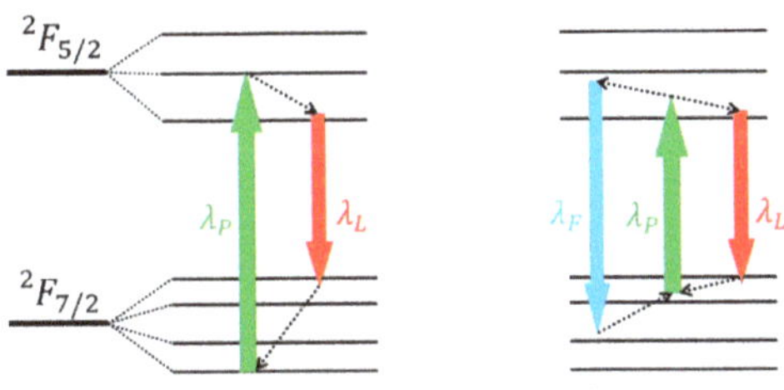

Figure 2. Energy diagrams of a traditional four-level laser (**a**) and a radiation-balanced laser (**b**). λ_P and λ_L are the pump and laser frequencies, respectively. λ_F is the mean fluorescence frequency. Solid and dushed arrows illustrate photon and phonon transitions, respectively.

As mentioned above, the quantum defect, η_q, is an unavoidable source of heat generation in lasers. The heat generation process is responsible for transverse-mode instability, thermal lensing, and catastrophic fiber damage. In RE-doped fiber lasers, heating can also cause a reduction in both the emission and absorption cross-sections of the RE ions, affecting the gain of the active fiber. Once the concentration of the RE ions exceeds a certain limit, cooperative effects such as an energy transfer process, a cross-relaxation process, nonradiative energy transfer, cooperative emission, and excitation become essential. Some of these processes can cause heat generation [27]. Cooperative effects limit the maximum ion concentration that can be doped into the fiber, resulting in the use of longer fibers. On the other hand, a longer fiber can serve as a source of nonlinear effects, for example, Raman and Brillouin scattering.

3.2. RE-Doped Radiation-Balanced Fiber Lasers

In 1999, Steven Bowman [28] proposed a radiation-balanced (athermal) RE-doped laser which operates without internal heating. In this laser, all photons generated in the laser cycle are annihilated in the cooling cycle, that is, the heat generated from the stimulated emission is offset by cooling from anti-Stokes emission (Figure 2b). The idea of optical cooling using anti-Stokes fluorescence was first proposed by Pringsheim in 1929 [29]. It is significant that in order to keep the radiation balance at each point in the laser medium, the pump intensity must be distributed properly along the length of the laser fiber. Any deviation from this distribution will result in heating or cooling in some parts of the laser medium. In radiation-balanced lasers and amplifiers, the power of the amplified signal increases only linearly with the length of the active medium. The theory of radiation-balanced (athermal) RE-doped fiber amplifiers was developed in [30]. To overcome the problem of linear signal power growth, fiber lasers and amplifiers with a cooler made from RE ions and integrated into the body of the device have been proposed and theoretically investigated [18,31,32]. The theory of mJ-level pulse amplifiers in Yb-doped double-clad optical fibers was developed in ref. [33].

In 2021, the first radiation-balanced silica fiber amplifier was demonstrated. An Yb^{3+}-doped silica fiber served as an active medium [34]. The core diameter of the silica fiber was 21 μm. Its numerical aperture was 0.13. The Yb^{3+} concentration was 2.52 wt.%. The fiber was co-doped with 2.00 wt.% Al to reduce concentration quenching. The wavelength of the pump was 1040 nm and the signal wavelength was 1064 nm. The mean fluorescence wavelength of Yb^{3+} was 1003.9 nm and the radiative and quenching lifetimes were 765 μs and 38 ms, respectively. The Yb^{3+} critical quenching concentration was 21.0 wt.%. The absorptive loss of the fiber was 18 dB/km. The temperature of the fiber decreased to ~130 mK below room temperature. Different schemes of athermal lasers and amplifiers are overviewed in ref. [35]. The latest advances in radiation-balanced fiber lasers and amplifiers are considered in ref. [36].

3.3. Power Evolution of RE-Doped Fiber Lasers

The rapid development of active RE-doped optical fibers has led to the development of high-power lasers with output powers of tens or hundreds of watts, sometimes even several kilowatts, from a single fiber. High-power laser sources are widely used in industrial precision processing. They can serve as a new platform for strong-field physics research using peak power over petawatt.

As noted above, Yb^{3+} ions are widely used in fiber lasers. Since they have only two manifolds, they are free from a number of cooperative effects. The quantum efficiency of Yb^{3+}-doped fiber lasers is close to 100% due to the low energy difference between the pump and laser photons. The Yb^{3+} ions possess a broadband absorption spectrum ranging from ~850 nm to ~1080 nm, with the peak in the range between ~915 nm and ~980 nm. The window of emission wavelengths from Yb^{3+}-doped fiber lasers is limited to a small band (1050–1160 nm). At this junction, all-fiber laser oscillators operating at ~1000 nm are well developed. Their power evolution throughout the past several years is presented in Figure 3 [37–48].

As can be seen in Figure 3, the maximum power ~8.5 kW was reached in 2020. Fiber lasers with wavelengths of 980 nm, 1007 nm, 1018 nm, and 1150 nm have attracted significant interest since they are very useful in nonlinear frequency conversion and efficient pumping of other RE-doped fibers which have absorption bands located near these wavelengths. A 1.3 kW, 1018 nm fiber laser with a beam quality M^2 less than 1.1 was reported by IPG Photonics [49].

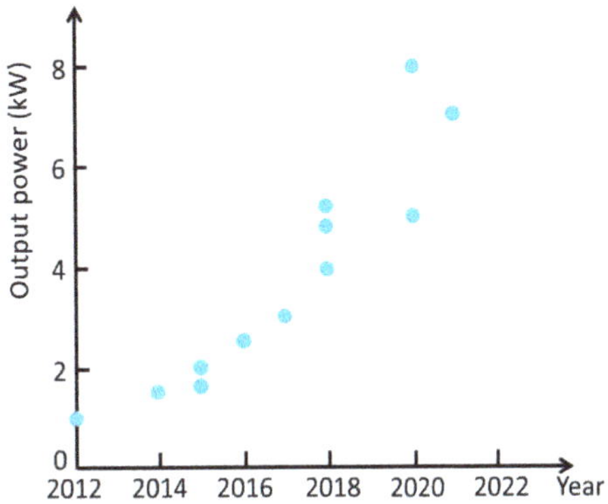

Figure 3. Power evolution of all-fiber laser oscillators operating at ~1000 nm [37–48].

The Er^{3+}-doped and Er^{3+}-Yb^{3+} co-doped fiber lasers partly complement the emission range of 1530–1590 nm. The power evolution of these fiber lasers at ~1500 nm throughout the past several years is presented in Figure 4 [50–61]. The highest power achieved with Er^{3+}-doped fibers with a relatively multimode operation (M^2 ~ 10.5) is 656 W [56].

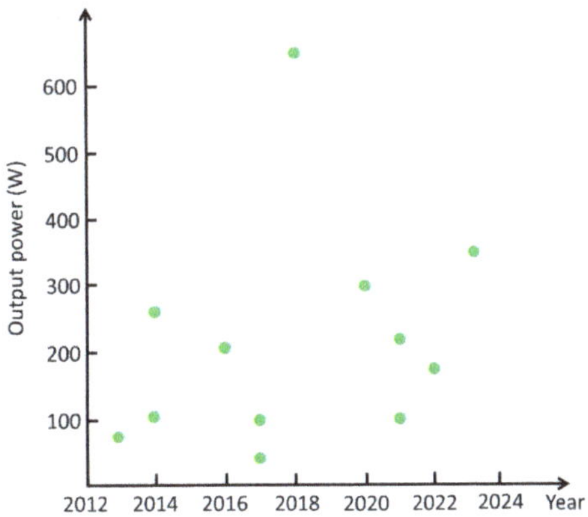

Figure 4. Power evolution of Er^{3+}-doped and Er^{3+}-Yb^{3+} co-doped fiber lasers operating at ~1.5 μm [50–61].

The Tm^{3+}-doped and Tm^{3+}-Ho^{3+} co-doped fiber lasers have an emission range between ~1900 nm and ~2100 nm. The output power ~1.2 kW at ~2000 nm was obtained with these lasers in 2018. Their power evolution is presented in Figure 5 [62–71].

Recent experiments have demonstrated the single-mode Ho-doped fiber laser providing 407 W of output power [72].

The Er^{3+}- and Tm^{3+}-doped lasers are significantly limited in power and efficiency compared to Yb^{3+}-doped fiber lasers. As one can see in Figures 4 and 5, the erbium lasers are limited to a few hundred watts; the thulium lasers have achieved power levels of around a kilowatt. Greater quantum defects, difficulty in increasing the ion concentration, excited state absorption, and cooperative effects prevent the erbium and thulium dopants from competing in the output power with ytterbium.

Although fiber laser systems exceeding 100 kW are commercially available [73], there is still a lot of room for further development of the RE-doped fiber lasers with excellent optical properties.

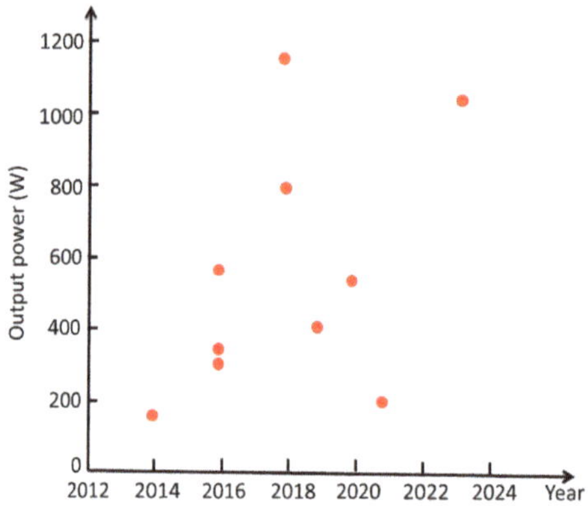

Figure 5. Power evolution of Tm^{3+}-doped and Tm^{3+}-Ho^{3+} co-doped fiber lasers operating at ~2 μm [62–71].

4. Raman Fiber Lasers

There are substantial white spaces in the wavelength spectrum where no RE-doped fiber lasers are available. Raman fiber lasers can cover these white spaces. In contrast to RE-doped lasers, Raman lasers are free from photodarkening.

4.1. Heat Mitigation in Raman Fiber Lasers

In 1963, Charles Townes introduced the concept of stimulated Raman scattering (SRS) [74]. In ref. [74], it has been shown that since the laser light incident on a medium is coherent, the resulting emission occurs through a parametric process due to the interaction between the coherent molecular oscillations (optical phonons) and the coherent laser light.

The lasing mechanism of Raman lasers is based on stimulated Stokes Raman scattering (SSRS) in a Raman medium (Figure 6a). An incoming pump photon with the energy $\hbar\omega_P$ scatters into a lower-energy Stokes photon $\hbar\omega_S$ and a phonon with the energy $\hbar\Omega_{op}$, which serves as a source of heat generation. As can be seen in Figure 6a, Raman lasers, like RE-doped lasers, suffer from heat dissipation inside the active medium caused by the quantum defect between the pump and lasing (Stokes) photons. Like in the case of RE-doped lasers, heat generation in Raman lasers can degrade their performance. Heat mitigation in Raman lasers can be achieved in different ways.

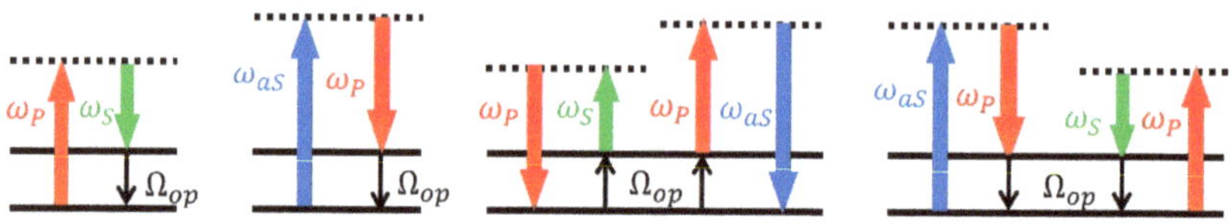

Figure 6. Processes that take place in a Raman laser: (**a**) stimulated Stokes Raman scattering (SSRS), (**b**) stimulated anti-Stokes Raman scattering (SARS), (**c**) coherent anti-Stokes Raman scattering (CARS) when it converts a Stokes photon and a pump photon to an anti-Stokes photon and a pump photon, annihilating two phonons, and (**d**) CARS when it converts an anti-Stokes photon and a pump photon to a Stokes photon and a pump photon, creating two phonons. ω_P, ω_S, ω_{aS}, and Ω_{op} are the pump, Stokes, anti-Stokes, and optical phonon frequencies, respectively.

One way to mitigate heat dissipation in Raman lasers has been proposed in [75–77]. It relies on three different Raman processes that can occur inside the medium of an optically pumped Raman laser illustrated in Figure 6. Besides SSRS, stimulated anti-Stokes Raman scattering (SARS) and coherent anti-Stokes Raman scattering (CARS) can take place in the

Raman medium. The maximum Raman gain point is considered in Figure 6, that is, all scattering processes are exactly at the Raman resonance. As stated above, the SSRS, which is responsible for lasing, is a source of heat generation. In common with SSRS, the SARS process generates quantum-defect heating too. In this process, an anti-Stokes photon is converted into a pump photon and a phonon (Figure 6b). CARS at the Raman resonance presented in Figure 6c is a four-wave mixing process. It converts a pump photon and a Stokes photon into a pump photon and an anti-Stokes photon, annihilating two phonons responsible for heating [78]. This process reduces the quantum-defect heating in the Raman medium. Phase matching is very important in this scheme. If it is not satisfied, the induced phase variation along the propagation direction of the amplified signal will cause the CARS process to alternate, reversing periodically at the coherence length (Figure 6c,d). This approach to Raman processes, which was developed by Bobbs and Warner in ref. [78], is different from the traditional one, where the CARS is seen as a mechanism that converts two pump photons into Stokes and anti-Stokes photons without exchanging energy with the Raman medium. By this means, the intrinsic heat-mitigation technique in Raman lasers relies on CARS (Figure 6c) instead of anti-Stokes fluorescence.

The harnessing of frequency-selective dissipative coupling to reduce the heat generated by CARS reversed cycles in active Raman media with phase mismatching was proposed in ref. [79] for waveguide systems. This approach can be applied to a fiber Raman system too. Heat mitigation in this approach is accomplished using a coupled waveguide structure, which includes the active Raman waveguide (RW), where the Stokes signal undergoes amplification via SSRS, and a dissipative waveguide (DW), which is tuned to the anti-Stokes wavelength so as to evacuate the corresponding anti-Stokes photons from the RW by coupling. The DW introduces optical loss that partially offsets the growth of the anti-Stokes signal in the RW and hence suppresses the reversed CARS cycles that would otherwise result in heat generation in the RW. It is shown that the frequency-selective dissipative coupling provided by the DW can reduce the heat in active Raman media by a factor of up to five when the CARS phase mismatch is compensated for by the optimum level of coupling between the RW and the DW.

The alternative approach to mitigating heat dissipation in Raman lasers is based on an ultralow quantum defect. In the majority of cases, high-power Raman fiber lasers are made of pure silica fibers or germanium-doped fibers. The lowest quantum defect for such a Raman laser pumped at 1064 nm and emitted at 1075 nm is 1.02% [80]. The pump power of the laser is 6.5 W and the output power is 1.1 W. It is difficult to achieve a high-power Raman output with such a low quantum defect, because the Raman gain in the pure silica fiber or germanium-doped fiber is relatively low at such a small frequency shift.

In 2020, the phosphosilicate-doped passive fiber was proposed for low quantum defect applications [81]. This fiber has a strong boson peak and a Raman gain peak at a frequency shift of ~40 THz in its Raman gain spectrum. The Raman gain peak is related to the asymmetric stretching vibrations supported by phosphorus–oxygen double bonds. This peak has been used for high power generation in Raman fiber lasers [82]. The boson peak in the phosphosilicate fiber is ascribed to an excess density of vibrational states [83]. In ref. [81], an ultralow quantum defect Raman laser based on the boson peak in phosphosilicate fibers has been reported for the first time. In this laser, the amplification process is based on a strong boson peak located at a frequency shift of 3.65 THz in its Raman gain spectrum. At a pump wavelength of 1066 nm and a pump power of 18.6 W, the Stokes output at the wavelength 1080 nm has reached 12.5 W. The quantum defect is 1.3%. For the increased pump wavelength, 1072 nm, the output power at 1080 nm is 10.7 W. In this case, the quantum defect is 0.74%. This is the lowest quantum defect ever reported for Raman fiber lasers.

4.2. Brightness Enhancement in Raman Fiber Lasers

In cladding-pumped RE-doped fiber lasers, the low-brightness pump light is coupled into the fiber cladding while the laser signal undergoes amplification in a fiber core, which supports only low-order modes with superior brightness relative to the pump. Brightness enhancement in the system can be estimated as follows:

$$BE = B_L / B_P \tag{16}$$

where B_L and B_P are the laser output brightness and the pump input brightness, respectively. Like RE-doped fiber lasers, Raman fiber lasers can provide brightness enhancement too. For example, one can use the same cladding-pumped approach (usually based on double cladding) to achieve brightness enhancement in Raman lasers.

Another way to achieve brightness enhancement in Raman fiber lasers consists of utilizing multimode graded-index (MM-GRIN) fibers. As in the case of the cladding-pumped approach, in the MM-GRIN approach, the Stokes laser signal is generated in the high-brightness core, while the pump propagates through a larger diameter of the fiber cladding. The first Raman fiber laser based on MM-GRIN fibers was demonstrated in 2018 [84]. A total of 135 W of the output power was obtained with 68% efficiency relative to launched pump power. Its brightness enhancement was $BE \approx 5.6$.

Much progress has been made in the last few years in both of these techniques. In 2018, the highest brightness enhancement $BE \approx 7$ for the cladding-pumped configuration was demonstrated at a record Raman fiber laser power of 1.2 kW with 85% efficiency relative to launched pump power [85]. At such high (over kW) output power levels, the beam quality $M^2 \approx 2.75$ was obtained.

In 2019, a Raman fiber laser power of 1002.3 W at 1060 nm with 84% efficiency was demonstrated using the MM-GRIN configuration [86]. The brightness enhancement was about 2.57 at the maximum output power. The beam quality was $M^2 \approx 5$.

5. Brillouin Fiber Lasers

Brillouin scattering is the interaction of optical waves with acoustic waves in a medium. It was described by Leon Brillouin in 1922 [87], and independently by Leonid Mandelstam in 1926 [88]. In the 1960s, Charles Townes, who introduced SRS in 1963, realized that parametric interactions could also occur between laser light and the acoustic branch of the phonon spectrum, that is, between laser light and acoustic oscillations (acoustic phonons), or sound waves. This process is known as stimulated Brillouin scattering (SBS). The first demonstration of SBS was reported in 1964 in quartz and sapphire crystals [89]. As a third-order nonlinear optical process, SBS requires intense radiation. Compared to typical Raman frequency shifts (50–1400 cm^{-1}), the Brillouin frequency shift (0.1–2 cm^{-1}) is extremely small. Brillouin scattering (both stimulated and spontaneous) manifests itself through the generation of backward or forward propagating waves shifted from the frequency of the incident pump wave by an amount determined by the frequency of the acoustic phonon. When the acoustic wave is moving away from the pump light beam, the Stokes shift takes place in the system (Figure 7a). When the acoustic wave is moving toward the pump light beam, the anti-Stokes shift takes place (Figure 7b).

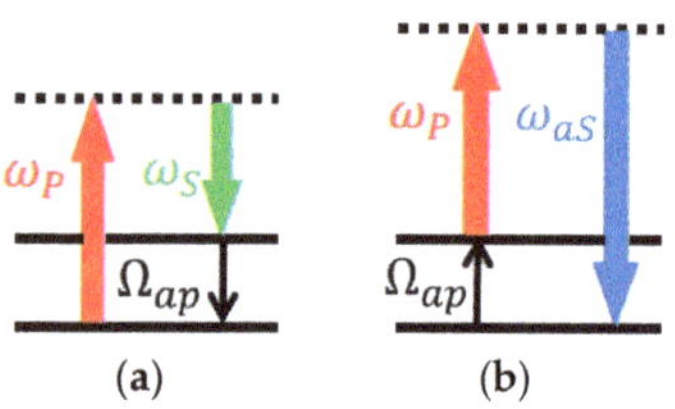

Figure 7. The energy level diagram of Brillouin scattering. (**a**) Brillouin Stokes scattering; (**b**) Brillouin anti-Stokes scattering. ω_P, ω_S, ω_{aS}, and Ω_{ap} are the pump, Stokes, anti-Stokes, and acoustic phonon frequencies, respectively.

The energy and momentum conservation requirements of Stokes and anti-Stokes Brillouin scattering have the forms

$$\omega_S = \omega_P - \Omega_{ap} \quad \text{and} \quad \mathbf{k_S} = \mathbf{k_P} - \mathbf{K_{ap}} \tag{17}$$

for Stokes scattering and

$$\omega_{aS} = \omega_P + \Omega_{ap} \quad \text{and} \quad \mathbf{k_{aS}} = \mathbf{k_P} - \mathbf{K_{ap}} \tag{18}$$

for anti-Stokes scattering. Here ω_P, ω_S, and ω_{aS} are the pump, Stokes, and anti-Stokes photon frequencies, respectively. Ω_{ap} is the acoustic wave frequency. $\mathbf{k_P}$, $\mathbf{k_S}$, $\mathbf{k_{aS}}$, and $\mathbf{K_{ap}}$ are the pump, Stokes, anti-Stokes, and acoustic phonon wave vectors, respectively. In the SBS process, $\Omega_{ap} \ll \omega_P, \omega_S, \omega_{aS}$. As for Raman scattering, the Stokes scattering process is accompanied by phonon generation (heating); the anti-Stokes scattering process is accompanied by phonon absorption (cooling).

SBS can be used for the development of Brillouin lasers. A typical Brillouin laser involves coupling a laser pump light source, whose spectral width is much less than the Brillouin gain bandwidth, to the optical fiber in order to produce SBS with the Stokes signal propagating in the backscattered path of the incident pump light. The first Brillouin fiber laser was presented in 1976 [90]. Since then, remarkable progress in the development of Brillouin fiber lasers has been made. Great effort went into the development of lasers with a relatively low pump threshold, high efficiency, small linewidth, and especially multiwavelength configurations.

Multiwavelength Fiber Lasers

SBS is the most attractive technique for the development of multiwavelength fiber lasers, which have the advantages of a narrow linewidth, low-intensity noise, stable operation at room temperature, and potential applications in optical communication, optical sensors, gas spectroscopy, and microwave photonics.

The cascaded SBS process occurs when the power of the first-order Stokes signal reaches the SBS generation threshold of the second-order Stokes (Figure 8). The first-order anti-Stokes signal can also be generated due to the four-wave mixing process between the pump and the first-order Stokes wave. The typical frequency shift in the Brillouin spectrum in most glass is about ~10 GHz. This narrow frequency shift is a challenge for the demultiplexing and signal filtering processes in optical communication applications. Several approaches have been developed to increase the frequency spacing range and to reduce the complexity of demultiplexing. These approaches include widely used double Brillouin frequency spacing of 0.17 nm [91–93] and a triple Stokes line spacing of 0.25 nm [94–97]. A quintuple wavelength spacing of 0.4 nm was presented in ref. [98], where four quintuple Brillouin Stokes signals with a high Stokes power of 10 dBm and a signal-to-noise ratio of 50 dB were achieved. The recorded Stokes signals were tuned over a wavelength range of 40 nm (1540–1580 nm).

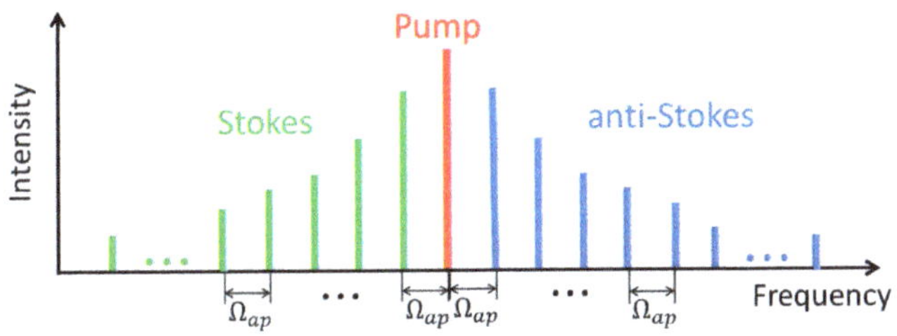

Figure 8. Illustration of the cascaded SBS process. It occurs via the interplay of SBS and four-wave mixing.

The sextuple Stokes signal 0.48 nm has been demonstrated in ref. [99], utilizing a triple unidirectional ring laser cavity. Three sextuple channels are achieved and tuned over a wavelength range of 15 nm (1560–1575 nm). In this work, up to three sextuple Stokes lines with a high signal power of 10 dBm and a large optical signal-to-noise ratio of more than 55 dB have been demonstrated. It has been shown that the generated Stokes lines can be tuned over a wavelength range of 15 nm from 1560 to 1575 nm.

A switchable multiwavelength Brillouin fiber laser has been proposed and demonstrated with channel spacing from 10 GHz up to even more than 100 GHz at a step of 10 GHz [100]. The structure comprises two nonlinear fiber loops, a feedback path, and a tunable optical bandpass filter for switchable operation. The multiwavelength Brillouin fiber laser is generated in the nonlinear fiber loop, where a dual-pump source produces dual Brillouin lasers that give rise to a wideband multiwavelength Brillouin fiber laser via the four-wave mixing effect. Some achievements of multiwavelength Brillouin fiber lasers across different spectral regions have been discussed in ref. [101].

Fiber lasers based on backward SBS have been studied and employed for over thirty years. As mentioned, SBS in fibers takes place in the forward direction as well, with amplification bandwidths which are narrower by two orders of magnitude. The first forward Brillouin fiber laser, using a bare off-the-shelf, panda-type PMF, has been reported in ref. [102]. Pump light in one principal axis provides Brillouin amplification for a co-propagating lasing signal of the orthogonal polarization. Feedback is provided by Bragg gratings at both ends of the optical fiber cavity. Single-mode, few-mode, and multimode regimes of operation have been observed. A laser threshold has been observed at a pump power level of 26.5 dBm. The output power of the laser was saturated at 250 μW due to the onset of backward SBS in the fiber cavity. The differential slope efficiency of the laser output power between the threshold and saturation was approximately 0.005. Forward SBS fiber lasers are extremely sensitive to changes in media outside the cladding. They can serve as highly coherent laser sources and ultraprecision forward SBS sensors.

6. Fiber Laser Applications

At present, fiber lasers are used widely in both science and industry. As mentioned above, in the 1990s, fiber lasers came into wide use in the realm of the telecommunications industry, providing high-quality single-mode low-power systems. Fiber lasers in fiber telecommunications are mainly used as light sources. Since fiber lasers have the characteristics of high efficiency, stability, and good modulation performance, they can provide stable and high-quality optical signals.

With the collapse of the telecommunications markets in the early 2000s, fiber laser producers shifted their focus to meeting the needs of industrial material processing, and of military and medical sectors. This shift to alternative markets required higher-power lasers. One particular advantage of fiber lasers is that they are inherently simple and can therefore provide easily manufacturable systems to meet a range of power levels and output quality.

6.1. Material Processing

The application of fiber lasers in the field of material processing mainly includes laser cutting, laser welding, and laser marking. High-speed and high-precision material processing can be achieved using high-power and high-beam-quality fiber lasers.

The use of fiber lasers has allowed, for the first time, for the generation of striation-free cuts in the case of oxygen-assisted laser cutting of 1 mm and 2 mm thick mild steel sheets [103]. Nowadays, fiber laser flexibility enables the cutting of a wide range of materials from steel and aluminum to nonmetal materials such as plastics and glass of different thicknesses. The fiber laser cutting process results in a high-quality cutting edge. High-power CW fiber lasers are commonly used for industrial applications. As the power output increases, the cutting speed and the ability to process thicker materials improve, making them ideal for high-volume production.

The excellent mode quality and high focusability of low-power single-mode fiber lasers are the most important features for achieving small spot sizes in the range of some microns. As a result, the welding of microparts with very high processing speeds, the lowest energy input, reduced heat-affected zones, and minimized distortion has been achieved [104]. At present, fiber laser welding creates strong welds and good-quality joints, even with dissimilar materials. With the benefit of consistent accuracy, laser welding is well suited to manufacturing processes with repetitive operations. Fiber lasers are also very useful in microprecision welding and laser spot welds.

Fiber lasers are perfect for laser marking. Laser marking is a process in which a laser beam is used to mark or engrave a surface at a high speed and low-operating cost.

6.2. Military Sector

Fiber lasers have high temperature and high corrosion resistance, which is especially suitable for military applications. Fiber lasers can serve as one of the main light sources of tactical laser weapons [105]. In fiber optic radar, fiber lasers can be used as laser sources to provide high-power and high-stability laser signals for target detection and tracking.

6.3. Medical Sector

With the continuous progress of science and technology, fiber lasers are increasingly widely used in laser therapy [106]. Laser energy can be delivered through optical fibers to accurately treat diseased tissues. In laser therapy, fiber lasers have the advantages of high energy density, high orientation, and minimally invasive treatment. The application of fiber lasers in laser therapy includes but is not limited to laser beauty treatment, laser eye therapy, laser tumor therapy, and laser endoscopic therapy.

6.4. Optical Frequency Combs

The invention of optical frequency combs (OFCs), also known as the rulers of light, has enabled numerous applications in the fields of optical metrology, spectroscopy, astronomy, and quantum and attosecond science [107].

Over the past two decades, various types of OFC sources have been demonstrated, including solid-state or fiber-based mode-locked combs, electro-optic combs, and microresonator-based combs. At present, mode-locked fiber combs are technically mature and have a deeper degree of commercialization. This is due to their low noise performance and long-term reliability. OFCs can be realized utilizing various mode-locking techniques, such as nonlinear polarization rotation, a real saturable absorber, and a nonlinear amplifying loop mirror. Fiber laser-based OFCs are particularly attractive for free-space time–frequency transfer, low-noise microwave generation, and gas molecule detection.

6.5. Quantum Applications

Quantum optics is based on lasers as light sources. Fiber lasers can be used as pump sources in single-photon emitters and in waveguide quantum electrodynamics [108]. Fiber lasers are actively engaged in the manipulation and readout of quantum bits. They are used widely in the trapping and cooling of atoms and ions as well as in quantum sensing and metrology.

7. Conclusions

The present work gives a brief overview of the latest achievements in fiber lasers including RE-doped fiber lasers, Raman fiber lasers, and Brillouin fiber lasers. In comparison to conventional bulk lasers, fiber lasers have a number of advantages.

- The inherently flexible structure of fibers enables the use of much longer (up to several kilometers) gain distances, which provides higher optical gains, compared to other lasers.
- The optical feedback in fiber lasers is usually provided by fiber Bragg gratings imprinted in the fiber. This integrated design provides a stable optical cavity.

- In fiber lasers, single-mode fibers typically offer the best beam performance, which is very important in a number of applications.
- In fiber lasers, the optical path is enclosed within protective cladding layers, and as a result, the laser beam is less susceptible to exterior disturbance compared to traditional lasers.

It has been shown that the Er^{3+}- and Tm^{3+}-doped lasers are significantly limited in power and efficiency compared to Yb^{3+}-doped fiber lasers, which can reach output powers of tens or hundreds of watts. Heat generation in traditional high-power fiber lasers is a serious problem.

- One solution to the heat generation problem is the development of radiation-balanced lasers based on cooling with anti-Stokes fluorescence. The efficiency of the laser cooling process can be improved by applying novel low-phonon fiber materials. The recent achievements in implementing new materials in high-power fiber lasers have been presented in ref. [109], where adverse interactions between the laser light and the host are discussed, and novel composition glass fiber designs and fabrication methodologies are presented.
- The alternative approach to mitigating heat dissipation in Raman lasers is based on an ultralow quantum defect.

As of now, the development of multiwavelength Brillouin fiber lasers with an increased number of Stokes channels that present high flatness and reasonable optical-to-signal ratio is a challenge that fiber laser technology still faces. Multiwavelength Brillouin fiber lasers operating at optical communication wavebands such as 1.5 μm have been widely reported. However, there is still a lot of room for further development of multiwavelength Brillouin fiber lasers in the 1.0 μm, 1.3 μm, and 2.0 μm spectral regions. In these spectral regions, multiwavelength Brillouin fiber lasers show a low optical-to-signal ratio and poor flatness. Finding the best setup parameters for such lasers can considerably improve their performance in these promising spectral ranges.

Although fiber lasers have already revolutionized the laser industry, they continue to be a very promising technique for different applications, including industrial material processing, in military and medical sectors, and quantum physics. Considerable room remains for their improvement.

- As already noted, the optimization of fiber materials is very important for the future development of high-power lasers. The absorption and radiation characteristics of fiber materials can be improved.
- Widely used pump technology such as laser diode pumps and fiber laser pumps can be improved in terms of pump efficiency and fiber absorption efficiency.
- The optical design of fiber lasers can be improved too. For example, specially designed fibers can be developed to achieve brightness enhancement in Raman lasers.

Funding: This research received no external funding.

Institutional Review Board Statement: Not applicable.

Informed Consent Statement: Not applicable.

Data Availability Statement: Not applicable.

Conflicts of Interest: The author declares no conflicts of interests.

References

1. Kepler, J. *Dioptrice;* Prague, Czech Republic, 1611.
2. Huygens, C. *Traité de la Lumière;* Gressner & Schramm: Leipzig, Germany, 1690.
3. Baid, J.L. An improved method of and means for producing optical images. *Br. Pat.* **1927**, *285*, 738.
4. van Heel, A.C.S. A new method of transporting optical images without aberrations. *Nature* **1954**, *173*, 39. [CrossRef]
5. Hirschowitz, B.I.; Curtiss, L.E.; Peters, C.W.; Pollard, H.M. Demonstration of a new gastroscope, the fiberscope. *Gastroenterology* **1958**, *35*, 50–53. [CrossRef] [PubMed]

6. Kapany, N.S. Fiber Optics. VI. Image Quality and Optical Insulation. *J. Opt. Soc. Am.* **1959**, *49*, 779–787. [CrossRef]
7. Kao, K.C.; Hockham, G.A. Dielectric-fibre surface waveguides for optical frequencies. *Proc. Inst. Electr. Eng.* **1966**, *133*, 1151–1158. [CrossRef]
8. Miya, T.; Terunuma, Y.; Hosaka, T.; Miyashita, T. Ultimate low-loss single-mode fibre at 1.55 μm. *Electron. Lett.* **1979**, *15*, 106–108. [CrossRef]
9. Stolen, R.H.; Ippen, E.P.; Tynes, A.R. Raman oscillation in glass optical waveguide. *Appl. Phys. Lett.* **1972**, *20*, 62–64. [CrossRef]
10. Ippen, E.P.; Stolen, R.H. Stimulated Brillouin scattering in optical fibers. *Appl. Phys. Lett.* **1972**, *21*, 539–541. [CrossRef]
11. Smith, R.G. Optical power handling capacity of low loss optical fibers as determined by stimulated Raman and Brillouin scattering. *Appl. Opt.* **1972**, *11*, 2489–2494. [CrossRef]
12. Hasegawa, A.; Tappert, F. Transmission of Stationary Nonlinear Optical Pulses in Dispersive Dielectric Fibers. I. Anomalous Dispersion. *Appl. Phys. Lett.* **1973**, *23*, 142–144. [CrossRef]
13. Snitzer, E. Optical maser action of Nd^{3+} in a barium crown glass. *Phys. Rev. Lett.* **1961**, *7*, 444–446. [CrossRef]
14. Desurvire, E. *Erbium-Doped Fiber Amplifiers: Principles and Applications*; John Wiley & Sons, Inc.: Hoboken, NJ, USA, 1994.
15. Yeh, P.; Yariv, A.; Marom, E. Theory of Bragg fiber. *J. Opt. Soc. Am.* **1978**, *68*, 1196–1201. [CrossRef]
16. Knight, J.C.; Birks, T.A.; Russell, P.S.J.; Atkin, D.M. All-silica single-mode optical fiber with photonic crystal cladding. *Opt. Lett.* **1996**, *21*, 1547–1549. [CrossRef]
17. Li, J.; Duan, K.; Wang, Y.; Cao, X.; Zhao, W.; Guo, Y.; Lin, X. Theoretical analysis of the heat dissipation mechanism in Yb^{3+}-doped double-clad fiber lasers. *J. Mod. Opt.* **2008**, *55*, 459–471. [CrossRef]
18. Nemova, G.; Kashyap, R. High-power fiber lasers with integrated rare-earth optical cooler. In Proceedings of the SPIE Laser Refrigeration of Solids III, San Francisco, CA, USA, 23–28 January 2010; Volume 7614, pp. 761406–761416.
19. Chen, Y.; Yao, T.; Xiao, H.; Leng, J.; Zhou, P. Theoretical analysis of heat distribution in Raman fiber lasers and amplifiers employing pure passive fiber. *IEEE Photonics J.* **2020**, *12*, 1504713. [CrossRef]
20. Broer, M.M.; Krol, D.M.; Di Giovanni, D.J. Highly nonlinear near-resonant photodarkening in a thulium-doped aluminosilicate glass fiber. *Opt. Lett.* **1993**, *18*, 799–801. [CrossRef] [PubMed]
21. Atkins, G.R.; Carter, A.L.G. Photodarkening in Tb^{3+}-doped phosphosilicate and germanosilicate optical fibers. *Opt. Lett.* **1994**, *19*, 874–876. [CrossRef] [PubMed]
22. Weber, M.J. Science and technology of laser glass. *J. Non-Cryst. Solids* **1990**, *123*, 208–222. [CrossRef]
23. Layne, C.B.; Lowdermilk, W.H.; Weber, M. Multiphonon relaxation of rare-earth ions in oxide glasses. *J. Phys. Rev. B* **1977**, *16*, 10–20. [CrossRef]
24. Reisfeld, R.; Boehm, L.; Spector, N. Multiphonon relaxation rates and fluorescence lifetimes for Tm^{3+} in four oxide glasses. *Chem. Phys. Lett.* **1977**, *49*, 251–254. [CrossRef]
25. Reisfeld, R.; Jørgensen, C.K. Excited state phenomena in vitreous materials. In *Handbook on the Physics and Chemistry of Rare Earths*; Gschneidner, K.A., Jr., Eyring, L., Eds.; Elsevier: Amsterdam, The Netherlands, 1987.
26. Nemova, G. *Field Guide to Light-Matter Interaction*; SPIE Press: Bellingham, WA, USA, 2022.
27. Nemova, G. *Field Guide to Laser Cooling Methods*; SPIE Press: Bellingham, WA, USA, 2019.
28. Bowman, S.R. Laser without internal heat generation. *IEEE J. Quantum Electron.* **1999**, *35*, 115–122. [CrossRef]
29. Pringsheim, P. Zwei Bemerkungen über den Unterschied von Lumineszenz- und Temperaturstrahlung. *Z. Phys.* **1929**, *57*, 739–746. [CrossRef]
30. Nemova, G.; Kashyap, R. Athermal continuous-wave fiber amplifier. *Opt. Commun.* **2009**, *282*, 2571–2575. [CrossRef]
31. Nemova, G.; Kashyap, R. Fiber amplifier with integrated optical cooler. *J. Opt. Soc. Am. B* **2009**, *26*, 2237–2241. [CrossRef]
32. Nemova, G.; Kashyap, R. Raman fiber amplifier with integrated cooler. *IEEE J. Light. Technol.* **2009**, *27*, 5597–5601. [CrossRef]
33. Yu, N.; Xiong, M.; Dragic, P.D. FDTD modeling of excitation-balanced, mJ-level pulse amplifiers in Yb-doped double-clad optical fibers. *Opt. Express* **2023**, *31*, 32404–32421. [CrossRef]
34. Knall, J.M.; Engholm, M.; Boilard, T.; Bernier, M.; Digonnet, M.J.F. A radiation-balanced silica fiber amplifier. *Phys. Rev. Lett.* **2021**, *127*, 013903. [CrossRef]
35. Nemova, G.; Kashyap, R. Laser cooling of solids. *Rep. Prog. Phys.* **2010**, *73*, 086501. [CrossRef]
36. Nemova, G. Radiation-Balanced Lasers: History, Status, Potential. *Appl. Sci.* **2021**, *11*, 7539. [CrossRef]
37. Xiao, Y.; Brunet, F.; Kanskar, M.; Faucher, M.; Wetter, A.; Holehouse, N. 1-kilowatt CW all-fiber laser oscillator pumped with wavelength-beam-combined diode stacks. *Opt. Express* **2012**, *20*, 3296. [CrossRef]
38. Yu, H.L.; Wang, X.L.; Tao, R.M.; Zhou, P.; Chen, J.B. 1.5 kW, near diffraction-limited, high-efficiency, single-end-pumped all-fiber-integrated laser oscillator. *Appl. Opt.* **2014**, *53*, 8055. [CrossRef] [PubMed]
39. Mashiko, Y.; Nguyen, H.K.; Kashiwagi, M.; Kitabayashi, T.; Shima, K.; Tanaka, D. 2 kW single-mode fiber laser with 20-m long delivery fiber and high SRS suppression. *Proc. SPIE* **2015**, *9728*, 29–34.
40. Shi, W.; Fang, Q.; Xu, Y.; Qin, Y.G.; Fan, J.L.; Meng, X.J.; Zhang, Q.H. 1.63 kW monolithic continuous-wave single-mode fiber laser oscillator. *J Optoelectron. Laser* **2015**, *26*, 662.
41. Yang, B.L.; Zhang, H.W.; Shi, C.; Wang, X.L.; Zhou, P.; Xu, X.J.; Chen, J.B.; Liu, Z.J.; Lu, Q.S. Mitigating transverse mode instability in all fiber laser oscillator and scaling power up to 2.5 kW employing bidirectional-pump scheme. *Opt. Express* **2016**, *24*, 27828. [CrossRef] [PubMed]

42. Ikoma, S.; Nguyen, H.K.; Kashiwagi, M.; Uchiyama, K.; Shima, K.; Tanaka, D. 3 kW single stage all-fiber Yb-doped single-mode fiber laser for highly reflective and highly thermal conductive materials processing. *Proc. SPIE* **2017**, *10083*, 45–50.
43. Kensuke, S.; Shinya, I.; Keisuke, U.; Yuya, T.; Masahiro, K.; Daiichiro, T. 5-kW single stage all-fber Yb-doped single-mode fiber laser for materials processing. *Proc. SPIE* **2018**, *10512*, 45–50.
44. Yang, B.L.; Zhang, H.W.; Shi, C.; Tao, R.M.; Su, R.T.; Ma, P.F.; Wang, X.L.; Zhou, P.; Xu, X.J.; Lu, Q.S. 3.05 kW monolithic fiber laser oscillator with simultaneous optimizations of stimulated Raman scattering and transverse mode instability. *J. Opt.* **2018**, *20*, 025802. [CrossRef]
45. Yang, B.L.; Zhang, H.W.; Ye, Q.; Pi, H.Y.; Shi, C.; Tao, R.M.; Wang, X.L.; Xu, X.J. 4.05 kW monolithic fiber laser oscillator based on home-made large mode area fiber Bragg gratings. *Chin. Opt. Lett.* **2018**, *16*, 031407. [CrossRef]
46. Krämer, R.G.; Möller, F.; Matzdorf, C.; Goebel, T.A.; Strecker, M.; Heck, M.; Richter, D.; Plötner, M.; Schreiber, T.; Tünnermann, A.; et al. Extremely robust femtosecond written fiber Bragg gratings for an ytterbium-doped fiber oscillator with 5 kW output power. *Opt. Lett.* **2020**, *45*, 1447. [CrossRef]
47. Wang, Y.; Kitahara, R.; Kiyoyama, W.; Shirakura, Y.; Kurihara, T.; Nakanish, Y.; Yamamoto, T.; Nakayama, M.; Ikoma, S.; Shima, K. 8-kW single-stage all-fiber Yb-doped fiber laser with a BPP of 0.50 mm-mrad. *Proc. SPIE* **2020**, *11260*, 273–278.
48. Xi, X.M.; Wang, P.; Yang, B.L.; Wang, X.L.; Zhang, H.W.; Ning, Y.; Han, K.; Wang, Z.F.; Zhou, P.; Xu, X.J.; et al. The output power of the all-fiber laser oscillator exceeds 7 kW. *Chin. J. Lasers* **2021**, *48*, 0116001.
49. Nikolai, P.; Oleg, S.; Valentin, F.; Daniil, M.; Roman, Y.; Anton, F.; Alexey, D.; Ivan, U.; Valentin, G. High-efficient kW-level single-mode ytterbium fiber lasers in all-fiber format with diffraction-limited beam at wavelengths in 1000–1030 nm spectral range. *Proc. SPIE* **2020**, *11260*, 1126003.
50. Kotov, L.V.; Likhachev, M.E.; Bubnov, M.M.; Medvedkov, O.I.; Yashkov, M.V.; Guryanov, A.N.; Lhermite, J.; Février, S.; Cormier, E. 75 W 40% efficiency single-mode all-fiber erbium-doped laser cladding pumped at 976 nm. *Opt. Lett.* **2013**, *38*, 2230. [CrossRef] [PubMed]
51. Jebali, M.A.; Maran, J.N.; Larochelle, S. 264 W output power at 1585 nm in Er-Yb co-doped fiber laser using in-band pumping. *Opt. Lett.* **2014**, *39*, 3974. [CrossRef] [PubMed]
52. Kotov, L.V.; Likhachev, M.E.; Bubnov, M.M.; Medvedkov, O.I.; Yashkov, M.V.; Guryanov, A.N.; Février, S.; Lhermite, J.; Cormier, E. Yb-free Er-doped all-fiber amplifier cladding-pumped at 976 nm with output power in excess of 100 W. *Proc. SPIE* **2014**, *8961*, 149–154.
53. Daniel, C.; Herman, P.; Julia, L.; Scott, D.S. Single frequency 1560 nm Er:Yb fiber amplifier with 207 W output power and 50.5% slope efficiency. *Proc. SPIE* **2016**, *9728*, 426–431.
54. De Varona, O.; Fittkau, W.; Booker, P.; Theeg, T.; Steinke, M.; Kracht, D.; Neumann, J.; Wessels, P. Single-frequency fiber amplifier at 1.5 µm with 100 W in the linearly-polarized TEM00 mode for next-generation gravitational wave detectors. *Opt. Express.* **2017**, *25*, 24880. [CrossRef] [PubMed]
55. Han, Q.; Yao, Y.Z.; Tang, X.Y.; Chen, Y.F.; Yan, W.C.; Liu, T.G.; Song, H.L. Highly efficient Er–Yb co-doped double-clad fiber amplifier with an Yb-band resonant cavity. *Laser Phys. Lett.* **2017**, *14*, 025105. [CrossRef]
56. Lin, H.Q.; Feng, Y.J.; Feng, Y.T.; Barua, P.; Sahu, J.K.; Nilsson, J. 656 W Er-doped, Yb-free large-core fiber laser. *Opt. Lett.* **2018**, *43*, 3080. [CrossRef]
57. Matniyaz, T.; Kong, F.; Kalichevsky-Dong, M.T.; Dong, L. 302 W single-mode power from an Er/Yb fber MOPA. *Opt. Lett.* **2020**, *45*, 2910. [CrossRef]
58. Michaud, L.-C.; Veilleux, C.; Bilodeau, G.; Paquet, O.; Lebel-Cormier, M.-A.; Lemieux-Tanguay, M.; Pelletier-Ouellet, S.; Paradis, P.; Bellec, M.; Grégoire, N.; et al. 100-W-level single-mode ytterbium-free erbium fiber laser. *Opt. Lett.* **2021**, *46*, 2553. [CrossRef]
59. Yu, W.L.; Xiao, Q.R.; Wang, L.L.; Zhao, Y.; Qi, T.C.; Yan, P.; Gong, M.L. 219.6 W large-mode-area Er: Yb co-doped fiber amplifier operating at 1600 nm pumped by 1018 nm fiber lasers. *Opt. Lett.* **2021**, *46*, 2192. [CrossRef]
60. Yu, W.L.; Yan, P.; Qi, T.C.; Wu, Y.L.; Li, D.; Xiao, Q.R.; Gong, M.L. Highpower and high-brightness Er: Yb co-doped fiber MOPA operating at 1535 nm. *Opt. Express* **2022**, *30*, 16837. [CrossRef]
61. Li, W.; Qiu, Q.; Yu, L.; Gu, Z.; He, L.; Liu, S.; Yin, X.; Zhao, X.; Peng, J.; Li, H.; et al. Er/Yb co-doped 345-W all-fiber laser at 1535 nm using hybrid fiber. *Opt. Express* **2023**, *48*, 3027–3030. [CrossRef]
62. Stutzki, F.; Gaida, C.; Gebhardt, M.; Jansen, F.; Wienke, A.; Zeitner, U.; Fuchs, F.; Jauregui, C.; Wandt, D.; Kracht, D.; et al. 152 W average power Tm-doped fiber CPA system. *Opt. Lett.* **2014**, *39*, 4671. [CrossRef] [PubMed]
63. Walbaum, T.; Heinzig, M.; Schreiber, T.; Eberhardt, R.; Tünnermann, A. Monolithic thulium fiber laser with 567 W output power at 1970 nm. *Opt. Lett.* **2016**, *41*, 2632. [CrossRef] [PubMed]
64. Yin, K.; Zhu, R.Z.; Zhang, B.; Liu, G.C.; Zhou, P.; Hou, J. 300 W-level, wavelength-widely-tunable, all-fber integrated thulium-doped fiber laser. *Opt. Express* **2016**, *24*, 11085. [CrossRef]
65. Jiang, L.; Chen, L.; Xing, S.H.; Pu, W. 342 W narrow-linewidth continuous-wave thulium-doped all-fber laser. *Acta. Phys. Sin.* **2016**, *65*, 194209. [CrossRef]
66. Gaida, C.; Gebhardt, M.; Heuermann, T.; Stutzki, F.; Jauregui, C.; Limpert, J. Ultrafast thulium fber laser system emitting more than 1 kW of average power. *Opt. Lett.* **2018**, *43*, 5853. [CrossRef] [PubMed]
67. Yao, W.C.; Shen, C.F.; Shao, Z.H.; Wang, J.L.; Wang, F.; Zhao, Y.G.; Shen, D.Y. 790 W incoherent beam combination of a Tm-doped fiber laser at 1941 nm using a 3 × 1 signal combiner. *Appl. Opt.* **2018**, *57*, 5574. [CrossRef] [PubMed]

68. Liu, Y.Z.; Cao, C.; Xing, Y.B.; Liao, L.; Cao, R.T.; Zhang, F.F.; Chen, Y.S.; Wang, Y.B.; Peng, J.G.; Li, H.Q.; et al. 406 W narrow-linewidth all-fiber amplifier with Tm-doped fiber fabricated by MCVD. *IEEE Photonics Technol. Lett.* **2019**, *31*, 1779. [CrossRef]
69. Zi, L.Y.; Bing, X.Y.; Lei, L.; Bo, W.Y.; Gang, P.J.; Qing, L.H.; Li, D.N.; Yan, L.J. 530 W all-fiber continuous-wave Tm-doped fiber laser. *Acta. Phys. Sin.* **2020**, *69*, 184209.
70. Motard, A.; Louot, C.; Robin, T.; Cadier, B.; Manek-Hönninger, I.; Dalloz, N.; Hildenbrand-Dhollande, A. Diffraction limited 195-W continuous wave laser emission at 2.09 μm from a Tm^{3+}, Ho^{3+} co-doped single-oscillator monolithic fiber laser. *Opt. Express* **2021**, *29*, 6599. [CrossRef]
71. Ren, C.; Shen, Y.; Zheng, Y.; Mao, Y.; Wang, F.; Shen, D.; Zhu, H. Widely-tunable all-fiber Tm doped MOPA with > 1 kW of output power. *Opt. Express* **2023**, *2931*, 22733–22739. [CrossRef] [PubMed]
72. Alexander, H.; Nikita, S.; John, H.; Adrian, C. High power resonantly pumped holmium-doped fiber sources. *Proc. SPIE* **2014**, *8982*, 898292.
73. IPG Photonics. "High Power CW Fiber Lasers". (IPG Photonics). Available online: https://www.ipgphotonics.com/en/products/lasers/highpower-cw-fiber-lasers (accessed on 8 February 2024).
74. Garmire, E.; Pandarese, F.; Townes, C.H. Coherently driven molecular vibrations and light modulation. *Phys. Rev. Lett.* **1963**, *11*, 160–163. [CrossRef]
75. Vermeulen, N.; Debaes, C.; Muys, P.; Theinpont, H. Mitigating heat dissipation in Raman lasers using coherent anti-Stokes Raman scattering. *Phys. Rev. Lett.* **2007**, *99*, 093903. [CrossRef]
76. Vermeulen, N.; Debaes, C.; Thienpont, H. Mitigating heat dissipation in near- and mid-infrared silicon-based Raman lasers using CARS: I. Theoretical analysis. *IEEE J. Sel. Top. Quantum Electron.* **2007**, *13*, 770–782. [CrossRef]
77. Vermeulen, N.; Debaes, C.; Thienpont, H. Mitigating heat dissipation in near- and mid-infrared silicon-based Raman lasers using CARS: II. Numerical demonstration. *IEEE J. Sel. Top. Quantum Electron.* **2007**, *13*, 783–788. [CrossRef]
78. Bobbs, B.; Warner, C. Raman-resonant four-wave mixing and energy transfer. *J. Opt. Soc. Am. B* **1990**, *7*, 234–238. [CrossRef]
79. Nemova, G.; Caloz, C. Heat Evacuation from Active Raman Media Using Frequency-Selective Dissipative Coupling. *Phys. Rev. Res.* **2021**, *3*, 013050. [CrossRef]
80. Belanger, E.; Bernier, M.; Faucher, D.; Cote, D.; Vallee, R. Highpower and widely tunable all-fiber Raman laser. *J. Light. Technol.* **2008**, *26*, 1696–1701. [CrossRef]
81. Zhang, Y.; Xu, J.; Ye, J.; Song, J.; Yao, T.; Zhou, P. Ultralow-quantum-defect Raman laser based on the boson peak in phosphosilicate fiber. *Photonics Res.* **2020**, *8*, 1155–1160. [CrossRef]
82. Dong, J.; Zhang, L.; Zhou, J.; Pan, W.; Gu, X.; Feng, Y. More than 200 W random Raman fiber laser with ultra-short cavity length based on phosphosilicate fiber. *Opt. Lett.* **2019**, *44*, 1801–1804. [CrossRef] [PubMed]
83. Tanaka, H.; Shintani, H. Universal link between the boson peak and transverse phonons in glass. *Nat. Mater.* **2008**, *7*, 870–877.
84. Glick, Y.; Shamir, Y.; Wolf, A.A.; Dostovalov, A.V.; Babin, S.A.; Pearl, S. Highly efficient all-fiber continuous-wave Raman graded-index fiber laser pumped by a fiber laser. *Opt. Lett.* **2018**, *43*, 1027–1030. [CrossRef]
85. Glick, Y.; Shamir, Y.; Aviel, M.; Sintov, Y.; Goldring, S.; Shafir, N.; Pearl, S. 1.2 kW clad pumped Raman all-passive-fiber laser with brightness enhancement. *Opt. Lett.* **2018**, *43*, 4755–4758. [CrossRef]
86. Chen, Y.; Leng, J.; Xiao, H.; Yao, T.; Zhou, P. Pure passive fiber enabled highly efficient Raman fiber amplifier with record kilowatt power. *IEEE Access* **2019**, *7*, 28334–28339. [CrossRef]
87. Brillouin, L. Diffusion de la lumière et des Rayons X par un corps transparent homogène. *Ann. Phys.* **1922**, *9*, 88–122. [CrossRef]
88. Mandelstam, L. Light scattering by inhomogeneous media. *Zh. Russ. Fiz-Khim.* **1926**, *58*, 381.
89. Chiao, R.; Townes, C.; Stoicheff, B. Stimulated Brillouin scattering and coherent generation of intense hypersonic waves. *Phys. Rev. Lett.* **1964**, *12*, 592–595. [CrossRef]
90. Hill, K.O.; Kawasaki, B.S.; Johnson, D.C. CW Brillouin laser. *Appl. Phys. Lett.* **1976**, *28*, 608–609. [CrossRef]
91. Shee, Y.G.; Al-Mansoori, M.H.; Yaakob, S.; Man, A.; Zamzuri, A.K.; Mahamd Adikan, F.R.; Mahdi, M.A. Millimeter wave carrier generation based on a double-Brillouin-frequency spaced fiber laser. *Opt. Express* **2012**, *20*, 13402–13408. [CrossRef]
92. Abass, A.K.; Al-Mansoori, M.H.; Jamaludin, M.Z.; Abdullah, F.; Al-Mashhadani, T.F.; Ali, M.H. L-band multi-wavelength brillouin-raman fber laser with 20-GHz channel spacing. *Fiber Integr. Opt.* **2014**, *33*, 56–67. [CrossRef]
93. Al-Mashhadani, T.F.; Al-Mashhadani, M.K.S.; Yucel, M.; Goktas, H.H. Influence of bidirectional cavity structure on the Brillouin Stokes signal characteristics in ring BFL. *Optik* **2019**, *185*, 359–363. [CrossRef]
94. Al-Mansoori, M.H.; Al-Sheriyani, A.; Al-Nassri, S.; Hasoon, F.N. Generation of efficient 33 GHz optical combs using cascaded stimulated Brillouin scattering effects in optical fiber. *Laser Phys.* **2017**, *27*, 65112. [CrossRef]
95. Al-Mansoori, M.H.; Al-Sheriyani, A.; Younis, M.A.A.; Mahdi, M.A. Widely tunable multiwavelength Brillouin-erbium fiber laser with triple Brillouin-shift wavelength spacing. *Opt. Fiber Technol.* **2018**, *41*, 21–26. [CrossRef]
96. Al-Mashhadani, T.F. Erbium gain effects on Stokes signal performance in a Fabry-Perot Brillouin Erbium fiber laser. *Opt. Quantum Electron.* **2019**, *51*, 189. [CrossRef]
97. Al-Mashhadani, T.F.; Al-Mashhadani, M.K.S.; Goktas, H.H.; Yucel, M.; Celebi, F.V. Widely triple Brillouin frequency shift multiwavelength Brillouin erbium fiber laser. *Opt. Quantum* **2020**, *52*, 288. [CrossRef]
98. Al-Mashhadani, M.K.S.; Al-Mashhadani, T.F.; Goktas, H.H. Tunable 50 GHz laser comb generation of multiwavelength Brillouin erbium fber laser. *Opt. Commun.* **2020**, *464*, 125542. [CrossRef]

99. Awsaj, M.K.; Al-Mashhadani, T.F.; Al-Mashhadani, M.K.S.; Ali, A.Y.; Zan, M.S.D.; Arsad, N. Multiwavelength fiber laser sources with 60 GHz Brillouin frequency shift. *Opt. Quantum Electron.* **2023**, *55*, 528. [CrossRef]
100. Shi, Y.; Wang, T.; Hao, Y.-Z.; Bai, H.-Y.; Yang, Y.-D.; Xiao, J.-L.; Chen, Y.-L.; Huang, Y.-Z. Wideband multiwavelength Brillouin fiber laser with switchable channel spacing. *Appl. Opt.* **2023**, *62*, 2130–2136. [CrossRef]
101. Silva, L.C.B.; Segatto, M.E.V. Advances in multi-wavelength Brillouin fiber lasers: An outlook across different spectral regions. *Opt. Fiber Technol.* **2023**, *76*, 103246. [CrossRef]
102. Bashan, G.; Diamandi, H.H.; Zehavi, E.; Sharma, K.; London, Y.; Zadok, A. A forward Brillouin fibre laser. *Nat Commun.* **2022**, *13*, 3554. [CrossRef]
103. Sobih, M.; Crouse, P.L.; Li, L. Elimination of striation in laser cutting of mild steel. *J. Phys. D Appl. Phys.* **2007**, *40*, 6908–6916. [CrossRef]
104. Wagner, F.; Grupp, M.; Vollertsen, F. Laser beam micro welding with single mode fibre lasers. In Proceedings of the Third International WLT-Conference on Lasers in Manufacturing (LIM), Munich, Germany, 13–16 June 2005; pp. 769–772.
105. Ludewigt, K.; Liem, A.; Stuhr, U.; Jung, M. High-power laser development for laser weapons. In *Hogh Power Lasers: Technology and Systems, Platforms, Effects III*; SPIE: Strasbourg, France, 2019; Volume 11162, pp. 46–53.
106. Baac, H.W.; Uribe-Patarroyo, N.; Bouma, B.E. High-energy pulsed Raman fiber laser for biological tissue coagulation. *Opt. Express* **2014**, *22*, 7113–7123. [CrossRef] [PubMed]
107. Yan, P.; Xu, W.; Hu, H.; Zhang, Z.; Li, Z.; Shu, R. Recent advances, applications, and perspectives in erbium-doped fiber combs. *Photonics* **2024**, *11*, 192. [CrossRef]
108. Sheremet, A.S.; Petrov, M.I.; Iorsh, I.V.; Poshakinskiy, A.P.; Poddubny, A.N. Waveguide quantum electrodynamics: Collective radiance and photon-photon correlations. *Rev. Mod. Phys.* **2023**, *95*, 015002. [CrossRef]
109. Dragic, P.D.; Cavillon, M.; Ballato, J. Materials for optical fiber lasers: A review. *Appl. Phys. Rev.* **2018**, *5*, 041301. [CrossRef]

applied sciences

MDPI

Review

Advances in High–Speed, High–Power Photodiodes: From Fundamentals to Applications

Qingtao Chen [1,2,*], Xiupu Zhang [3,*], Mohammad S. Sharawi [4] and Raman Kashyap [1,2,5]

1 Fabulas Laboratory, Department of Electrical Engineering, Polytechnique Montreal, Montreal, QC H3T 1J4, Canada; raman.kashyap@polymtl.ca
2 Fabulas Laboratory, Poly–Grames Research Center, Polytechnique Montreal, Montreal, QC H3T 1J4, Canada
3 Department of Electrical and Computer Engineering, Concordia University, Montreal, QC H3G 1M8, Canada
4 Blue Origin LLC, Kent, WA 98032, USA; m.sharawi@ieee.org
5 Fabulas Laboratory, Department of Engineering Physics, Polytechnique Montreal, Montreal, QC H3T 1J4, Canada
* Correspondence: qingtao.chen@polymtl.ca (Q.C.); xzhang@ece.concordia.ca (X.Z.)

Abstract: High–speed, high–power photodiodes play a key role in wireless communication systems for the generation of millimeter wave (MMW) and terahertz (THz) waves based on photonics–based techniques. Uni–traveling–photodiode (UTC–PD) is an excellent candidate, not only meeting the above–mentioned requirements of broadband (3 GHz~1 THz) and high–frequency operation, but also exhibiting the high output power over mW–level at the 300 GHz band. This paper reviews the fundamentals of high–speed, high–power photodiodes, mirror–reflected photodiodes, microstructure photodiodes, photodiode–integrated devices, the related equivalent circuits, and design considerations. Those characteristics of photodiodes and the related photonic–based devices are analyzed and reviewed with comparisons in detail, which provides a new path for these devices with applications in short–range wireless communications in 6G and beyond.

Keywords: high–speed photodiodes; high–power photodiodes; millimeter–wave; terahertz wave; photodiode–integrated devices; wireless communication

1. Introduction

High data rate and high–quality transmission, large–capacity data storage, wireless short–range links, sensing, and imaging bring new challenges for current communication systems. Parts of the of microwave (MW), millimeter wave (MMW), and terahertz (THz) bands, which cover the frequency from 20 GHz to 3 THz, as shown in Figure 1, provide an appropriate choice for the above–mentioned challenges [1–4]. The exciting MMW and THz bands (20 GHz~1 THz), which are located between the radio waves and light waves, combine both their qualities and advantages, including large available bandwidth, high information capacity, transparent for some kinds of materials, high–spatial resolution, smaller beam scattering than infrared and visible light, and greater safety for the human body than x–rays and gamma rays [5–7]. Based on the above–mentioned advantages of the frequency band, it would bring about many potential applications [7], such as high–capacity wireless links, short–range wireless communication systems, and inspection system, high–resolution imaging transmission, spectroscopy, remote gas sensing and detection, security system and astronomical radio telescope system. More–over, this frequency band would support 6G technology for a high data rate (100~100 Gbit/s), helping towards a fully connected world in the future.

However, the large atmospheric attenuation (Figure 1) [8,9] in wireless communication systems needs to be considered (i.e., the attenuation caused by the rain and fog, free–space propagation loss (FSPL), gas loss, and signal loss from fiber to chip [8–11]) with the exception of focusing on the two key figures, i.e., the data rates or information capacity and transmission distance. To compensate for the FSPL limitation, high–gain and high–output power antennas or phased array antennas for beam forming/beam steering [12,13] should be considered for both transmitter and receiver systems. For the transmitter in communication systems, using large–scale photonic integrated circuit (PIC) techniques has an unparalleled impact on the continually reducing footprint, coupling loss, low power consumption and cost, while gradually increasing stability, reliability, and functionality [14]. Photonics–based technology is not only used for the development of MMW to THz communication techniques in systems with lower loss in optical fibers, large bandwidth, and miniaturized device/system [6–9,15–25] but it can also reduce some drawbacks which are brought forth by electronics–based technology, including high propagation loss in coaxial cables, narrow bandwidth, incompatibility with system operations at high–frequency bands and bulky feeding methods [26–29].

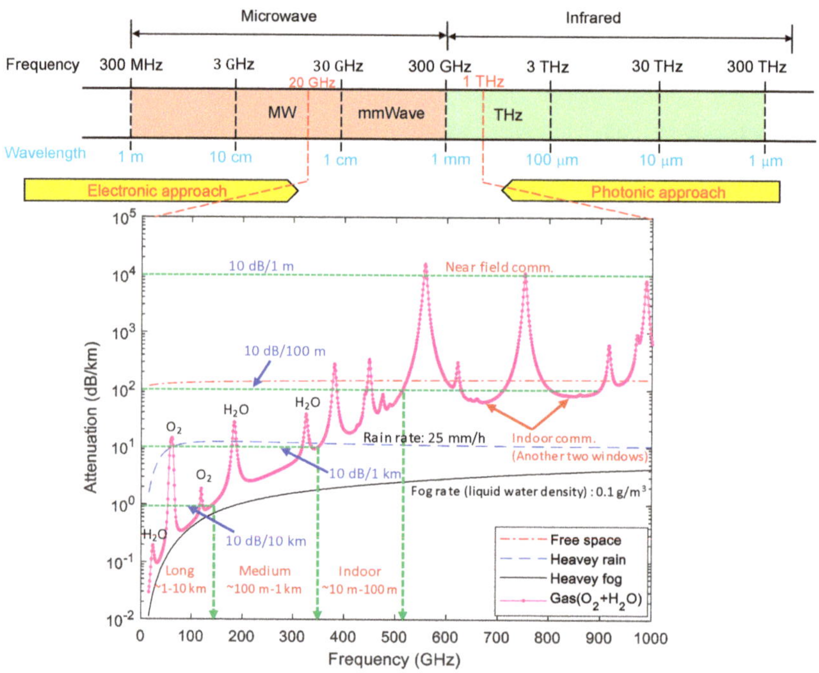

Figure 1. Features of microwave, millimeter–wave, and terahertz spectrum [8–11].

In this scheme, photonic–based techniques are derived from optical to RF signal conversions utilizing the photomixing method [30]. Since there is an output power limitation using a single photomixer, the InP–based semiconductor devices have been employed and led by a uni–traveling–carrier photodetector (UTC–PD) [31,32], operating at a long fiber–optics communication band (1.31~1.55 μm), which can be used to further enhance the output power up to a mW level around 300 GHz [9]. However, the new requirements of high data rate transmission in future 6G poses another challenge for THz systems. In addition, due to the lack of appropriate power amplifiers in the THz system transmitting terminally along with the difficulty of miniaturizing the monolithically integrated system employing Si–lens, the combination of high–speed, high–power UTC–PD (or UTC–PD array) and high–gain, high–output power antenna and/or arrays would be a better choice to tackle the power limitation problem. Furthermore, the optical–to–electrical (O/E) conversion efficiency and heat dissipation at a high bias voltage also needs to be considered. We could use resonant cavity enhanced (RCE) structure [33] employing distributed Bragg reflector (DBR) [34], subwavelength grating (SWG) mirrors [35] as bottom reflectors and dielectric layers as top reflectors to increase O/E conversion efficiency, while using high conductivity AlN [36,37], diamond [38] or SiC [39] substrates to reduce heat dissipation. Moreover, the slow light effect resulting from Bragg grating structures [40,41] with a remarkably low group velocity might offer a possible and promising solution to successfully compress optical signals and enhance light–matter interactions, and the enhanced O/E conversion efficiency in PDs could be possible while the joule heat problem at a higher bias voltage, device footprint reduction, and low power consumption could also be solved in the future.

This review work is structured as follows. Section 2 presents the overall design considerations for photodiodes (PDs), including the principles of PIN–PD and UTC–PD, 3–dB bandwidth analysis and discussion, equivalent circuit models, and saturation current and RF output power. Section 3 shows the solutions of bandwidth–responsivity trade–off, which are integrated or quasi–integrated with high–reflected mirrors for III–V and group–IV photodiodes, and microstructure photodiodes with microholes for enhancing light–matter interactions. Section 4 narrates the photodiode photonic–integrated applications, which relate to the photodiodes integrated with short stubs for output power improvement as well as the integrations with planar antennas for transmitters/emitters from MMW to THz bands. The discussion and conclusions are given in Sections 5 and 6.

2. Overall Design Considerations for Photodiodes

2.1. Principles of PIN–PD and UTC–PD

The band diagrams for PIN–PD [42,43] and UTC–PD [5,31,32,44,45] are shown in Figure 2. PIN–PD has a simple three–layer structure, i.e., wide–bandgap P– and N–layers, and a depleted absorber. Both holes and electrons in the depleted absorber of PIN–PD contribute to the response, but the output response mainly depends on the low–velocity carrier holes, whose transport velocity is an order of magnitude lower than that of electrons, which limits the photoresponse to some extent.

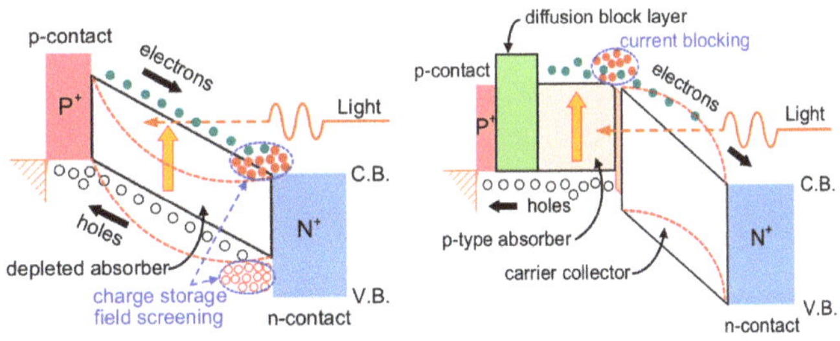

Figure 2. Band diagrams of (**a**) PIN–PD and (**b**) UTC–PD.

The UTC–PD band structure, by dividing the PIN–PD absorber into two layers, is composed of a p–type neutral absorber and a wide–bandgap (depleted) carrier collector [31,32]. By employing a bandgap grading or/and doping grading in the absorber, the quasi–field is formed to reduce the electron traveling time effectively. In the meantime, the photogenerated majority holes in the InGaAs absorber respond quickly within several picoseconds due to the high concentration of electrons, leading to a negligible effect on the photoresponse; therefore, the high–velocity electrons dominate the photoresponse of UTC–PDs. Furthermore, the velocity of electrons ($3 \sim 5 \times 10^7$ cm/s) is usually 6~10 bigger than that of holes (5×10^6 cm/s) [5], together with the quasi–ballistic transport (overshoot velocity) [46] in the InP collector, which leads to a quite short total delay time in the whole UTC–PD. For PIN–PD and UTC–PD with the same absorber (W_a), the UTC–PD ($W_a \cong W_c$) represents a superior frequency response with a shorter traveling time in the collector due to the large difference in carrier velocities, though the carriers in UTC–PD travels ~three times the average distance than that in PIN–PD [44].

During high optical input, the charge intensity increases together with the increasing photocurrent in the depletion, producing a space charge effect due to the charge accumulation, while the field and potential profiles are modulated resulting in output saturation and a nonlinear response. For PIN–PD, as shown in Figure 2a, the charge storage happens at high injection conditions, resulting in the band bending and field screening, which finally causes output current saturation. However, in UTC–PD, shown in Figure 2b, only electrons participate in the space charge effect, whose velocity overshoot, even facing the decreased electrical field. Finally, the saturation occurs because of the current blocking produced from the heterojunction interface between the absorber and collector; while this state will be slower than that in PIN–PD, it is because UTC–PD can withstand a charge density an order of magnitude higher than PIN–PD [5,44]. In addition, the space charge in UTC–PD can not only be compensated by doping the spacer layer at the absorber/collector (InGaAs/InP) heterojunction interface but also can be reduced by employing a heavily n–doped InP charge (cliff) layer to increase the electrical field of the spacer layer [47]. Therefore, UTC–PD is a typical representative of high–speed response and high–output current.

In a typical UTC–PD structure, the bandgap discontinuity (the abrupt conduction band barrier) at the InGaAs absorber and InP collector interface causes the current blocking effect, thus leading to the device output current saturation. To smooth this discontinuity and suppress the blocking effect, the compositional graded quaternary system material $In_{1-x}Ga_xAs_yP_{1-y}$ (for spacer layers) is usually employed to improve device performance [48]. Therefore, the design of $In_{1-x}Ga_xAs_yP_{1-y}$ lattice–matched to InP [49] must be considered, while the cutoff wavelength–related refractive index and permittivity is shown in Figure 3a. However, the introduction of quaternary system material (Figure 3b) might bring certain difficulties for epitaxial growth and the device fabrication process. Therefore, a dipole–doped regime, as shown in Figure 3b is incorporated to reduce the conduction band barrier, which uses a lower dipole–doping concentration ($\leq 5 \times 10^{18}$ cm^{-3} to avoid out–diffusion) to fully suppress the conduction band offset only left with a spike–like barrier for electrons to tunnel through [50–52].

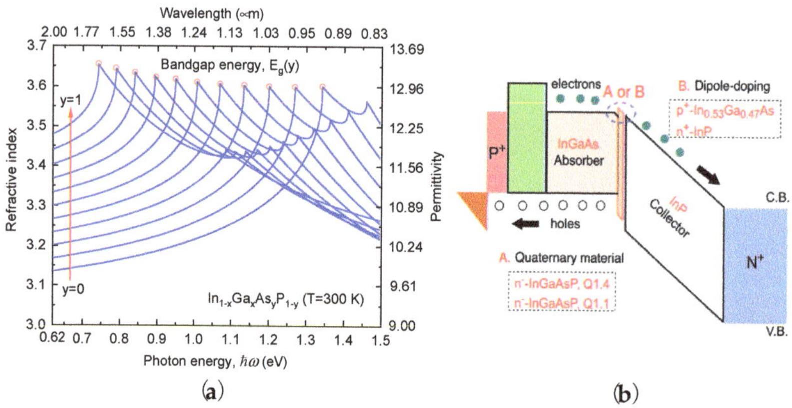

Figure 3. (**a**) The cutoff wavelength–related refractive index and permittivity for $In_{1-x}Ga_xAs_yP_{1-y}$ with the As–composition changed from 0 to 1. (**b**) Solutions for suppressing the bandgap discontinuity by compositional graded quaternary material and dipole–doping regimes.

2.2. 3–dB Bandwidth Analysis and Discussion

Response time is a physical quantity that characterizes how quickly a PD responds to an incident light signal, that is, it represents the total time it takes for a PD to convert an optical signal into an electrical signal.

The total response time of PD generally consists of three parts: (a) diffusion time of electrons in the absorption layer, τ_a, (b) circuit time constant, τ_{RC}, determined by junction capacitance and load resistance, and (c) drift time of electrons in the collection layer, τ_c [44].

PIN–PD is usually made up of parts (a) and (b), while (c) belongs to UTC–PD [44,53,54]. Therefore, the total response time for general UTC–PD can be given as

$$\tau = \sqrt{\tau_a^2 + \tau_c^2 + \tau_{RC}^2}, \quad (1)$$

while the PIN–PD can be dealt with as a special case, just having a single intrinsic/depletion layer, W_a. The RC–delay time–limited bandwidth, f_{RC}, and transit–time–limited bandwidth, f_{RC}, usually consist of the main 3–dB bandwidth of a PD, which here the diffusion is neglected at the high–frequency response. We assume that a PD has one single intrinsic layer (Figure 4) with the i–layer thickness of d and the area of A, so the f_{RC} can be expressed as

$$f_{RC} = \frac{1}{2\pi RC} = \frac{1}{2\pi R}\frac{d}{\varepsilon A}, \quad (2)$$

where the junction capacitance $C = \varepsilon A/d$ approximately equals PD capacitance (in ideal cases), the ε is the relative permittivity of i–layer, and R is the resistance and equals the sum of diode series resistance, R_S, and the load resistance, R_L, has a typical value of 50 Ω.

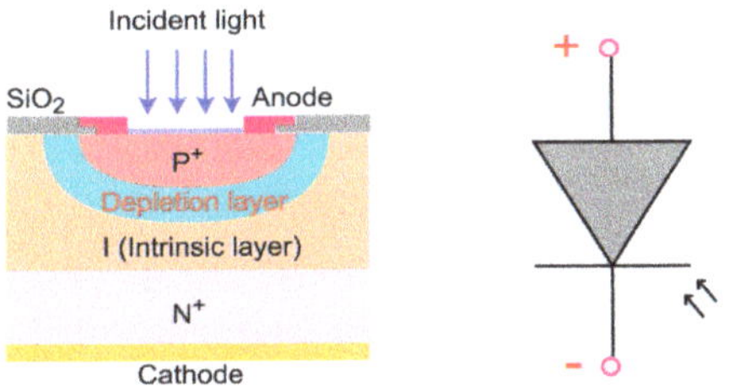

Figure 4. Cross–sectional diagram of PIN–PD and PD symbol.

It needs to be noted that Equation (2) will be changed if a PD has multiple dielectric/depletion layers and the corresponding values will be replaced by the relative permittivity, ε_{req}, and the total depletion thickness, d_T, respectively, which are shown as follows [55]:

$$\varepsilon_{req} = \left[\sum_{m=1}^{n} \frac{d_m}{d_T \varepsilon_{rm}}\right]^{-1}, \quad (3)$$

and

$$d_T = d_1 + d_2 + \cdots + d_n. \quad (4)$$

Therefore, the total capacitance, C_T, and the RC–limited 3–dB bandwidth, $f_{3dB,RC}$, can be rewritten as

$$C_T = \frac{\varepsilon_0 \varepsilon_{req} A}{d_T} = 8.854 \times 10^{-6} \frac{\varepsilon_{req} A}{d_T} \ (\mathrm{pF}), \quad (5)$$

and

$$f_{3dB,RC} = \frac{1}{2\pi RC_T} = \frac{1}{2\pi R}\frac{d_T}{\varepsilon_0 \varepsilon_{req} A}, \quad (6)$$

where ε_0 is the vacuum permittivity, with a value of 8.854×10^{-12}F/m.

For a UTC–PD with the same thickness of the absorber and collector, assuming a pure diffusion case without quasi–field, the 3–dB bandwidth can be given as [44,54]

$$f_{3dB,tr} \cong \frac{1}{2\pi\tau_a} = \frac{1}{2\pi\left(W_a^2/3D_e + W_a/v_{th}\right)}, \quad (7)$$

where D_e is the electron diffusion constant, v_{th} is the thermionic velocity and the related equations and values are shown in Table 1.

Table 1. The explanations for parameters [44] used for the 3–dB bandwidth calculation.

Symbol	Value and Units	Explanations
k_B	1.38×10^{-23} J/K	Boltzmann constant
q	1.6×10^{-19} C	Electron charge
T	300 K	Temperature
m_0	9.11×10^{-31} kg	Electron mass in vacuum
μ_e	5000 cm^2/Vs	Electron mobility in p–InGaAs
μ_h	150 cm^2/Vs	Hole mobility in p–InGaAs
$v_{e,os}$, v_{ns}	4×10^7 cm/s	Electron overshoot velocity in InGaAs and InP
v_{hs}	4.5×10^6 cm/s	Hole saturation velocity
D_e	$2k_BT\mu_e/q$	Electron diffusion constant
m^*	$0.043m_0$	Electron effective mass at the band edge
v_{th}	$\sqrt{2k_BT/\pi m^*}$	Thermionic–emission velocity
$v_{e,dri}$	$\sim \mu_e(\phi/W_a)$	Electron drift velocity in absorber
$v_{e,diff}$	$\sim 3D_e/W_a$	Electron diffusion velocity in absorber
τ_a	$W_a^2/3D_e + W_a/v_{th}$	Carrier traveling time by electron diffusion
τ_c	$W_c/v_{e,os}$	Electron drift time in collector
k_B	1.38×10^{-23} J/K	Boltzmann constant

In addition, for the electron drift/delay time–limited bandwidth in the collector of UTC–PD, $f_{3dB,co}$, can be considered as

$$f_{3dB,co} \cong \frac{1}{2\pi\tau_c}, \tag{8}$$

where the electron overshoot velocity in the InP collector is $v_{e,os} = 4 \times 10^7$ cm/s, while the corresponding electron drift time $\tau_c = W_c/v_{e,os}$ which is around 0.5~1 ps for a collector with a thickness of 200~400 nm. However, the electron drift time, τ_c, usually can be ignored for a designed UTC–PD which has a similar thickness for the absorber and collector due to the relatively greater carrier traveling time (τ_a) in the absorber. The τ_a can be separately calculated to be 3.74 ps and 11.2 ps for the 200 nm and 600 nm absorber by [44,54]

$$\tau_a = \frac{W_a^2}{3D_e} + \frac{W_a}{v_{th}}. \tag{9}$$

It is quite clear that we can ignore the electron drift time when the thickness of the absorber approximately equals the collector.

Therefore, the total bandwidth, $f_{3dB,UTC}$, as a general case, can be written as

$$f_{3dB,UTC} = \frac{1}{2\pi\tau} = \frac{1}{2\pi\sqrt{\tau_a^2 + \tau_c^2 + \tau_{RC}^2}}, \tag{10}$$

while the similar values for the absorber and collector, i.e., $W_a \cong W_c$, so Equation (10) can be changed to

$$f_{3dB,UTC} = \frac{1}{2\pi\tau} \cong \frac{1}{2\pi\sqrt{\tau_a^2 + \tau_{RC}^2}}, \tag{11}$$

and the $f_{3dB,UTC}$ also has another form, as shown in Equation (12), for treating PD with a small junction area less than 10 μm^2.

$$f_{3dB,UTC} = \frac{1}{2\pi\tau} \cong \frac{1}{2\pi\sqrt{\tau_a^2}} = \frac{1}{2\pi\tau_a}. \tag{12}$$

For PIN–PD, both of the holes and electrons are in the single intrinsic/depletion layer, W_a, while the hole and electron saturation velocity are v_{hs} and v_{ns}, so the 3–dB bandwidth for a PIN–PD with a thin absorber can be given as [43,54]

$$f_{3dB,PIN} \approx \frac{3.5}{2\pi\tau_a}, \tag{13}$$

where the τ_a equals $W_a/\overline{v}$, and the average velocity formula is $1/\overline{v}^4 = 0.5\left(1/\overline{v}_{ns}^4 + 1/\overline{v}_{hs}^4\right)$. Meanwhile, a thinner absorber can make PIN–PD obtain a larger transit–time–limited bandwidth, but it also causes a decrease in the RC–limited bandwidth by Equation (2), while the drawback can be overcome in UTC–PD, just by independently designing the absorber and collector. Therefore, a thinner absorber can be used in UTC–PD for a larger transit–time–limited bandwidth without sacrificing the RC–limited bandwidth by well–designing a relatively thick collector. Additionally, some references show the detailed theoretical and experimental analysis of the characteristics for PIN–PD and UTC–PD, including energy band, electrical field, doping mechanism, optimization of absorber and collector, bandwidth, responsivity, saturation, and output power [56–62] as well as transient response [63], dark current [64,65], and P–contact shapes [66]. According to the above–mentioned bandwidth equations and the values from Table 1, the 3–dB bandwidths for PIN–PD and UTC–PD can be calculated as shown in Figure 5a, including the assumed values during the calculation.

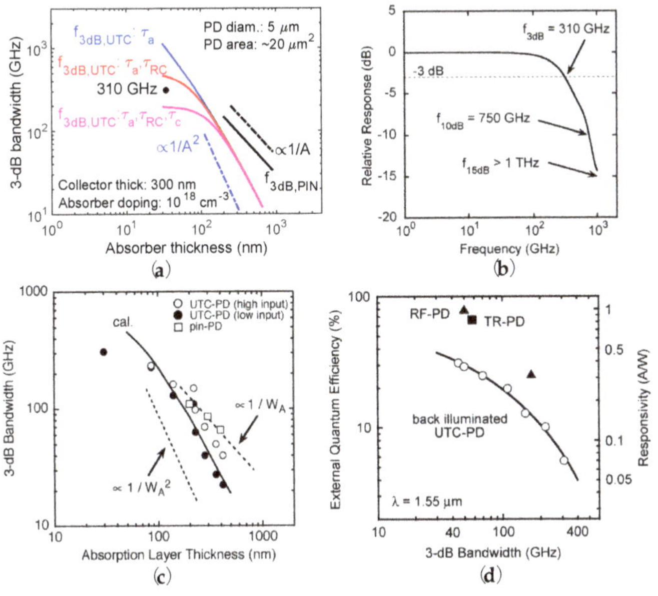

Figure 5. (**a**)The calculated 3–dB bandwidth versus the absorber thickness for UTC–PD and PIN–PD. (**b**) The 310 GHz bandwidth [67] is used to indicate the calculated accuracy in (**a**). (**c**) Measured and calculated 3–dB bandwidth for UTC–PD (at high and low input) and PIN–PD [44]. (**d**) Quantum efficiency, responsivity versus 3–dB bandwidth for back–illuminated UTC–PD [44], refracting–facet UTC–PD (RF–PD) [68], and total–reflection UTC–PD (TR–PD) [69].

It can be seen that the bandwidth of UTC–PD is proportional to $1/A^2$ [44,54] when the absorber thickness is larger than 100 nm, while it is proportional to $1/A$ for PIN–PD [44,54]. Figure 5c is from [44] as a reference. The well–designed UTC–PD has improved 3–dB bandwidths over 150 GHz, 220 GHz [70], while the maximum value is up to 310 GHz (Figure 5b), and the 10–dB and 15–dB bandwidths are separately 750 GHz and >1 THz, but the responsivity is only 0.07 A/W [67]. To improve the low responsivity of UTC–PD, the refracting–facet UTC–PD (RF–PD) [68] and total reflection UTC–PD (TR–PD) [69] are proposed to increase the light propagation length in the absorber, where RF–PD uses an angled edge to reflect the light while TR–PD employs a V–grooved mirror integrated with two PD to totally reflect the back–illuminated light. A high responsivity of 1 A/W with 50 GHz 3–dB bandwidth (0.32 A/W and 170 GHz) is achieved for RF–PD, while the TR–PD shows 0.83 A/W responsivity and a 58 GHz 3–dB bandwidth, as shown in Figure 5d.

Up to now, many PIN–PD and UTC–PD structures are proposed to improve the device performance, including frequency response, output power, saturation current, and responsivity, while the common band diagrams, including simple p–i–n type (Figure 2a), partially depleted absorber (PDA) type, dual depletion type, general UTC type, hybrid absorber type, charge–compensated (CC) hybrid absorber type and near–ballistic (NB)/NB–CC type [31,32,38,44,47,71–80], are summarized and shown in Figure 6.

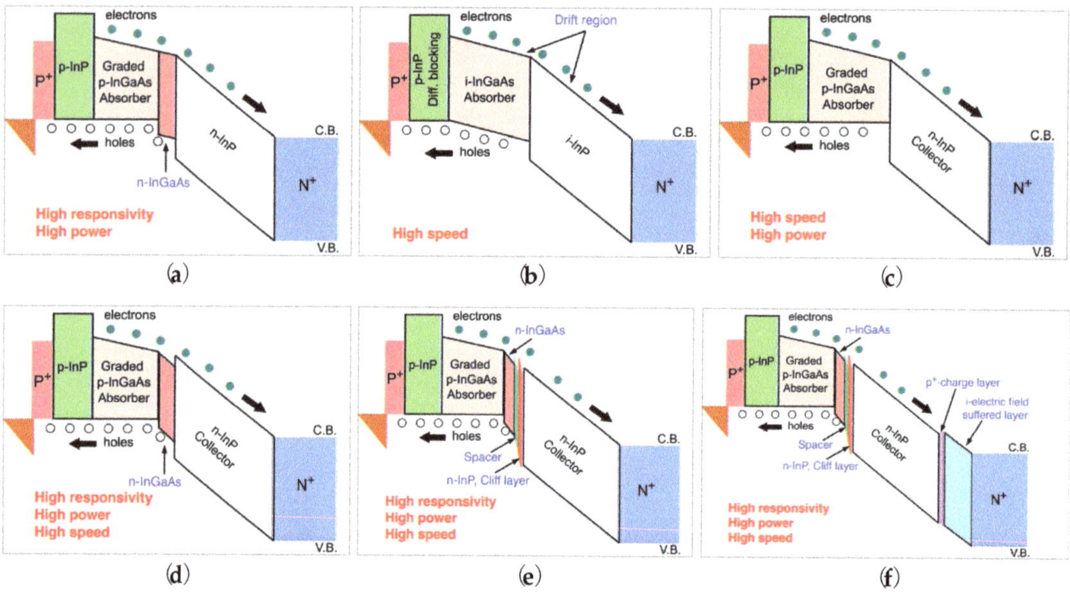

Figure 6. Band diagrams for designing different kinds of PDs to realize various performances, including high responsivity, high speed, and high power [71,72]. (**a**) Partially depleted absorber PD (PDA–PD) band diagram [73,74,81]; (**b**) dual depletion PD band diagram [75,76]; (**c**) UTC–PD band diagram [31,32,44]; (**d**) hybrid absorber structure PD band diagram [47,71,72,77]; (**e**) CC–MUTC–PD band diagram [38,47,78]; and (**f**) near–ballistic (NB) CC–MUTC–PD band diagram [79,80].

In practice for UTC–PD, we need to choose one of them for adequate design (such as device size, bandwidth, absorber and collector layers, doping, electrical field, electron mobility, diffusivity, and diffusion time, etc.) according to the requirements and applications. We take the UTC–PD design as an example and assume three types of p–doping in the absorber with the values of 2.5×10^{17} cm^{-3}, 1×10^{18} cm^{-3}, and 2.5×10^{18} cm^{-3}, corresponding the electron mobility (μ_e) of 6000, 5000 and 4000 cm^2/Vs [44]. Using equations shown in Part 2.2, the electron traveling time in the absorber (τ_a), the electron drift time in the collector (τ_c), and the electron diffusion constant (i.e., electron diffusivity, D_e) can be calculated as given in Figure 7a–c and summarized in Figure 7d. From Figure 7d, we find that the τ_a is much bigger than τ_c and indicates the electron diffusion time in the absorber which determines the total carrier transit time [82], so it only needs to consider τ_a and use Equation (7), i.e., $f_{3dB,tr}$, to evaluate the device's 3–dB bandwidth at MMW and THz frequency bands when the absorber thickness is smaller than 100 μm.

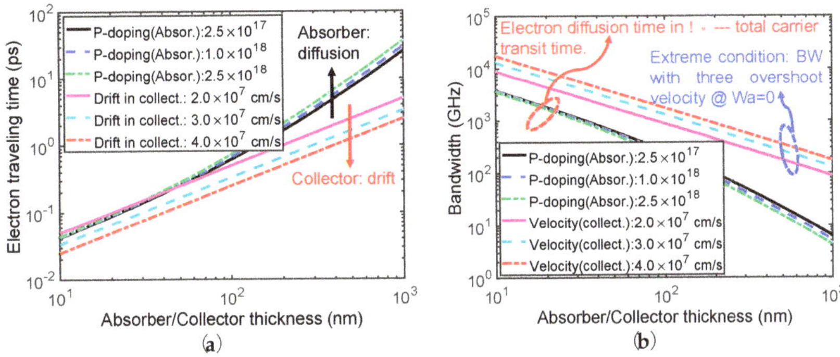

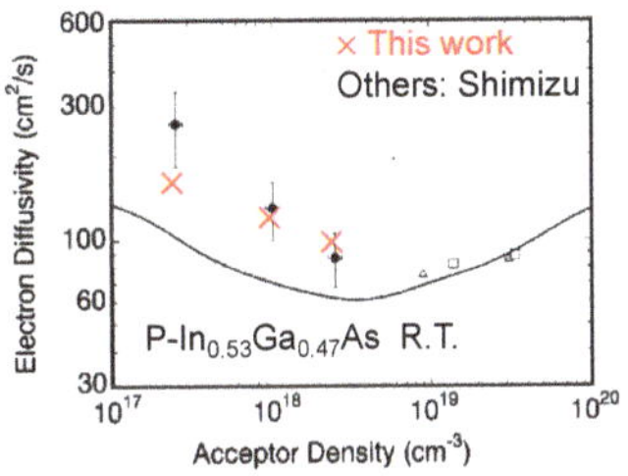

Absorber				Collector
P-doping (cm^{-3})	Electron mobility (cm^2/Vs) μ_e	Electron diffusivity (cm^2/s) D_e	Electron diffusion time τ_a (ps) @ 100 nm absorber	Electron diffusion time τ_c (ps) with $v_{e,os}$ @ 100 nm collector
2.5×10^{17}	6000	155	0.62	0.25~0.5
1.0×10^{18}	5000	129	0.66	0.25~0.5
2.5×10^{18}	4000	104	0.72	0.25~0.5

(d)

Figure 7. The calculated electron traveling time in the absorber and collector on 3–dB bandwidth (BW) [44,82,83]. (**a**) The electron traveling time versus the absorber and collector thickness (assume $W_a \cong W_c$); (**b**) relationship between the 3–dB bandwidth and absorber/collector thickness; (c) comparisons of the calculated electron diffusivity versus p–doping concentration in the absorber [83]; and (**d**) summary of three p–dopings in the absorber with the calculated values.

2.3. Equivalent Circuit Models

The time–delay equivalent circuit model for air–bridge type waveguide PIN–PD was thoroughly analyzed by Wang et al. [84–86], as shown in Figure 8. Fundamentally, a basic air–bridge typed–PD is a current generator, and the corresponding equivalent circuit is composed of four parts, as shown in Figure 9, including the PD itself, the air bridge, the coplanar waveguide (CPW), and the 50 Ω load [86,87]. The photocurrent can be obtained and simulated using a current source, I_{ph}, which is in parallel with the junction capacitance, C_j, and the large junction shunt resistance (megohms, usually $\geq$ 100 kΩ), R_j, and in series with a small series resistance, R_s (or R_c) (related to P^+, N^+ layers and ohmic contact resistance of metallic connection), as well as the air–bridge inductance from p–contact to p–electrode pad, L_b (or L_s), determined by the bridge geometry. The C_{dx} is the capacitance between the air bridge and the n–electrode pad. The characteristic impedance, Z_c, of the CPW, which can be measured with open and short circuit yielding a p–electrode pad capacitance, C_p and an n–electrode pad–induced pad inductance, L_p (or L_G), is calculated by $Z_c = \sqrt{L_p/C_p}$ with a common value of 50 Ω (e.g., $L_p = 55$ pH, $C_p = 22$ fF, so $Z_c = 50$ Ω in [87]). One thing to note is that some PDs do not use air bridge during the device fabrication process, so the parasitic capacitance between the p–electrode and n–electrode, C_{dx} and air–bridge inductance, L_b (or L_s) could not be used in the equivalent circuit model, as shown in Figure 10.

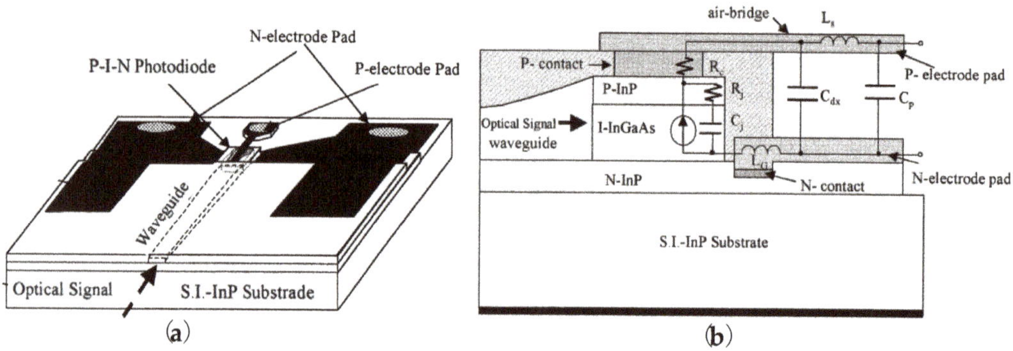

Figure 8. (**a**) Waveguide PIN–PD [84,86]; (**b**) equivalent circuit model [85,86].

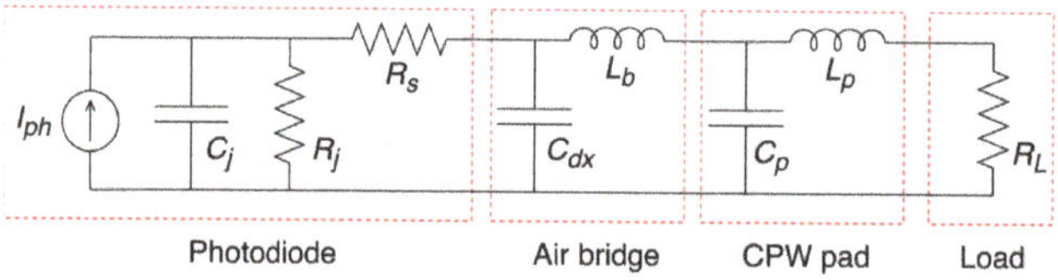

Figure 9. PD equivalent circuit model with air bridge, coplanar waveguide (CPW) pad, and load resistance.

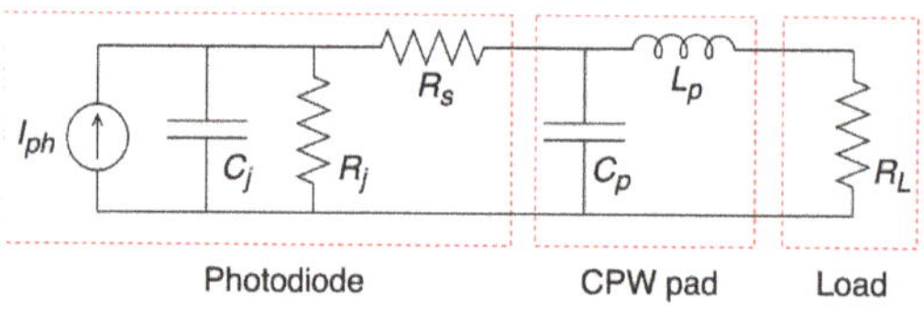

Figure 10. PD equivalent circuit model with coplanar waveguide (CPW) pad and load resistance.

To evaluate the opto–microwave conversion properties of high–speed PDs by using S–parameters (such as reflection coefficients S22 and optoelectronic conversion properties S21), a small–signal radio frequency (RF) equivalent circuit for PD is reported considering synthetically both the analysis of the carrier transit effect [88] and the external parasitic resistance–capacitance (RC) delay–time constant [85]. The small–signal RF equivalent circuit is shown in Figure 11, which uses the voltage (RF voltage, V_{ac})–controlled current source (VCCS, $I_{ac} = g_m V_{ac}$, g_m is denoted as transconductance representing optical–to–RF conversion quantum efficiency, i.e., responsivity) to replace of the current source, I_{ph} in Figures 9 and 10, while the overall frequency response (O–E 3–dB bandwidth) of PD is determined by both of the transit time and the RC–delay time constant, i.e., $1/f_{3dB}^2 = 1/f_{tr}^2 + 1/f_{RC}^2$. The transit–time bandwidth, $f_{tr} = 1/(2\pi\tau_{tr})$ of ac–photocurrent, I_{ac} is determined by the carrier transit–time effect ($\tau_{tr} = R_t C_t$), which is indirectly affected by selecting the values of input circuit elements of C_t, R_t, and g_m. The signal source impedance is R_0. The RC–delay time constant bandwidth can be calculated using Equation (6). In addition, the elements in an external circuit, such as in the air bridge, CPW pad (and P–contact shapes [66]), and load, also have an effect on the overall frequency response of PDs.

The similar and/or improved equivalent circuit models and RF small–signal equivalent circuit models, which are given from some specific examples for both PIN–PD and UTC–PD, with and without air bridge, can be found in previous works [25,85–87,89–119]. Some typical improved RC–delay time models are the modified UTC–PD (MUTC−PD) (Figure 12a) [89] and the dipole–doped UTC–PD (Figure 12b) [50], which are exploited as follows. For the MUTC–PD [89] (Figure 12a) and dual–drifting layer UTC–PD (DDL–UTC–PD) [100], Li et al. focus on the analysis of the absorbing depleted region (resistance, R_u, and capacitance, C_u) and the non–absorbing depleted region (resistance, R_j, and capacitance, C_j). Based on the RF small–signal equivalent circuit model without considering the air–bridge circuit, as shown in Figure 13a, the improved equivalent circuit model of the RC–delay time is shown in Figure 13b, while the further simplified one without considering R_u and C_u is reported by Han et al. [90], as shown in Figure 13c. In the above–mentioned RC–delay time models, the bulk material resistance is R_{11}, while R_{22} represents P– and N–contact resistance, $C_{p\text{–}elect}$ is the P–electrode parasitic capacitance. Based on this model, the experimental and circuit analysis exhibit the bandwidth of an MUTC–PD up to 40 GHz, while the affected factors of the device performance, such as space charge screening, self–induced electrical field and over–shoot effects, are also discussed in detail [89].

In particular, the dipole–doped structure [50], as shown in Figure 12b, has been proposed by Wang et al. to tackle the bandgap discontinuity between the InGaAs absorber and the InP collector interface, and also to simplify the difficulty of epitaxy growth. The corresponding equivalent circuit for the RC–delay time [96] proposed by Meng et al. is shown in Figure 13d, where the resistance and capacitance for the dipole–doped part are denoted as R_{dd} and C_{dd}. R_{aborb} and $R_{collect}$ represent the resistance of the absorber and collector, while the junction capacitance is C_j. Through carefully choosing the values for circuit components, the simulated bandwidth matches the measured value very well from 10 MHz to 20 GHz and can be estimated to be ~62.5 GHz at a −5 V bias voltage.

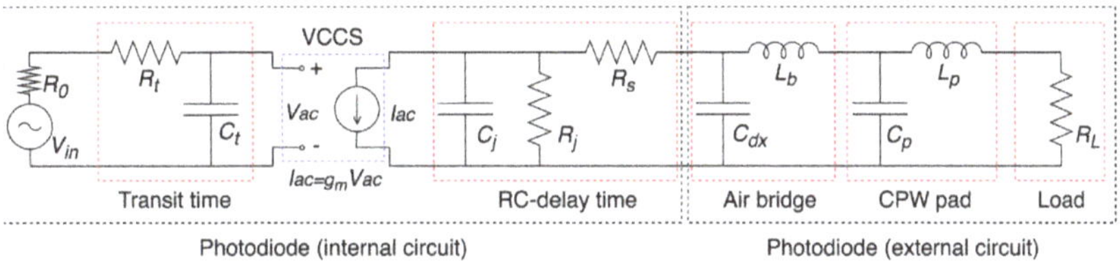

Figure 11. RF small–signal equivalent circuit model of PD, including internal circuit and external circuits which consist of air bridge, coplanar waveguide (CPW) pad, and load resistance.

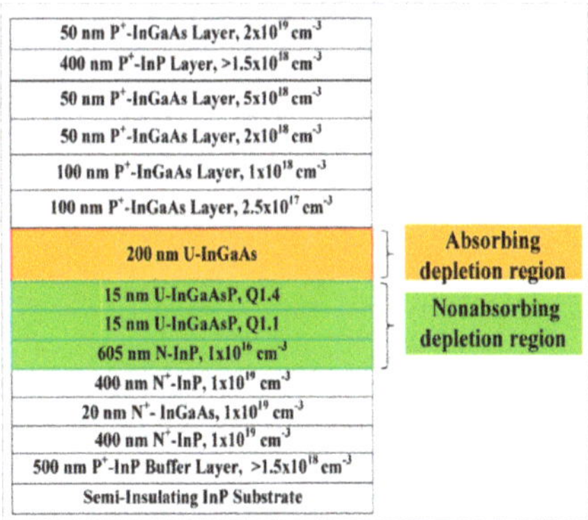

(**a**)

Dipole-doping UTC-PD

Material	Layer	Doping
$In_{0.6}Ga_{0.4}As$	Cap	p-type 1×10^{19} cm^{-3}
InP	Barrier	p-type 1×10^{19} cm^{-3}
$In_{0.53}Ga_{0.47}As$	Absorber	p-type 2×10^{18} cm^{-3}
$In_{0.53}Ga_{0.47}As$		p-type 1×10^{18} cm^{-3}
$In_{0.53}Ga_{0.47}As$		p-type 5×10^{17} cm^{-3}
$In_{0.53}Ga_{0.47}As$		undoped
$In_{0.53}Ga_{0.47}As$	Dipole-doping	p-type 1×10^{18} cm^{-3}
InP		n-type 1×10^{18} cm^{-3}
InP	Collector	n-type 1×10^{17} cm^{-3}
InP		n-type 5×10^{16} cm^{-3}
InP	Sub-collector	n-type 5×10^{18} cm^{-3}
S. I. – InP substrate		

(**b**)

Figure 12. Epitaxial structures of (**a**) modified UTC–PD [89]; and (**b**) dipole–doped UTC–PD [50].

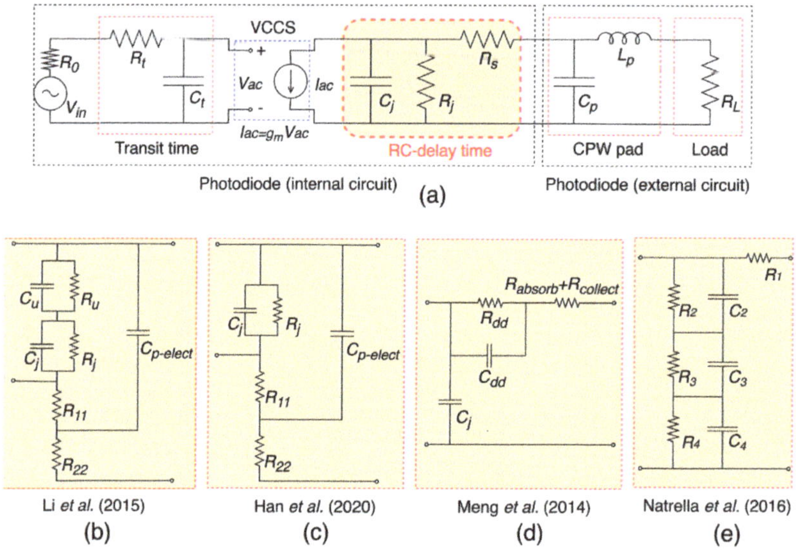

Figure 13. (**a**) RF small–signal equivalent circuit model of PD without air–bridge; (**b**–**e**) RC–delay time models designed by Li [89], Han [107], Meng [96] and Natrella [99,120].

To carefully analyze and depict UTC–PD impedance and photoresponse behavior, an accurate circuit model is proposed [99]. This model not only provides detailed explanations for inconsistent values at low frequencies (here, from DC to 30 GHz, the CPW parasitic effects are negligible) between the classical circuit models of PDs and measurements but even presents a comprehensive study for both the measured device impedance and S11 up to 110 GHz, while showing the values of frequency photoresponse up to 67 GHz. The RC–delay time of this new model is shown in Figure 13e, where two RC parallel circuits (R_2C_2 and R_3C_3) are used to model the two spacers (or transition layers) in UTC–PD, while the R_4C_4 parallel circuit denotes the carrier collector, and the R_1 represents the series resistance with a small value (related to P^+, N^+ layers and ohmic contact resistance of metallic connection/doped materials). In addition, the heavily p–doped neutral thick absorber usually can be ignored due to a negligible resistive effect [99]. It is noted that the parasitic capacitance, C_p, and inductance, L_4 of the CPW pad cannot be ignored if the designed device photoresponse is above 75 GHz. This model provides useful and accurate information to optimize the UTC–PD complex impedance over the frequency range, thus providing a method for power maximization from UTC–PD to an antenna and then radiating to the air, by employing complex conjugate impedance matching between them [99,120]. Based on this new accurate model, some UTC–PD structures are designed and integrated with antennas [22,25,121]. In Figure 14, the matching network is employed to optimize the impedance matching to increase the RF output power, while stray capacitance is introduced in model–2 (without in model–1) to represent the parasitic effect resulting from the flip–chip bonding [25]. Compared with Figure 14a, the accurate equivalent circuit model is used for UTC–PD in Figure 14b, and the extra capacitance is also employed [22].

Moreover, by optimizing the impedance transmission line of the signal pad in CPW, a new transmission line with a characteristic impedance of 96 Ω is introduced forming a new low–inductive peaking CPW [36,37,101]. Subsequently, the charge–compensated MUTC–PD (CC–MUTC–PD) [78] is flip–bonded with this new CPW on aluminum nitride (AlN) (or diamond submounts for thermal dissipation [38,122]) die for further increasing the bandwidth and decreasing the output power roll–off at >120 GHz frequency [121]. Similarly, an inductive 115 Ω characteristic impedance CPW, as shown in Figure 15, is proposed and fabricated to improve the PD bandwidth from 28 GHz to 37.5 GHz [90], while further optimization is applied for MUTC–PD to obtain higher 3–dB bandwidth (and RF output power) from 156 GHz (−0.16 dBm @ 170 GHz [111]) up to 220 GHz (−4.94 dBm @ 230 GHz [113]). Note that some special examples with detailed parameters have been reported in the above–mentioned literatures cited in this paper, so here they are not given.

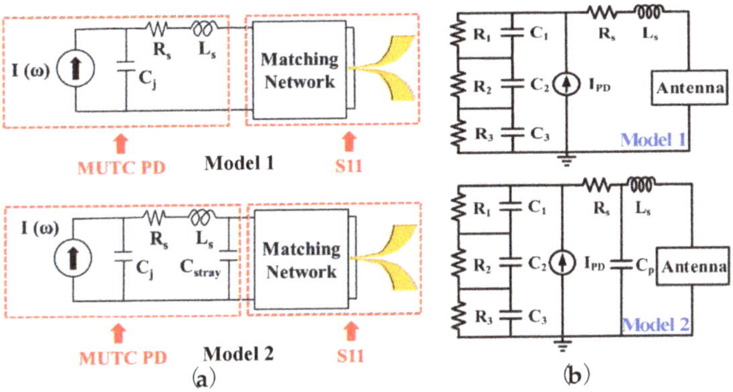

Figure 14. Equivalent circuit models for MUTC–PD integrated with antennas. (**a**) Classical equivalent circuit models of MUTC–PD with Vivaldi antenna and matching network [25]; (**b**) improved equivalent circuit models of CC–MUTC–PD with planer antennas [22].

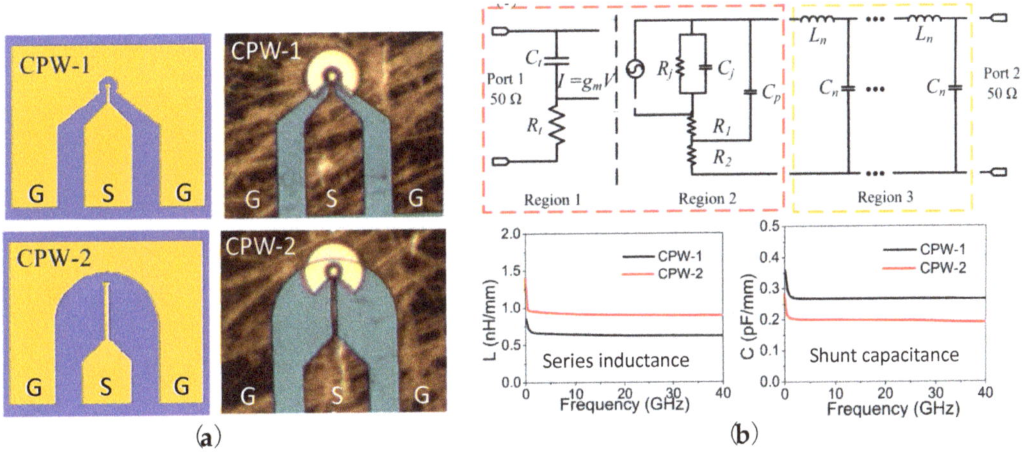

Figure 15. Two CPW schematic diagrams, equivalent circuit models, and simulated results [107]. (**a**) Conventional CPW–1 with 50 Ω impedance and the new CPW–2 with 115 Ω impedance, and the fabricated CPW–1, and CPW2; (**b**) PD equivalent circuit model (Region 1: transit time; Region 2: RC–delay time; Region 3: CPW pad) and the simulated series inductance and shunt capacitance.

2.4. Saturation Current and Output Power

High–power PDs with a broadband frequency response are key components serving as microwave photonic antenna systems [123–125], phased array antennas, MMW imaging and sensing systems [5], and analog fiber–optic links wireless communications [8,126–129]. The high–speed and high–power PD is a promising candidate, such as the UTC–PD which can be monolithically integrated with a matching circuit using a coplanar waveguide short stub to achieve saturation output power, obtaining a value of −13.2 dBm (20.8 mW) at 100 GHz with a −3 V bias voltage [90].

Furthermore, narrow bandwidth high–power also serves as an analog photonic emitter and oscillator systems. The charge–compensated MUTC–PD (CC–MUTC–PD) with a cliff layer [78] is an excellent candidate to achieve high power and high frequency which are realized by the flip–chip techniques on AlN [36,37] or diamond [38] submounts, as shown in Figure 16a, for thermal dissipation. Figure 16b shows that the dissipated power of the device bonded on the diamond submount is faster than that of the devices bonded both on AlN and no bonding. Similarly, the simulated results indicate that the lowest junction temperature is produced from the device bonded on the diamond submount when using the same input power, as shown in Figure 16c.

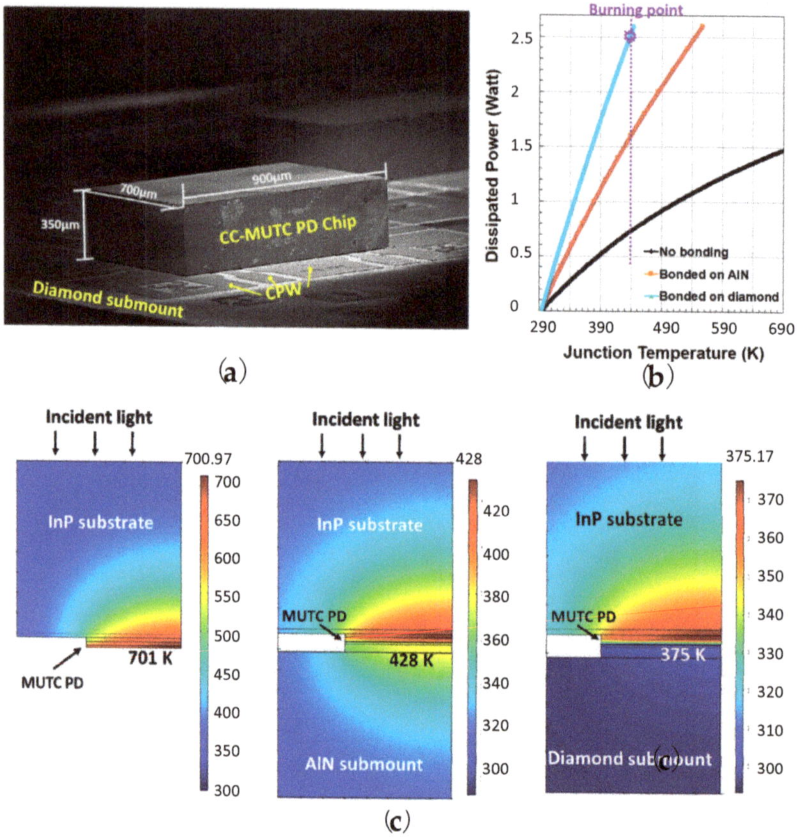

Figure 16. (**a**) Flip–chip bonded CC–MUTC–PD; (**b**) PD junction temperature vs. dissipated power; and (**c**) junction temperature comparisons for PD with and without submounts [38].

Considering to use the well–designed high impedance CPW signal transmission line, bonding on AlN submount, a 9 μm CC–MUTC–PD with 70 GHz bandwidth achieves an RF output power of −2.6 dBm at 160 GHz [121]. To further improve the RF output power, high thermal conductivity substrates, including AlN and diamond, are employed bonding with CC–MUTC–PDs using flip–chip techniques [114], which shows 3–dB bandwidth of 105 GHz, 117 GHz, 129 GHz, and 150 GHz for PDs with diameters of 10 μm, 8 μm, 6 μm and 4 μm at a −3 V bias voltage, achieving an RF output power of −3 dBm at 150 GHz and −5.7 dBm at 165 GHz. Using the same methods for heat dissipation, the record RF output powers for gated modulation at 1 GHz and 10 GHz are separately 41.5 dBm (14.2 V) and 40 dBm (10 W) for 40 μm CC–MUTC–PD at a −36 V bias voltage [122], as shown in Figure 17.

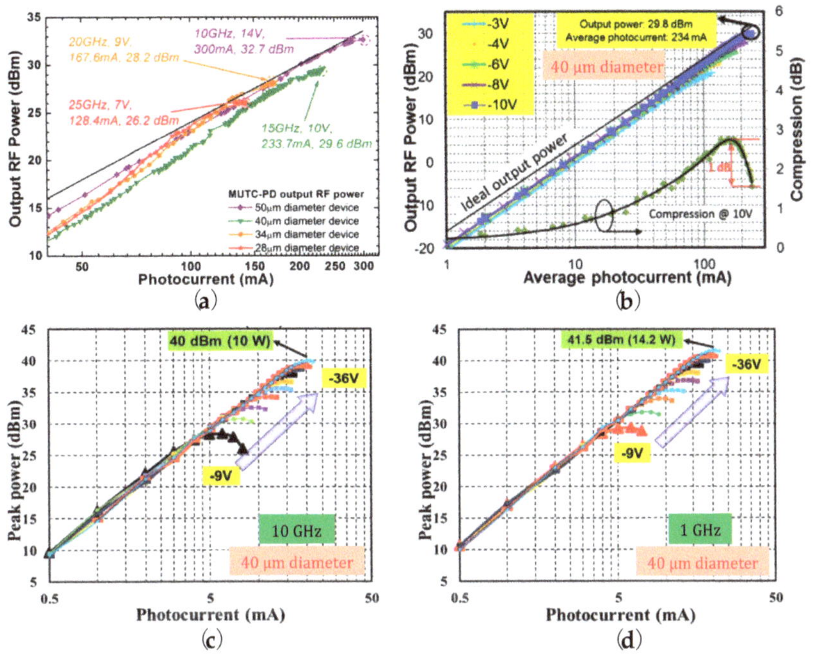

Figure 17. CC–MUTC–PD RF output power and peak power at different photocurrents and bias voltages. (**a**) RF power vs. photocurrent with different device diameters and bias voltages [130]; (**b**) RF power vs. photocurrent with different bias voltages for a 40 μm diameter device [122]; (**c**) peak power at 10 GHz pulse signal vs. photocurrent with different bias voltages [122]; and (**d**) peak power at 1 GHz pulse signal vs. photocurrent with different bias voltages [122].

In brief, the PD saturation current and RF output power is affected by several factors, including device area, PD types, bias voltage, fabrication quality, CPW design, junction heat with high–bias voltages, short microwave stub [131], submounts, measurements and so on, which need to be considered comprehensively. Furthermore, the superior conductivity of SiC substrate also is an excellent candidate for heat dissipation. Using the wafer–bonding technique, InP–based UTC–PD bonded on the SiC substrate [39] can increase the RF output power to a mW–level, i.e., 0.22 dBm (1.05 mW) at 300 GHz with 15.5 mA photocurrent and 2.01 dBm (1.59 mW) with the optimized bias at 18 mA photocurrent [132,133], while the output power is up to 4.04 dBm (2.53 mW) at 273 GHz with the optimized bias voltage at 18 mA photocurrent when this PD integrates with a taper slot antenna [134,135], as shown in Figure 18. The comparisons of 3–dB bandwidth, responsivity, saturation current or 1–dB compression saturation current and RF output power for different PDs [67,94,95,97,101,111,113–115,119,121–124,130,131,134,136–152] are shown in Table 2.

Table 2. The 3–dB bandwidth (BW), saturation current (I_{Sat} or I_{ph}), responsivity (R), bias voltage (V_{bias}) and output power (P_{out}) comparisons for PDs.

PD Type *	BW (GHz) **	R (A/W)	I_{Sat} (mA)	P_{out}(dBm), V_{bias}	Ref.
UTC–PD	310	0.07	NA	11 (12.59 mW) @ 100 GHz, −0.5 V	[67]
NBUTC–PD	120 (64 μm^2, −3.0 V)	0.15	24.6	6.28 (4.25 mW) @ 110 GHz, −5 V	[94]
NBUTC–PD	270 (3.5 μm, −3.0 V)	0.12	17	6.12 (4.093 mW) @ 170 GHz, −2 V	[95,146]
NBUTC–PD	225 (3 μm, −1.5 V)	0.08	13	−1.74 (0.67 mW) @ 260 GHz, −2 V	[97]

Table 2. *Cont.*

PD Type *	BW (GHz) **	R (A/W)	I_{Sat} (mA)	P_{out}(dBm), V_{bias}	Ref.
CC–MUTC–PD	>110 (4 μm)	0.17	27	7.80 (6.03 mW) @ 110 GHz (6 μm), −4 V	[101]
MUTC–PD	156 (4.5 μm, −3 V)	0.165	14.6	−0.53 (0.885 mW) @ 170 GHz, −3.5 V	[111]
MUTC–PD	150 (4.5 μm, −3 V)	0.165	14.6	−0.16 (0.964 mW) @ 170 GHz, −3.5 V	[111]
MUTC–PD	200 (3.6 μm)	0.07	8.4	−2.14 (0.611 mW) @ 200 GHz, −2 V	[113]
MUTC–PD	230 (3 μm)	0.07	5.85	−4.94 (0.321 mW) @ 220 GHz, −2 V	[113]
CC–MUTC–PD	105 (10 μm, −5.0 V)	0.15	>30	14 (25.12 mW) @ 80 GHz (10 μm), −2 V	[114]
CC–MUTC–PD	117 (8 μm, −3.0 V)	0.15	>20	11.3 (13.49 mW) @ 100 GHz (8 μm), −2 V	[114]
CC–MUTC–PD	117 (8 μm, −3.0 V)	0.15	>20	8.50 (7.08 mW) @ 110 GHz (8 μm), −2.5 V	[114]
CC–MUTC–PD	150 (4 μm, −3.0 V)	0.15	8.0	−3.0 (0.5 mW) @ 150 GHz (4 μm), −2.5 V	[114]
CC–MUTC–PD	100 (7 μm, −1.1 V)	0.11	4.9	−19.3 (11.75 μW) @ 300 GHz (7 μm), 0 V	[115]
CC–MUTC–PD	125 (5 μm, −1.1 V)	0.11	25.3	−10.5 (89.13 μW) @ 300 GHz (10 μm), −1.8 V	[115]
WG–MUTC–PD	165 (2 × 15 μm², −1.2 V)	0.312	10.6	-1.69 (0.678 mW) @ 215 GHz, −2 V	[119]
CC–MUTC–PD	70 (9 μm)	0.2	40	-2.60 (0.55 mW) @ 160 GHz, −5 V	[121]
CC–MUTC–PD	13 (40 μm, −6 V)	0.7	234 (−10 V)	41.5 (14.2 W) @ 1 GHz, −36 V	[122]
CC–MUTC–PD	13 (40 μm, −6 V)	0.7	234 (−10 V)	40 (10 W) @ 10 GHz, −36 V	[122]
CC–MUTC–PD	13 (40 μm, −6 V)	0.7	234 (−10 V)	29.8 (0.955 W) @ 15 GHz, −10 V	[122]
UTC–PD	120 (8 μm)	NA	21	0 (1 mW) @ 120 GHz, −3V	[123,124]
CC–MUTC–PD	10 (50 μm)	0.75	300	32.7 (1.86 W) @ 10 GHz, −14 V	[130]
CC–MUTC–PD	11 (50 μm)	0.75	228	30.1 (1.02 W) @ 10 GHz, −11 V	[131]
UTC–PD on SiC	NA	NA	15.5	0.22 (1.05 mW) 300 GHz, −1 V	[134]
UTC–PD on SiC	NA	NA	18	2.01 (1.59 mW) 300 GHz, bias optimized	[134]
PIN–PD	1 (90 μm)	NA	620	26.5 (0.447 W) @ 1 GHz, −5 V	[136]
PIN–PD	> 6 (34 μm)	NA	260	26 (0.398 W) @ 6 GHz, −5 V	[136]
PIN–PD	> 5 (50 μm)	0.55	120	25.8 (0.38 W) @ 5 GHz	[137]
PIN–PD	7 (70 μm)	0.6	200	29 (0.794 W) @ 5 GHz, −5.8 V	[138]
PIN–PD	40 ∼ 50 (10 μm)	~0.7	30	14 (25.12 mW) @ 50 GHz, −2 V	[139]
MUTC–PD array	23.8 (40 μm)	0.438	86.8	16 (39.81 mW) @ 12 GHz, −3 V	[140]
MUTC–PD	34 (20 μm, −4 V)	0.41	~70	17.1 (51.29 mW) @ 30 GHz, −4 V	[141]
MUTC–PD	∼ 45 (15 μm, −4 V)	0.41	38	14.6 (28.84 mW) @ 40 GHz, −4 V	[141]
WG–UTC–PD	> 67 (3 × 10 μm², −3 V)	0.19	13	1.1 (1.29 mW) @ 100 GHz, −3 V	[142]
TTR–UTC–PD	> 110 (3 × 20 μm²)	0.5	16	0 (1 mW) @ 110 GHz, −8 V	[143]
NBUTC–PD	290 (64 μm², −3.0 V)	0.15	13.6	1.85 (1.53 mW) @ 110 GHz, −3 V	[144]
NBUTC–PD	280 (64 μm², −3.0 V)	0.15	18	4.12 (2.58 mW) @ 110 GHz, −3 V	[144]
NBUTC–PD	200 (64 μm², −3.0 V)	0.15	29	8.10 (6.46 mW) @ 110 GHz, −3 V	[144]
NBUTC–PD	180 (64 μm², −3.0 V)	0.15	37	10.7 (11.75 mW) @ 110 GHz, −3 V	[144]
Type–II UTC–PD	140 (3.2 μm, 0 V)	0.09	8	−13.9 (40.74 μW) @ 160 GHz, 0 V	[145]
NBUTC–PD	315 (3.5 μm, −0.5 V)	0.1	13	0.174 (1.04 mW) @ 280 GHz, −1.2 V	[147,148]
RCE–UTC–PD	300 (3 × 3 μm²)	0.12	9.8	-1.25 (0.75 mW) @ 300 GHz, −1 V	[149]
Two UTC–PD	70 per PD	0.17	20	0.79 (1.2 mW) @ 300 GHz, −3.9 V	[150]
Type–II UTC–PD	330 (3 μm, −0.5 V)	0.11	13	-3 (0.5 mW) @ 320 GHz, −1 V	[151]
MUTC–PD	~100	0.22	20	-2.7 (0.54 mW) @ 350 GHz, −2.5 V	[152]

* Waveguide–type PDs are denoted as 'WG', while others are vertical double–mesa PDs. ** PD diameter, area, and bias voltage are shown in brackets.

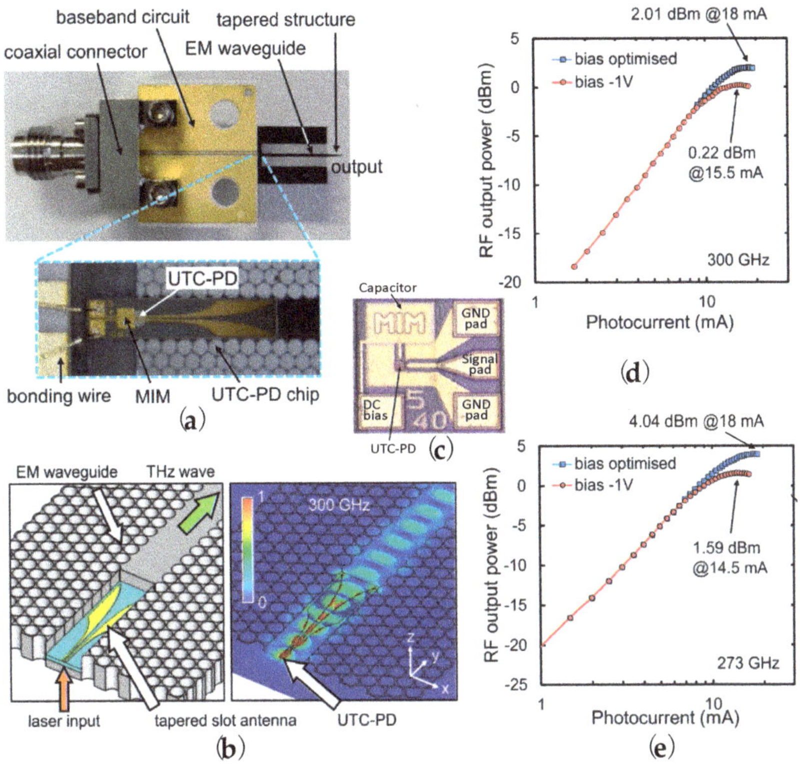

Figure 18. Two mW RF output power UTC–PD on SiC substrate in 300 GHz band [134]. (**a**) UTC–PD chip and the packaged module; (**b**) schematic of UTC–PD and effective–medium (EM) waveguide, while showing the electrical field distribution; (**c**) UTC–PD chip on SiC, where a short–stub matching circuit and a metal–insulator–metal (MIM) capacitor is used to increase the output power; (**d**) UTC–PD chip RF output power and photocurrent at 300 GHz; and (**e**) RF output power and photocurrent at 273 GHz for the packaged module (UTC–PD chip integrated with tapered slot antenna).

3. Solutions of Bandwidth–Responsivity Trade–Off

3.1. High–Reflected Mirrors for Photodiodes

High–performance PD based on III–V compound semiconductors serves as a key component and has already been demonstrated in various applications in wireless communications and radio–over–fiber and antenna systems [5]. The key parameters of PDs, which are responsivity, bandwidth, and saturation power, have been well designed for increasing one of them. For tackling the trade–off between bandwidth and responsivity in PD, several kinds of structures have been reported. The first structure is the resonant cavity enhanced (RCE) structure [33]. By selecting distributed Bragg reflectors (DBRs) as a bottom mirror in the PD, the incident light is reflected to the active layer, to achieve secondary absorption, as shown in Figure 19a. The detailed requirements for DBR design, which is also amenable to the RCE scheme, i.e., many number of different semiconductor material combinations, have been already reported in a review article [153].

Based on the reflection theory of RCE, many kinds of reflected mirror models, as shown in Figure 19b–f, which are used in vertical–mesa and waveguide types of PDs to improve the responsivity without sacrificing the bandwidth, have been proposed in the last 30 years. Most of the fabrication of reflected mirrors are the material combinations of DBRs (Figure 19b) [154–165], while periodic (or non–periodic) strip or concentric circular subwavelength gratings (SWGs), two–dimensional square bulk SWGs

(Figure 19c) [140,161,166–174], Au mirrors [149], and even the scattering structure formed by a periodic cylindrical particle array [175], also can be employed as reflected ones. Those types of mirrors are usually either on the PD bottom (i.e., bottom mirrors) realized by direct epitaxial growth by metal–organic chemical vapor Deposition (MOCVD) or quasi–monolithically integrated using micron–level–thick benzocyclobutene (BCB) bonding (or integrated [69]) techniques, or on the top of the PD (i.e., top mirrors) implemented by employing the metal contacts (or strip gratings) by magnetic sputtering (or electron beam evaporation) or the flip–chip method with chemical etching [158,176–178].

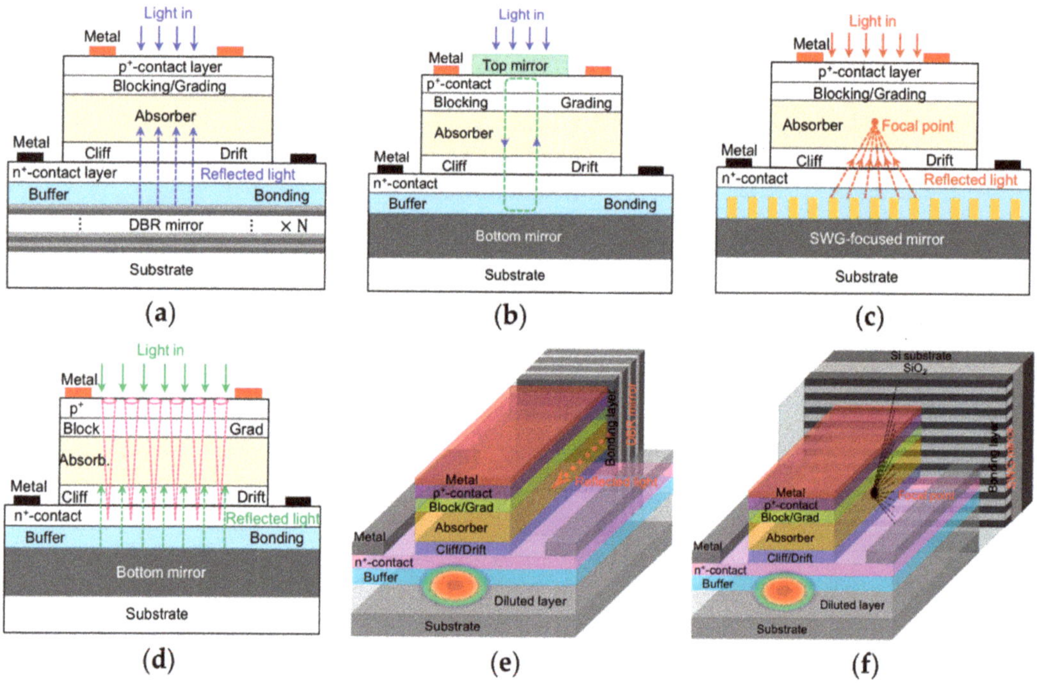

Figure 19. Photodiodes (PIN–PDs or UTC–PDs) are integrated or quasi–integrated with different kinds of bottom mirrors or reflectors. (**a**) Conventional RCE photodiode model, i.e., photodiode with DBR mirror; (**b**) photodiode with dual mirrors, i.e., bottom mirrors (DBR or subwavelength grating (SWG) mirrors) and top mirrors (dielectric layers, or SWG or 'metal' mirrors); (**c**) photodiode with SWG–focused mirror; (**d**) photodiode with tapered holes and bottom mirror (DBR or SWG mirrors); (**e**) waveguide photodiode with DBR mirror; and (**f**) waveguide photodiode with SWG–focused mirror (strip SWG or concentric circular SWG mirrors).

Figure 20 shows PIN–PD and UTC–PDs integrated or quasi–integrated (BCB bonding) with two kinds of bottom mirrors [34,35,140,157,179] It can be seen from Figure 20a that the GaAs–based PIN–PD is integrated with 4 DBRs (three GaAs/AlGaAs DBRs and one Si/SiO_2 DBRs) forming a four–mirror and three–cavity(M4C3) structure operating at 1550 nm [157]. The fabricated M4C3 structure achieves 70% peak quantum efficiency, 36 GHz 3–dB bandwidth, and quite a narrow spectrum linewidth (full–width at half–maximum (FWHM)) of 0.75 nm which is well–suited to high–density wavelength division multiplexing (WDM) communication systems [155]. The HR–UTC–PD [34], as shown in Figure 20b, is quasi–integrated with high–reflectivity (HR) DBR mirrors, which makes the responsivity increase by 23.2% without sacrificing the 3–dB bandwidth. As depicted in Figure 20c, the FR–UTC–PD [35] is quasi–integrated with a focusing reflection (FR) mirror, which is designed with a non–periodic concentric circular high–contrast grating (NP–CC–

HCG) on the silicon substrate. Based on the 84.59% FR–efficiency, the integrated device achieves 36.5% improvement for the responsivity, without altering the 3–dB bandwidth. Figure 20d shows a symmetric–connected UTC–PD array (SC–PDA) integrated with a two–dimensional (2D) SWG beam–splitter [140], where the integrated device obtains an RF output power of 16 dBm at 12 GHz with a 87.9 mA saturation current and 0.438 A/W responsivity, showing stronger high–power handling capability than that of a single photodetector (~11 dBm@12 GHz and 46.7 mA, 0.179 A/W), without complex coupling, phase matching as well as any cooling techniques. Therefore, these kinds of mirrors perfectly realize the decoupling between the responsivity and bandwidth for photodiodes, while the improvement of responsivity and/or bandwidth can be further enhanced if using the well–optimized bonding process in the future.

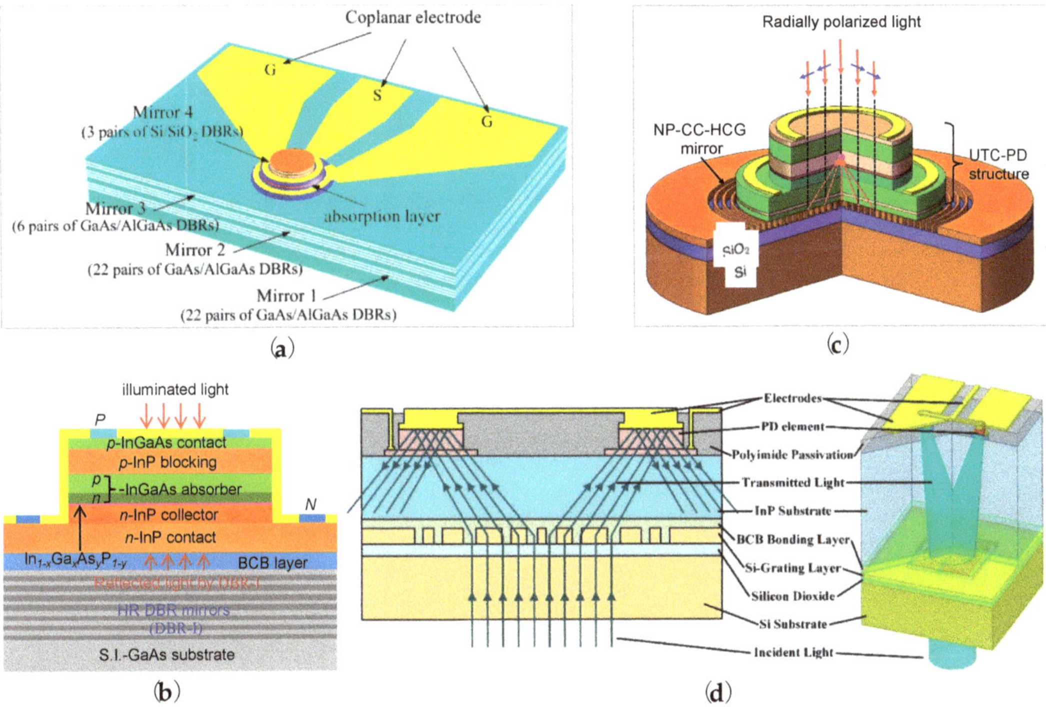

Figure 20. UTC–PDs integrated with different kinds of mirrors. (**a**) PIN–PD integrated with 4 DBRs forming a four–mirror and three–cavity (M4C3) structure [157]; (**b**) UTC–PD integrated with high–reflectivity (HR) DBR mirrors, using HR–UTC–PD for short [34]; (**c**) UTC–PD integrated with non–periodic concentric circular high–contrast grating (NP–CC–HCG, i.e., focusing reflection (FR) mirror), referred to it as FR–UTC–PD [35]; and (**d**) symmetric–connected UTC–PD array integrated with 2D–SWG beam–splitter [140].

3.2. Microhole Design for Photodiodes

Moreover, the application of micro–holes [180–185] etched from the top until through the active regions is also an effective way to assist the light–matter interactions. Figure 21 shows PDs with different kinds of micro– or/and nano–holes in active layers for enhancing light–matter interactions to balance the frequency response and responsivity [180,184]. Figure 21(a1,a2) reports a novel design that uses the etched periodic or photon–trapping micro– and nanoholes (600~1700 nm diameters) with square (Figure 21(a3)) or hexagonal (Figure 21(a4)) shapes that penetrate from the top mesa to the bottom contact layer in the Si photodiode [180]. Such a structure reveals that the lateral propagating slow and stationary optical modes can be generated only using normal incident light, which promotes a well–rounded interaction of light with Si that is an order of magnitude greater compared with the same thick Si–film absorber while ensuring an ultrafast carrier transport. This design enables the Si–based PDs to monolithically integrate with CMOS and extends its applications for optical data links in data communications and computer networks [180].

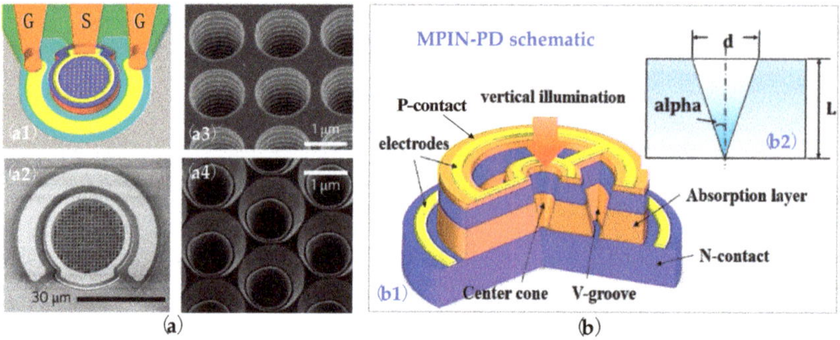

Figure 21. Photodiodes with different kinds of micro–holes for enhancing light–matter interactions. (**a**) Schematic of Si–based photodiode with micro–holes [180]: (**a1**) PD with GSG–pad; (**a2**) scanning electron micrograph (SEM) of PD active region; (**a3**) SEM of square holes; (**a4**) SEM of hexagonal holes. (**b**) Microstructure PIN photodiode (MPIN–PD) [184]: (**b1**) schematic diagram; (**b2**) design of V–groove and center cone.

Given the 1550 nm wavelength application, microstructure PIN–PD (MPIN–PD) is proposed and analyzed numerically by employing an InP substrate, as shown in Figure 21b [184]. The well–designed central obconical shape surrounded by a V–groove trench not only greatly facilitates the interaction of light and InGaAs absorber, but also increases the device responsivity and bandwidth simultaneously. With the optimized dimensions of the V–groove and center cone, together with the optimized circular electrodes, the efficiency and bandwidth of MPIN–PD are separately increased by 61% and 21%. In addition, the back–to–back type of PD [186], waveguide–coupled PD [187], dual–absorber PD [58,161,169] and near–ballistic PD [79] also can be used to achieve high responsivity.

With the evolution of the RCE scheme and tapered–holes structures, a novel model for vertical–mesa PD, as shown in Figure 19d, which is realized based on tapered holes and bottom mirrors, could be used to solve the trade–off between bandwidth and responsivity. For waveguide PDs, the responsivity could be improved by exploiting an SWG high–focused mirror which is made by periodic (or non–periodic) strip or concentric circular SWG bonded at the back of the device (Figure 19f), while the usage of DBR mirrors (Figure 19e) have already been simulated [188] and achieved experimentally [189] showing similar performance for the trade–off.

3.3. High–Reflected Mirrors for Group–IV Photodiodes

However, the PDs based on III–V compound semiconductors are not easy for CMOS compatibility, so the group–IV PDs are gradually paid more attention by researchers, which more easily enables electronic photonic integrated circuits (EPICs) with low–cost, compact, and large–scale integrations [190]. It is to be noticed that the Si–based PD cannot be directly used for the C–band and L–band fiber–optic telecommunication regimes due with the bandgap of 1.12 eV only resulting in a ~1107 nm cutoff wavelength.

Therefore, germanium (Ge) will play an important role for the above–mentioned telecommunication bands owing to a relatively small bandgap (less than 0.7 eV at room temperature), though its optical response rolls off rapidly beyond 1500 nm [191] while this defect can be solved by depositing the Ge–layer on the Ge–on–insulator (GOI) platform [190]. By introducing 0.16% tensile strain to the Ge active layer, the operation wavelength is extended to cover the entire range of telecommunication C– and L–bands (1530–1620 nm). The high refractive index Ge absorber with the low refractive indices of the bottom insulator and the deposited SiO_2 top layer separately serve as two mirrors which finally constitute a RCE–Ge PD on the GOI platform [190] to increase the responsivity, similar to the model shown in Figure 19b. Such RCE–Ge PD on GOI (i.e., GOI–PD) indicates that the responsivity can be greatly increased up to 0.15 A/W at 1550 nm while covering the entire C– and L–bands due to the resonant cavity effect.

Meanwhile, germanium–tin (GeSn) PD on Si is another typical group–IV candidate for SWIR applications which only needs to incorporate Sn into Ge shrinking the direct bandgap and causing redshift of the absorption edge from C–band to the mid–infrared range and beyond [192,193]. GeSn–based (2.5% Sn) PIN–RCE–PD on SOI use the buried oxide (BOX) and the deposited SiO_2 as the bottom and top mirrors forming multiple reflections to enhance light–matter interaction, which shows ~0.4 A/W responsivity at 0 V bias voltage in the C–band, which is much higher than that of the conventional GeSn PIN–PDs with the values of 0.1~0.3 A/W [194]. The introduction of 10% Sn extends the operation wavelength to a 2 μm band, where the metal–semiconductor–metal (MSM) RCE–GeSn–PD can work at 1550 nm and 2 μm wavelengths by optimizing the cavity length. Based on the RCE resonance effects, a record responsivity of 0.43 A/W is obtained with a −3 V bias voltage at 2 μm wavelength at room temperature, revealing that it is promising for CMOS–compatible photonic–integrated circuit applications covering the entire telecommunication bands in a 2 μm range [195].

4. Photodiode Photonic–Integrated Applications

4.1. Photodiodes Integrated with Short–Stubs

With the development of MMW to THz communication techniques, the photonics–based technology provides lower loss in the optical fiber and a large bandwidth while needing the PDs with a high output power and broadband high–frequency characteristics.

Furthermore, we could not ignore the connected compact rectangular waveguide (WR) output port for practical use in a package module [90,152,196–200]. Figure 22 shows two different equivalent circuits and the fabricated micrographs for the photonic MMW generator (PWG), where each PWG includes a UTC–PD, a short–stub matching circuit (impedance transformer), and a metal–insulator–metal (MIM) capacitor, C_{MIM} (i.e., C in Figure 22d,e) [196,199]. The calculated relative output power curves are given for comparisons as shown in Figure 22c, in which the higher power is obtained by PMG–A at ~100 GHz (W–band). Meanwhile, the PMG works at the J–band as shown in Figure 22f, where a WR–3 waveguide output port is used for practical use. It is noticed that a typical case is a coupler integrated with UTC–PD packaged with WR–1.5, which attains a record output power of −19 dBm at 500 GHz with a 9 mA photocurrent while showing a wide range 3–dB bandwidth of 340 GHz, i.e., from 470 GHz to 810 GHz [201]. The detailed results of PDs integrated with WR ports and short stubs [90,197,198,200,201] are shown in Table 3.

Table 3. The comparisons of PDs integrated with WR ports and short stubs.

Device Type	Coupling Type	Si Lens	I_{ph} (mA)	Output Power (dBm)	Ref.
UTC–PD	WR10 port (W–band)	No	25	13.2 (20.8 mW) @ 100 GHz	[90]
UTC–PD	WR3 port (J–band)	No	26	−8.73 (134 μW) @ 264 GHz	[190]
UTC–PD	WR8 port (F–band)	No	25	12.3 (17 mW) @ 120 GHz	[197,198]
UTC–PD	WR6 port (D–band)	No	7	> 10 (> 10 mW) @ 104~185 GHz	[200]
Coupler integrated UTC–PD	WR1.5 port	No	9	−18.96 (12.7 μW) @ 500 GHz	[201]
Coupler integrated UTC–PD	WR1.5 port	No	9	> −22 (> 6.31 μW) @ 470~810 GHz	[201]
Coupler integrated UTC–PD	WR1.5 port	No	6	−22 (6.31 μW) @ 500 GHz	[201]
Coupler integrated UTC–PD	WR1.5 port	No	6	> −25 (> 3.16 μW) @ 470~810 GHz	[201]

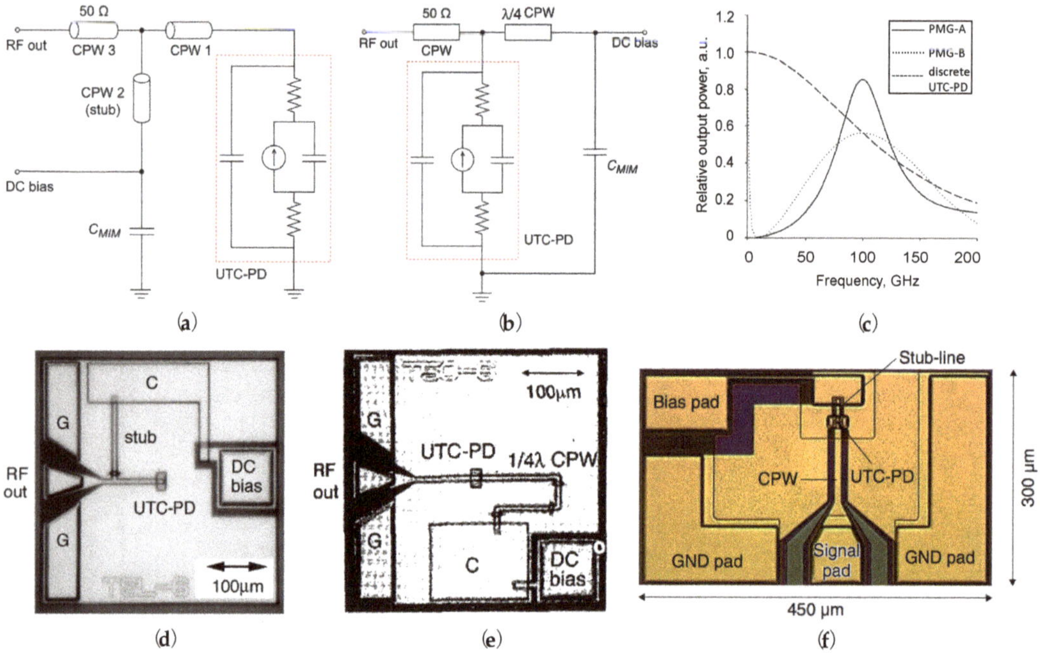

Figure 22. Photonic MMW generator (PMG) equivalent circuits and fabricated micrographs [196]. (**a**,**d**) PMG–A works at W–band; (**b**,**e**) PMG–B works at W–band; (**c**) relative output power comparisons for PMG–A, PMG–B, and UTC–PD without PMG; and (**f**) PMG works at J–band [168].

4.2. Photodiodes Integrated with Antennas

Another photonic–integrated device is PD with antennas for increasing output power for practical use. The general schematic diagram of the photonics–based transmitter and receiver is shown in Figure 23. One of the unique performances of UTC–PDs [44,54] is both a large bandwidth and high saturation output power, which can be used as an MMW signal emitter [123] for wireless applications.

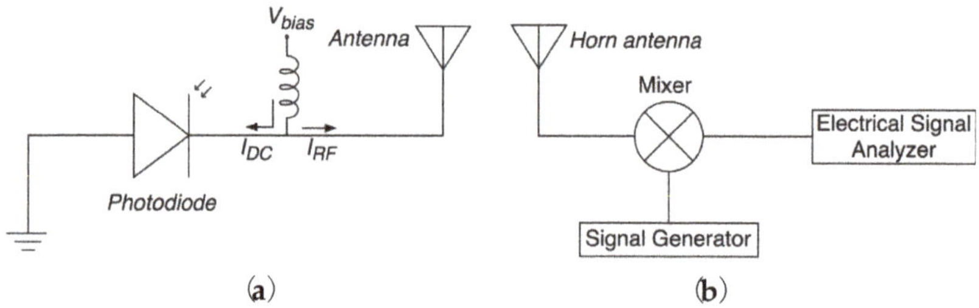

Figure 23. General schematic diagram of (**a**) PD photonic–integrated emitter and (**b**) receiver.

To obtain V–band–to–G–band (50 GHz~220 GHz) photonically driven emitters and extend applications to MMW and THz signal wireless links, high–power photonic–integrated emitters with three types of planar antennas have been proposed [22,202], as shown in Figure 24. The high–speed and high–power broadband CC–MUTC–PD [78,101] with 0.2 A/W responsivity and 12 dBm output power is used for the MMW and THz emitter design. The fabricated emitter with the sinuous antenna and 8 μm PD shows a record output power of 20 dBm at 90 GHz with a photocurrent of 20 mA while achieving an output power of

9.9 dBm at 70 GHz, 11.5 dBm at 100 GHz, all with 7 mA photocurrent, for log–periodic and spiral antennas, respectively [22].

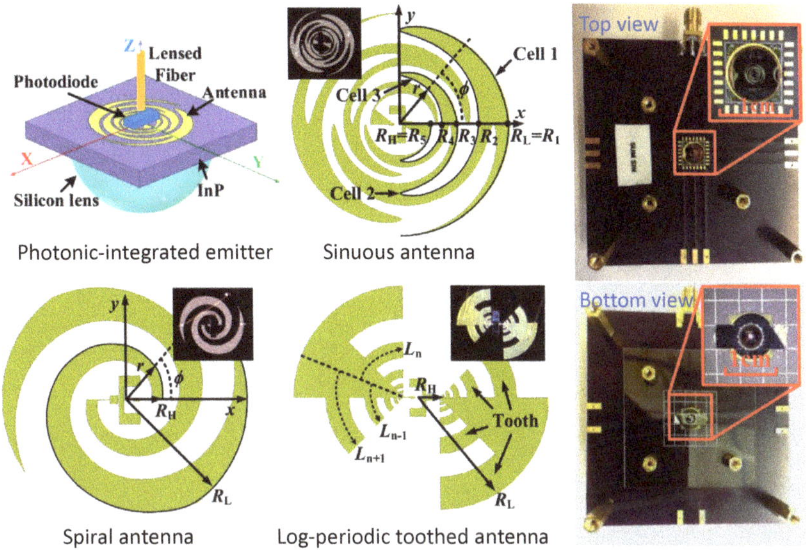

Figure 24. Schematic diagram of CC–UTC–PD photonic–integrated emitter with Si–lens, the three different planar antennas, and the top/bottom views of the fabricated emitter [22].

Using flip–chip technology, UTC–PD can be mounted on a planar slot antenna on the Si substrate [123,124,203], as shown in Figure 25a, and then boned on a hemispherical Si–lens to collimate the MMW signal direction, which generates a larger than 0.5 mW output power from the PD chip while radiating a bigger than 0.3 mW power into the free space, at a 120 GHz bandwidth [123,124]. Meanwhile, this design indicates the potential applications in future wireless communication, sensing, radar, measurement, and imaging systems based on photonic–assisted MMW generation and transmission techniques. Beyond the F–band, the electrical transmission line will bring loss and reflection, so the miniaturized PD–integrated antenna with a quasi–optical configuration (Figure 25b) is necessary for sub–MMW signal handling [204]. As shown in Figure 25c, a UTC–PD integrated with a log–periodic antenna achieves a saturation output power of 300 µW at a photocurrent of 20 mA (Figure 25d) [205]. Using a similar configuration, the UTC–PD integrated with a log–periodic antenna exhibits a high output power of 2.6 µW at 1.04 THz with a saturation current of 13 mA [204].

However, a single PD integrated with an antenna suffers some limitations, including the limited emitted power resulting from the saturation of PD and the limited antenna directivity due to the finite effective aperture of the planar antenna [206]. Here, four InP–based photodiodes operating at 300 GHz, which are integrated with a linear bowtie antenna array mounting on the Si–lens, are proposed, as shown in Figure 26, to increase the radiation power. Compared to the straight radiation of a single emitter, the array emitter realizes a 10.6 dB higher power (theoretical gain: 12 dB), while the 3–dB beamwidth is reduced by 8.5° and up to 22.5°, which achieves the output power increase and higher directivity. These photonic–integrated antenna arrays provide new solutions to wireless communications beyond 5G and will further exploit combinations of other photonic devices for application to future complex MMW and THz transceivers.

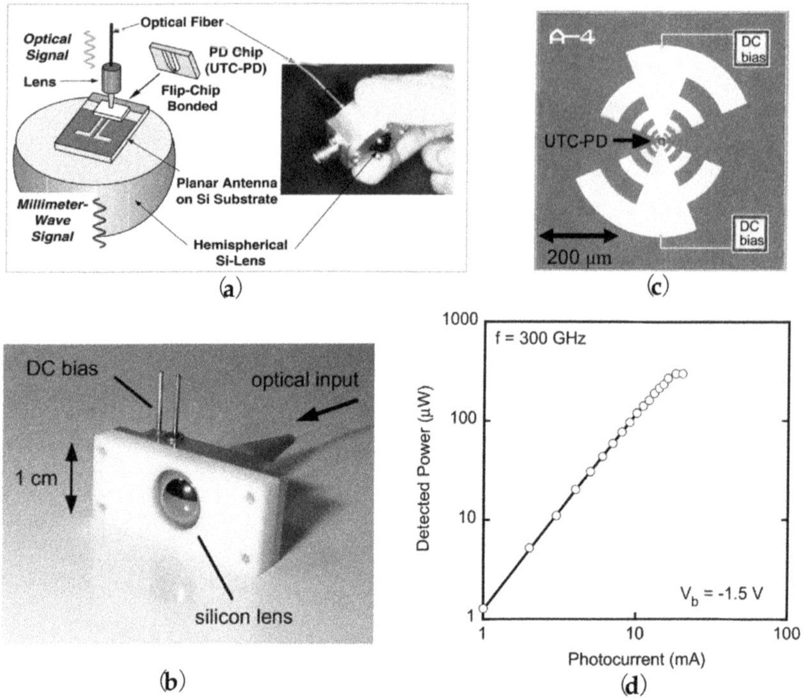

Figure 25. (**a**) UTC–PD integrated with slot antennas used as an MMW signal emitter for free–space applications [123,124]. (**b**) The fabricated quasi–optical UTC–PD with Si–lens module for practical use [204]. (**c**,**d**) UTC–PD integrated with log–periodic antenna (micrograph as shown in the inset) attains 300 μW output power and 20 mA photocurrent at a 1.5 V bias voltage [205].

In addition, a resonant antenna integrated with UTC–PD is another excellent candidate to be employed for output power radiation in the THz bands [207–209]. Figure 27a shows the THz UTC–PD chip integrated with a twin–diploe antenna [209] using an impedance matching circuit [90] to enlarge the output power and the resonant behaviors are found exhibiting the peak output powers of 5.9, 3.8, and 1.1 μW at 0.78, 1.04 and 1.53 THz (Figure 27b), while generating a record output power of 10.9 μW at 1 THz with a photocurrent of 14 mA [209].

Moreover, the THz wave and THz beam steering provide great opportunities for free–space applications, including imaging, sensing, radar, and wireless communications in the 6G era and beyond [15,135,210]. Figure 27c shows a 4 × 1 UTC–PD array integrated with a 4×4 unidirectional slot antenna to generate THz signals based on coupled fiber beats techniques [211]. The adaptive genetic algorithm is used to control the produced THz beam steering and the experiment achieves a precise pointing towards the intended direction of the 300 GHz beam, which greatly improves the THz power efficiency by beam steering and paves the way for THz wireless communication using large–scale UTC–PD array [211–213].

With the development of communication techniques, the frequency range (MMW to THz: 30 GHz~3 THz) between microwave and far–infrared is set to meet the growing demand. Exploiting and making the most of the MMW and THz wireless signals [7,214,215] with high–radiated power have long been studied and generated by the electronic system [26–28] or photonic system [6,8,9,15–23] or hybrid electronic–photonic systems [216,217], which have shown a very diverse range of application in information and communications, security screening, food and agriculture, biometrics, medical and medicine, drug inspection, semiconductor wafer inspection, and air pollution [218–220].

Given the photonic–generated THz signal, ultrafast broadband, and high–power PDs, such as UTC–PDs [44,54] and traveling–wave PD (TW–UTC–PD) [221], are necessary. By combining the TW–PD structure for high responsivity and the UTC–PD structure for high output power, a new TW–UTC–PD is proposed for integration with antennas for MMW and THz signal radiation [21,222]. The TW–UTC–PD separately integrates with the resonant, bowtie, and log–periodic antennas for generating THz signals, covering frequencies up to 1 THz. The radiated power at different photocurrent levels is shown in Figure 28, where the $4 \times 15\ \mu m^2$ PD integrated with a resonant antenna achieves an RF output power of 148 μW at 457 GHz and 24 μW at 914 GHz with a 10 mA photocurrent (Figure 28a). Figure 28b shows that the RF output power of 105 μW at 255 GHz and 10 μW at 612 GHz with a 13 mA photocurrent are attained by a $2.5 \times 50 \mu m^2$ PD integrated with a bowtie antenna while exhibiting 70 μW at 150 GHz with a 9 mA photocurrent for integrating with a log–periodic antenna.

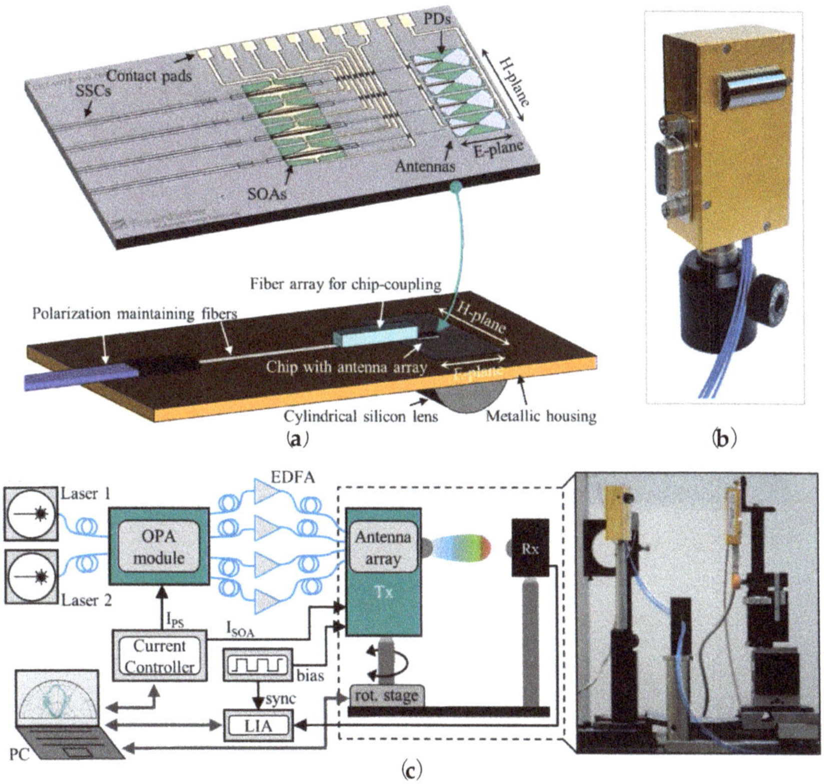

Figure 26. InP–based PIN–PD integrated with four bowtie antennas and mounted on Si–lens to increase emitted power and high directivity [206]. (**a**) The fabricated InP chip and the schematic for packaging of the antenna array; (**b**) the emitter chip module; and (**c**) measurement setup.

Another type of PD for THz applications is related to the design, usage, and optimization of the hybrid absorber UTC–PD (i.e., modified UTC–PD, MUTC–PD) [72], as shown in Figure 29. The band diagram for this type of MUTC–PD is shown in Figure 29a, where the absorber consists of a non–depleted absorber with p–doping and a depleted absorber without doping. The maximum 3–dB bandwidth can be calculated while just comprehensively analyzing the numerical relation of thickness among depletion (W_{dep}), non–depleted (W_{an}), and depleted (W_{ad}) absorbers. Figure 29d shows the frequency response at different quasi–fields in the non–depleted absorber, while the W_{an}, W_{ad} and W_{dep} are separately 0.12 μm, 0.06 μm, and 0.18 μm. The 3–dB bandwidth is enlarged from 603 to 1087 GHz with the increase in quasi–field from 10 to 30 kV/cm.

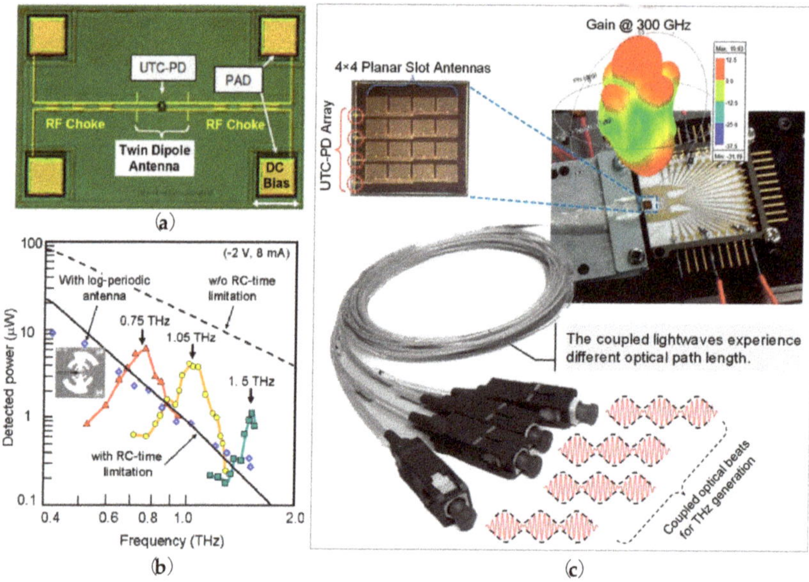

Figure 27. (**a**) THz UTC–PD chip integrated with twin–diploe antenna using impedance matching circuit and (**b**) the output power versus different frequency [209]. (**c**) Adaptive THz beam steering in fiber–coupled UTC–PD array chip to generate THz wave [211–213].

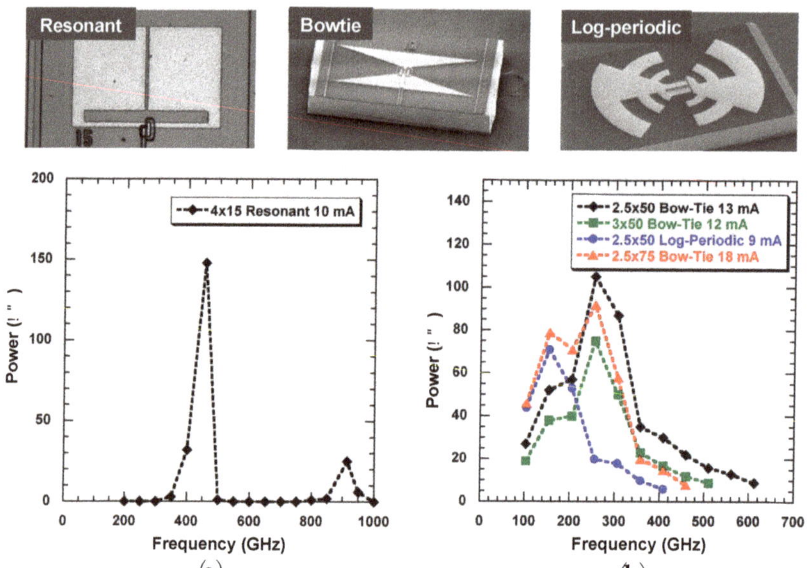

Figure 28. Traveling–wave UTC–PD (TW–UTC–PD) integrated with resonant, bowtie, and log–periodic antennas, and the radiated power at different photocurrent levels for resonant antenna (**a**), and for bowtie and log–periodic antennas (**b**), where the device dimensions are in μm^2 [21,222].

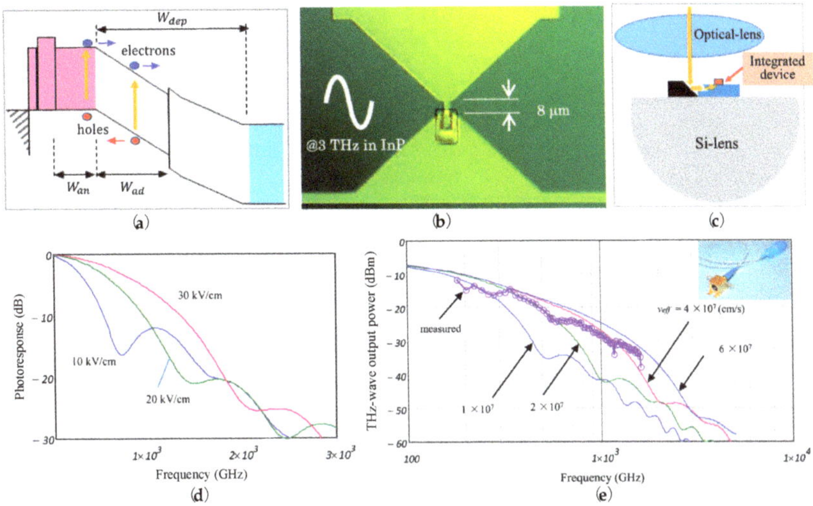

Figure 29. Hybrid absorber UTC–PD (modified UTC–PD, MUTC–PD) integrated with bowtie antenna [72]. (**a**) Band diagram of MUTC–PD; (**b**) MUTC–PD integrated with bowtie antenna; (**c**) incident light coupling to MUTC–PD by a refracting–facet plane and 45° mirror; (**d**) frequency response of MUTC–PD with three quasi–field values; (**e**) comparisons of the measured THz power (open circles) and calculated values of the integrated device using different electron effective velocity, v_{eff}. Note: The inset is the packaged device module (i.e., photomixer) with Si–lens and fiber pigtail.

The MUTC–PD is integrated with a bowtie antenna shown in Figure 29b, where Si–lens (Figure 29c) is employed to increase the output power. Being accompanied by a well–designed matching circuit and the rectangular waveguide output port, the quasi–optical antenna–integrated photomixer module can be completed as shown in the inset of Figure 29e. It can be seen that the measured output power gradually decreases with the increase in the frequency while the typical values are larger than −20 dBm at 300 GHz, −30 dBm at 1 THz, and approximately −40 dBm at 2 THz. Furthermore, the calculated results which changed with the different effective electron velocities are also shown in Figure 20e, as comparisons with the measured values. Parts of PDs photonic–integrated antennas [18,21,22,25,72,91,98,123,124,134,142,204–206,208,209,223–227] are summarized in Table 4.

Table 4. The comparisons of emitted power for PDs integrated with antennas.

Device Type *	Coupling Type	Si Lens	I_{ph} (mA)	Output Power (dBm)	Ref.
NBUTC–PD	Yagi antenna	No	30	–14.1 (25.7 mW) @ 100 GHz	[18,223]
TW–UTC–PD	Log–periodic antenna	No	9	−11.55 (70 μW) @ 150 GHz	[21]
TW–UTC–PD	Bowtie antenna	No	13	−9.79 (105 μW) @ 255 GHz	[21]
TW–UTC–PD	Resonant antenna	No	10	−8.3 (148 μW) @ 457 GHz	[21]
TW–UTC–PD	Bowtie antenna	No	13	−20 (10 μW) @ 612 GHz	[21]
TW–UTC–PD	Resonant antenna	No	10	−16.2 (24 μW) @ 914 GHz	[21]
CC–MUTC–PD	Log–periodic antenna	Yes	7	9.9 (9.77 mW) @ 70 GHz	[22]
CC–MUTC–PD	Sinuous antenna	Yes	7	11.5 (14.13 mW) @ 90 GHz	[22]
CC–MUTC–PD	Sinuous antenna	Yes	20	20 (100 mW) @ 90 GHz	[22]
CC–MUTC–PD	Spiral antenna	Yes	7	8.8 (7.59 mW) @ 100 GHz	[22]
MUTC–PD	Vivaldi antennas	No	10	5 (3.2 mW) @ 110 GHz	[25]
Hybrid absorber UTC–PD	Bowtie antenna	Yes	6	−20 (10 μW) @ 300 GHz	[72]

Table 4. Cont.

Device Type *	Coupling Type	Si Lens	I_{ph} (mA)	Output Power (dBm)	Ref.
Hybrid absorber UTC–PD	Bowtie antenna	Yes	6	−30 (1 μW) @ 1 THz	[72]
Hybrid absorber UTC–PD	Bowtie antenna	Yes	6	−35.2 (0.3 μW) @ 1.5 THz	[72]
Hybrid absorber UTC–PD	Bowtie antenna	Yes	6	−40 (0.1 μW) @ 2 THz	[72]
Hybrid absorber UTC–PD	Bowtie antenna	Yes	6	−44 (0.04 μW) @ 2.5 THz	[72]
MUTC–PD	Patch antenna	No	45	20 (100 mW) @ 60 GHz	[98]
TWPD	Bowtie antenna	Yes	6.3	−33 (0.5 μW) @ 40 GHz	[91]
UTC–PD	Slot antenna	Yes	>20	-5.23 (0.3 mW) @ 120 GHz	[123,124]
UTC–PD on SiC	Tapered slot antenna	No	14.5	1.59 (1.44 mW) @ 273 GHz, −1 V	[134]
UTC–PD on SiC	Tapered slot antenna	No	18	4.04 (2.53 mW) @ 273 GHz, bias optimized	[134]
WG–UTC–PD	Bowtie antenna	Yes	13.5	−12.22 (60 μW) @ 250 GHz	[142]
UTC–PD	Log–periodic antenna	Yes	13	−25.9 (2.6 μW) @ 1.04 THz	[204]
UTC–PD	Log–periodic antenna	Yes	20	−5.23 (300 μW) @ 300 GHz	[205]
PIN–PD array	Bowtie antenna	Yes	NA	>10 dB vs. an emitter @ 300 GHz	[206]
UTC–PD	Log–periodic antenna	Yes	15	−11.2 (75 μW) @ 414 GHz	[208]
UTC–PD	Log–periodic antenna	Yes	15	−17.4 (18 μW) @ 609 GHz	[208]
UTC–PD	Log–periodic antenna	Yes	15	−20.8 (8.3 μW) @ 804 GHz	[208]
UTC–PD	Twin dipole antenna	Yes	14	−19.6 (10.9 μW) @ 1.04 THz	[209]
UTC–PD	Log–periodic antenna	Yes	20	−9.2 (120 μW) @ 300 GHz	[224]
Type–II UTC–PD	Dual–ridge horn antenna	No	NA	−15 (31.6 μW) @ 240 GHz	[225]
UTC–PD	TEM–horn antenna	Yes	2.75	−29.5 (1.13 μW) @ 940 GHz	[226]
UTC–PD	TEM–horn antenna	Yes	2.5	−33.4 (0.46 μW) @ 1.04 THz	[226]
UTC–PD	TEM–horn antenna	Yes	2.88	−35.1 (0.31 μW) @ 1.365 THz	[226]
UTC–PD	slot antenna	Yes	10	−15.53 (28 μW) @ 700 GHz	[227]
UTC–PD	slot antenna	Yes	10	−24.56 (3.5 μW) @ 1.25 THz	[227]

* Waveguide– and traveling–wave–type PDs are denoted as 'WGPD' and TWPD, while others are vertical–mesa PDs.

5. Discussion

During the design of high–speed and high–output power photodiodes, the compatible photodiode–integrated devices, the measurement setups, and applications for communication systems, there are lots of considerations and challenges that need to be noticed. They are mainly reflected as follows: (1) The efficient optical coupling and precise alignment between the photodiode and other optical components. (2) The wider bandwidth with high response for photodiodes. (3) The wider dynamic range and linearity for higher output power without small saturation or signal distortion. (4) The effects resulting from temperature and heat from high–power conditions. (5) The reliability and robustness of photodiodes, and the scalability and modularity of photodiode–integrated devices. (6) The potential limitations at the 300 GHz band, such as the coverage range with high propagation loss, the penetrating obstacles from the walls, buildings, and vegetation, and the complexity and cost of exploiting and deploying photodiodes to the MMW and THz bands. (7) The design of corresponding other components matching with photodiodes in THz links and systems.

Though achieving high–speed, high–power photodiodes and photodiode–integrated devices with over 2 mW output power in the THz band has been realized, it is still a big challenge to apply photodiodes to 6G technology. Compared with existing communication systems, 6G requires ultra–high data rates (such as 200~1000 Gbit/s) and also requires the matching terahertz sources, photodiodes, and components in THz links and systems. However, considering the wider applications of photodiodes in transparent materials, including cell phone screens, smart windows, and automobile/aircraft win screens, is also a big challenge. A possible solution could be to implement the integration of photodiodes with waveguide–based Bragg grating devices in glass inscribed using femtosecond laser in the future.

6. Conclusions

This paper reviews the advances in high–speed, high–power photodiodes from fundamentals to applications. (1) For the overall design and consideration of photodiodes, this paper discusses the principles of PIN–PD and UTC–PD, while reporting the detailed calculation of the 3–dB bandwidth for the two typical photodiodes and summarizes the energy band diagrams for several kinds of photodiodes. Given the equivalent circuits, this paper shows the general ones and RF small–signal ones together, and also considers the different designs for the RC–delay time part and the external circuit parts, including the air bridge, CPW pad, and load. Additionally, the DC saturation and RF output power are also analyzed in detail and a table for comparison is also included. (2) The solutions of bandwidth–responsivity trade–off for photodiodes are discussed in detail, which range from the use of high–reflected mirrors to the making of microholes and extending the reflected mirrors for group–IV photodiodes. (3) The photodiode photonic–integrated applications are presented, including the integration with short stubs and various planar antennas for high output power at high–speed conditions, and showing the parameter comparisons in tables, which provides a reference for the practice use of photonics–based devices operating at the system–level with the radiation power up to the mW–level and beyond in the THz band.

Author Contributions: Conceptualization, Q.C.; methodology, software, Q.C.; validation, Q.C.; formal analysis, Q.C.; investigation, Q.C.; resources, Q.C.; data curation, Q.C.; writing—original draft preparation, Q.C.; writing—review and editing, Q.C. and X.Z.; visualization, Q.C.; supervision, X.Z., M.S.S. and R.K.; project administration, M.S.S. and R.K.; funding acquisition, M.S.S. and R.K. All authors have read and agreed to the published version of the manuscript.

Funding: This work was funded by the Natural Sciences and Engineering Research Council of Canada (NSERC), through the New Frontiers in Research Funding Fund (NFRF) with project number NFRFE–2020–00355.

Institutional Review Board Statement: Not applicable.

Informed Consent Statement: Not applicable.

Data Availability Statement: Not applicable.

Conflicts of Interest: Author Mohammad S. Sharawi was employed by the company Blue Origin LLC. The remaining authors declare that the research was conducted in the absence of any commercial or financial relationships that could be construed as a potential conflict of interest.

References

1. Giordani, M.; Polese, M.; Mezzavilla, M.; Rangan, S.; Zorzi, M. Toward 6G Networks: Use Cases and Technologies. *IEEE Commun. Mag.* **2020**, *58*, 55–61. [CrossRef]
2. Rajatheva, N.; Atzeni, I.; Bjornson, E.; Bourdoux, A.; Buzzi, S.; Dore, J.-B.; Erkucuk, S.; Fuentes, M.; Guan, K.; Hu, Y.; et al. White Paper on Broadband Connectivity in 6G. *arXiv* **2020**, arXiv:2004.14247.
3. Xiaohu, Y.; Cheng-xiang, W.; Jie, H.; Xiqi, G.; Michael, W.; Yongming, H.; Chuan, Z.; Yanxiang, J.; Min, Z.; Dongming, W.; et al. Towards 6G Wireless Communication Networks: Vision, Enabling Technologies, and New Paradigm Shifts. *Sci. China Inf. Sci.* **2020**, *64*, 110301.
4. Koenig, S.; Lopez-Diaz, D.; Antes, J.; Boes, F.; Henneberger, R.; Leuther, A.; Tessmann, A.; Schmogrow, R.; Hillerkuss, D.; Palmer, R.; et al. Wireless Sub-THz Communication System with High Data Rate. *Nat. Photonics* **2013**, *7*, 977–981. [CrossRef]
5. Nagatsuma, T.; Ito, H.; Ishibashi, T. High-Power RF Photodiodes and Their Applications. *Laser Photonics Rev.* **2009**, *3*, 123–137. [CrossRef]
6. Renaud, C.C.; Natrella, M.; Graham, C.; Seddon, J.; Van Dijk, F.; Seeds, A.J. Antenna Integrated THz Uni-Traveling Carrier Photodiodes. *IEEE J. Sel. Top. Quantum Electron.* **2018**, *24*, 8500111. [CrossRef]
7. Wakatsuki, A.; Muramoto, Y.; Ishibashi, T. Development of Terahertz-Wave Photomixer Module Using a Uni-Traveling-Carrier Photodiode. *NTT Tech. Rev.* **2012**, *10*, 1–7.
8. Seeds, A.J.; Shams, H.; Fice, M.J.; Renaud, C.C. TeraHertz Photonics for Wireless Communications. *J. Light. Technol.* **2015**, *33*, 579–587. [CrossRef]

9. Nagatsuma, T.; Ducournau, G.; Renaud, C.C. Advances in Terahertz Communications Accelerated by Photonics. *Nat. Photonics* **2016**, *10*, 371–379. [CrossRef]
10. ITU. *International Telecommunication Union ITU-R P.840-3: Attenuation Due to Clouds and Fog*; ITU: Geneva, Switzerland, 1999.
11. ITU. *International Telecommunication Union ITU-R P.676-6: Attenuation by Atmospheric Gases*; ITU: Geneva, Switzerland, 2005.
12. Han, S.; Chih-Lin, I.; Xu, Z.; Rowell, C. Large-Scale Antenna Systems with Hybrid Analog and Digital Beamforming for Millimeter Wave 5G. *IEEE Commun. Mag.* **2015**, *53*, 186–194. [CrossRef]
13. Omam, Z.R.; Abdel-wahab, W.M.; Gigoyan, S.; Safavi-naeini, S. High Gain 4 × 4 SIW Passive Phased Array Antenna. In Proceedings of the 2020 IEEE International Symposium on Antennas and Propagation and North American Radio Science Meeting, Toronto, ON, Canada, 5–10 July 2020; pp. 45–46.
14. Nagarajan, R.; Member, S.; Joyner, C.H.; Schneider, R.P.; Bostak, J.S.; Butrie, T.; Dentai, A.G.; Dominic, V.G.; Evans, P.W.; Kato, M.; et al. Large-Scale Photonic Integrated Circuits. *IEEE J. Sel. Top. Quantum Electron. Electron.* **2005**, *11*, 50–65. [CrossRef]
15. Song, H.J.; Nagatsuma, T. Present and Future of Terahertz Communications. *IEEE Trans. Terahertz Sci. Technol.* **2011**, *1*, 256–263. [CrossRef]
16. Harter, T.; Ummethala, S.; Blaicher, M.; Muehlbrandt, S.; Wolf, S.; Weber, M.; Adib, M.M.H.; Kemal, J.N.; Merboldt, M.; Boes, F.; et al. Wireless THz Link with Optoelectronic Transmitter and Receiver. *Optica* **2019**, *6*, 1063–1070. [CrossRef]
17. Bowers, S.M.; Abiri, B.; Aflatouni, F.; Hajimiri, A. A Compact Optically Driven Travelling-Wave Radiating Source. In Proceedings of the Optical Fiber Communication Conference 2014, San Francisco, CA, USA, 9–13 March 2014; pp. 18–20.
18. Shi, J.-W.; Kuo, F.-M.; Wu, Y.-S.; Chen, N.-W.; Shih, P.-T.; Lin, C.-T.; Jiang, W., Jr.; Wong, E.-Z.; Chen, J.; Chi, S. A W-Band Photonic Transmitter-Mixer Based on High-Power Near-Ballistic Uni-Traveling-Carrier Photodiodes for BPSK and QPSK Data Transmission Under Bias Modulation. *IEEE Photonics Technol. Lett.* **2009**, *21*, 1039–1041. [CrossRef]
19. Costanzo, R.; Yang, Z.; Beling, A.; Bowers, S.M. Wideband Balanced Photoreceivers with InP-Based Photodiodes and 65 Nm CMOS TIAs for Use in Optical Frequency Synthesis Systems. *J. Light. Technol.* **2019**, *37*, 5833–5839. [CrossRef]
20. Ferguson, B.; Zhang, X.-C. Materials for Terahertz Science and Technology. *Nat. Mater.* **2002**, *1*, 26–33. [CrossRef] [PubMed]
21. Rouvalis, E.; Renaud, C.C.; Moodie, D.G.; Robertson, M.J.; Seeds, A.J. Traveling-Wave Uni-Traveling Carrier Photodiodes for Continuous Wave THz Generation. *Opt. Express* **2010**, *18*, 11105–11110. [CrossRef] [PubMed]
22. Shen, X.; Morgan, J.; Costanzo, R.; Sun, K.; Woodson, M.; Estrella, S.B.; Beling, A.; Bowers, S.M. High-Power V-Band-to-G-Band Photonically Driven Electromagnetic Emitters. *IEEE Trans. Microw. Theory Tech.* **2021**, *69*, 1474–1487. [CrossRef]
23. Sun, K.; Beling, A. High-Speed Photodetectors for Microwave Photonics. *Appl. Sci.* **2019**, *9*, 623. [CrossRef]
24. Nagatsuma, T.; Horiguchi, S.; Minamikata, Y.; Yoshimizu, Y.; Hisatake, S.; Kuwano, S.; Yoshimoto, N.; Terada, J.; Takahashi, H. Terahertz Wireless Communications Based on Photonics Technologies. *Opt. Express* **2013**, *21*, 23736. [CrossRef]
25. Sun, K.; Moody, J.; Li, Q.; Bowers, S.M.; Beling, A. High Power Integrated Photonic W-Band Emitter. *IEEE Trans. Microw. Theory Tech.* **2018**, *66*, 1668–1677. [CrossRef]
26. Assefzadeh, M.M.; Babakhani, A. Broadband Oscillator-Free THz Pulse Generation and Radiation Based on Direct Digital-to-Impulse Architecture. *IEEE J. Solid-State Circuits* **2017**, *52*, 2905–2919. [CrossRef]
27. Wang, C.; Lu, B.; Lin, C.; Chen, Q.; Miao, L.; Deng, X.; Zhang, J. 0.34-Thz Wireless Link Based on High-Order Modulation for Future Wireless Local Area Network Applications. *IEEE Trans. Terahertz Sci. Technol.* **2014**, *4*, 75–85. [CrossRef]
28. Razavian, S.; Babakhani, A. A THz Pulse Radiator Based on PIN Diode Reverse Recovery. In Proceedings of the 2019 IEEE BiCMOS and Compound semiconductor Integrated Circuits and Technology Symposium (BCICTS), Nashville, TN, USA, 3–6 November 2019; IEEE: New York, NY, USA, 2019; pp. 1–4.
29. Deng, J.; Burasa, P.; Wu, K. Joint Multiband Linear Interferometric Receiver for Integrated Microwave and Terahertz Sensing and Communication Systems. *IEEE Trans. Microw. Theory Tech.* **2024**, 1–13. [CrossRef]
30. Ueda, A.; Noguchi, T.; Iwashita, H.; Sekimoto, Y.; Ishiguro, M.; Takano, S.; Nagatsuma, T.; Ito, H.; Hirata, A.; Ishibashi, T. W-Band Waveguide Photomixer Using a Uni-Traveling-Carrier Photodiode with 2-MW Output. *IEEE Trans. Microw. Theory Tech.* **2003**, *51*, 1455–1459. [CrossRef]
31. Ishibashi, T.; Shimizu, N.; Kodama, S.; ITO, H.; Nagatuma, T.; Furuta, T. Uni-Traveling-Carrier Photodiodes. In Proceedings of the Ultrafast Electronics and Optoelectronics 1997, Incline Village, NA, USA, 17–21 March 1997; pp. 83–87.
32. Ishibashi, T.; Ito, H. Uni-Traveling-Carrier Photodiodes. *J. Appl. Phys.* **2020**, *127*, 031101. [CrossRef]
33. Kishino, K.; Unlu, M.S.; Chyi, J.-I.; Reed, J.; Arsenault, L.; Morkoc, H. Resonant Cavity-Enhanced (RCE) Photodetectors. *IEEE J. Quantum Electron.* **1991**, *27*, 2025–2034. [CrossRef]
34. Chen, Q.; Huang, Y.; Zhang, X.; Duan, X.; Fei, J.; Ma, X.; Liu, T.; Wu, G.; Liu, K.; Ren, X. Uni-Traveling-Carrier Photodetector with High-Reflectivity DBR Mirrors. *IEEE Photonics Technol. Lett.* **2017**, *29*, 1203–1206. [CrossRef]
35. Chen, Q.; Fang, W.; Huang, Y.; Duan, X.; Liu, K.; Sharawi, M.S.; Ren, X. Uni-Traveling-Carrier Photodetector with High-Contrast Grating Focusing-Reflection Mirrors. *Appl. Phys. Express* **2020**, *13*, 016503. [CrossRef]
36. Zhou, Q.; Cross, A.; Fu, Y.; Beling, A.; Campbell, J.C. High-Power High-Bandwidth Flip-Chip Bonded Modified Uni-Traveling Carrier Photodiodes. In Proceedings of the IEEE Photonics Conference 2012, Burlingame, CA, USA, 23–27 September 2012; IEEE: New York, NY, USA, 2012; Volume 2, pp. 306–307.
37. Li, Z.; Fu, Y.; Piels, M.; Pan, H.; Beling, A.; Bowers, J.E.; Campbell, J.C. High-Power High-Linearity Flip-Chip Bonded Modified Uni-Traveling Carrier Photodiode. *Opt. Express* **2011**, *19*, B385–B390. [CrossRef]

38. Xie, X.; Zhou, Q.; Li, K.; Shen, Y.; Li, Q.; Yang, Z.; Beling, A.; Campbell, J.C. Improved Power Conversion Efficiency in High-Performance Photodiodes by Flip-Chip Bonding on Diamond. *Optica* **2014**, *1*, 429–435. [CrossRef]
39. Ito, H.; Shibata, N.; Nagatsuma, T.; Ishibashi, T. Terahertz-Wave Detector on Silicon Carbide Platform. *Appl. Phys. Express* **2022**, *15*, 026501. [CrossRef]
40. Kashyap, R. *Fiber Bragg Gratings*, 2nd ed.; Academic Press: Cambridge, MA, USA, 2010.
41. Chen, Q.; Boisvert, J.-S.; Sharawi, M.S.; Kashyap, R. Bragg Gratings with Novel Waveguide Models Fabricated in Bulk Glass via Fs-Laser Writing and Their Slow-Light Effects. *Opt. Express* **2024**, *32*, 188–204. [CrossRef]
42. Bowers, J.; Burrus, C. Ultrawide-Band Long-Wavelength p-i-n Photodetectors. *J. Light. Technol.* **1987**, *5*, 1339–1350. [CrossRef]
43. Kato, K. Ultrawide-Band/High-Frequency Photodetectors. *IEEE Trans. Microw. Theory Tech.* **1999**, *47*, 1265–1281. [CrossRef]
44. Ishibashi, T.; Furuta, T.; Fushimi, H.; Kodama, S.; Ito, H.; Nagatsuma, T.; Shimizu, N.; Miyamoto, Y. InP/InGaAs Uni-Traveling-Carrier Photodiodes. *IEICE Trans. Electron.* **2000**, *E83-C*, 938–949.
45. Ishibashi, T.; Ito, H. Uni-Traveling Carrier Photodiodes: Development and Prospects. *IEEE J. Sel. Top. Quantum Electron.* **2022**, *28*, 3803006. [CrossRef]
46. Kurishima, K.; Nakajima, H.; Kobayashi, T.; Ishibashi, T.; Matsuoka, Y. Fabrication and Characterization of High-Performance InP/InGaAs Double-Heterojunction Bipolar Transistors. *IEEE Trans. Electron Devices* **1994**, *41*, 1319–1326. [CrossRef]
47. Li, N.; Li, X.; Demiguel, S.; Zheng, X.; Campbell, J.C.; Tulchinsky, D.A.; Williams, K.J.; Isshiki, T.D.; Kinsey, G.S.; Sudharsansan, R. High-Saturation-Current Charge-Compensated InGaAs–InP Uni-Traveling-Carrier Photodiode. *IEEE Photonics Technol. Lett.* **2004**, *16*, 864–866. [CrossRef]
48. Wang, X.; Duan, N.; Chen, H.; Campbell, J.C. InGaAs–InP Photodiodes With High Responsivity and High Saturation Power. *IEEE Photonics Technol. Lett.* **2007**, *19*, 1272–1274. [CrossRef]
49. Adachi, S. Refractive Indices of III-V Compounds: Key Properties of InGaAsP Relevant to Device Design. *J. Appl. Phys.* **1982**, *53*, 5863–5869. [CrossRef]
50. Wang, H.; Mao, S. High Speed InP/InGaAs Uni-Traveling-Carrier Photodiodes with Dipole-Doped InGaAs/InP Absorber-Collector Interface. In Proceedings of the IPRM 2011—23rd International Conference on Indium Phosphide and Related Materials, Berlin, Germany, 22–26 May 2011; pp. 1–3.
51. Meng, Q.Q.; Wang, H.; Liu, C.Y.; Ang, K.S.; Guo, X.; Gao, B.; Tian, Y.; Kumar, C.M.M.; Gao, J. High-Photocurrent and Wide-Bandwidth UTC Photodiodes with Dipole-Doped Structure. *IEEE Photonics Technol. Lett.* **2014**, *26*, 1952–1955. [CrossRef]
52. Meng, Q.; Wang, H.; Liu, C.; Guo, X.; Gao, J.; Ang, K.S. High-Speed and High-Responsivity InP-Based Uni-Traveling-Carrier Photodiodes. *IEEE J. Electron Devices Soc.* **2017**, *5*, 40–44. [CrossRef]
53. Kato, K.; Hata, S.; Kawano, K.; Kozen, A. Design of Ultrawide-Band, High-Sensitivity p–i–n-Photodetectors. *IEICE Trans. Electron.* **1993**, *E76-C*, 214–221.
54. Ito, H.; Kodama, S.; Muramoto, Y.; Furuta, T.; Nagatsuma, T.; Ishibashi, T. High-Speed and High-Output InP–InGaAs Unitraveling-Carrier Photodiodes. *IEEE J. Sel. Top. Quantum Electron.* **2004**, *10*, 709–727. [CrossRef]
55. Bahl, I. Multilayer Dielectric Capacitor. In *Lumped Elements for RF and Microwave Circuits*; Artech House: London, UK, 2003; pp. 171–172.
56. Liu, F.; Huang, Y.; Kang, C.; Chen, Q.; Duan, X.; Liu, K.; Wang, Q.; Zhang, X.; Wang, J.; Ren, X. Optimization of High Speed and High Saturation Uni-Traveling-Carrier Photodiode. In Proceedings of the Asia Communications and Photonics Conference (ACPC), Hong Kong, China, 19–23 November 2015; p. ASu2A.11.
57. Wang, J.; Duan, X.; Huang, Y.; Liu, K.; Fei, J.; Chen, Q.; Ren, X. Novel 1.55 Mm Mushroom-Type Vertical-Illumination Photodiode. In Proceedings of the Asia Communications and Photonics Conference (ACPC), Wuhan, China, 2–5 November 2016; p. AF2A.68.2.
58. Liu, F.; Huang, Y.; Kang, C.; Chen, Q.; Duan, X.; Ren, X. High Speed and High Responsivity Dual-Absorption InGaAs/InP UTC-PDs. In Proceedings of the Opto-Electronics and Communications Conference (OECC), Shanghai, China, 28 June–2 July 2015; pp. 1–3.
59. Kang, C.; Huang, Y.; Liu, F.; Fei, J.; Chen, Q.; Liu, K.; Duan, X.; Wang, Q.; Wang, J.; Zhang, X.; et al. A Mushroom Dual-Absorption Partially Depleted Absorber Photodetector. In Proceedings of the Asia Communications and Photonics Conference, Hong Kong, China, 19–23 November 2015; p. ASu2A.1.
60. Liu, L.; Huang, Y.; Ma, X.; Fei, J.; Chen, Q.; Liu, K.; Duan, X.; Ren, X. Study of the Effect of Gaussian Doping in Absorber on High Speed Performance of Uni-Traveling-Carrier Photodetector. In Proceedings of the Asia Communications and Photonics Conference (ACPC), Wuhan, China, 2–5 November 2016; p. AF3F.7.
61. Chen, Q.; Huang, Y.; Duan, X.; Liu, F.; Kang, C.; Wang, Q.; Wang, J.; Zhang, X.; Ren, X. High-Speed Uni-Traveling-Carrier Photodetector with the New Design of Absorber and Collector. In Proceedings of the Opto-Electronics and Communications Conference (OECC), Shanghai, China, 28 June–2 July 2015; IEEE: New York, NY, USA, 2017.
62. Chen, Q.-T.; Huang, Y.-Q.; Fei, J.-R.; Duan, X.-F.; Liu, K.; Liu, F.; Kang, C.; Wang, J.-C.; Fang, W.-J.; Ren, X.-M. Fabrication and Characterization of Novel High-Speed InGaAs/InP Uni-Traveling-Carrier Photodetector for High Responsivity. *Chinese Phys. B* **2015**, *24*, 108506. [CrossRef]
63. Liu, T.; Huang, Y.; Chen, Q.; Fei, J.; Liu, K.; Duan, X.; Ren, X. Transient Simulation of UTC-PD Using Drift-Diffusion Model. In Proceedings of the International Conference on Numerical Simulation of Optoelectronic Devices (NUSOD), Copenhagen, Denmark, 24–28 July 2017; IEEE: New York, NY, USA, 2017; pp. 139–140.

64. Ma, X.; Huang, Y.; Fei, J.; Chen, Q.; Liu, T.; Liu, K.; Duan, X.; Yan, X.; Ren, X. Analysis of Dark Current Considering Trap-Assisted Tunneling Mechanism for InGaAs PIN Photodetectors. *Opt. Quantum Electron.* **2017**, *49*, 407. [CrossRef]
65. Ma, X.; Huang, Y.; Fei, J.; Chen, Q.; Liu, T.; Li, G.; Duan, X.; Liu, K.; Cai, S.; Ren, X. InP-Based Photodetectors with Wide Spectral Range and Low Dark Current for Optical Interconnection. In Proceedings of the Asia Communications and Photonics Conference (ACPC), Guangzhou, China, 10–13 November 2017; OSA: Washington, DC, USA, 2017; p. Su4K.6.
66. Liu, T.; Huang, Y.; Fei, J.; Chen, Q.; Ma, X.; Duan, X.; Liu, K.; Ren, X. Influences of Contact Electrode Shape and Incidence Direction on P-i-n Photodiodes. *IET Optoelectron.* **2019**, *13*, 151–154. [CrossRef]
67. Ito, H.; Furuta, T.; Kodama, S.; Ishibashi, T. InP/InGaAs Uni-Travelling-Carrier Photodiode with 310 GHz Bandwidth. *Electron. Lett.* **2000**, *36*, 1809–1810. [CrossRef]
68. Fukano, H.; Muramoto, Y.; Takahata, K.; Matsuoka, Y. High Efficiency Edge-Illuminated Uni-Travelling-Carrier-Structure Refracting-Facet Photodiode. *Electron. Lett.* **1999**, *35*, 1664. [CrossRef]
69. Ito, H.; Furuta, T.; Kodama, S.; Ishibashi, T. High-Efficiency Unitraveling-Carrier Photodiode with an Integrated Total-Reflection Mirror. *J. Light. Technol.* **2000**, *18*, 384–387. [CrossRef]
70. Ito, H.; Furuta, T.; Kodama, S.; Watanabe, N.; Ishibashi, T. InP/InGaAs Uni-Travelling-Carrier Photodiode with 220 GHz Bandwidth. *Electron. Lett.* **1999**, *35*, 1556–1557. [CrossRef]
71. Beling, A.; Campbell, J.C.; Pan, H.; Chen, H.; Bach, H.-G.; Mekonnen, G.G.; Schmidt, D. InP-Based High-Speed Photonic Devices. In Proceedings of the OFC/NFOEC 2008—2008 Conference on Optical Fiber Communication/National Fiber Optic Engineers Conference, San Diego, CA, USA, 24–28 February 2008; IEEE: New York, NY, USA, 2008; pp. 1–27.
72. Ishibashi, T.; Muramoto, Y.; Yoshimatsu, T.; Ito, H. Unitraveling-Carrier Photodiodes for Terahertz Applications. *IEEE J. Sel. Top. Quantum Electron.* **2014**, *20*, 79–88. [CrossRef]
73. Li, X.; Li, N.; Demiguel, S.; Zheng, X.; Campbell, J.C.; Tan, H.H.; Jagadish, C. A Partially Depleted Absorber Photodiode With Graded Doping Injection Regions. *IEEE Photonics Technol. Lett.* **2004**, *16*, 2326–2328. [CrossRef]
74. Li, X.; Li, N.; Demiguel, S.; Campbell, J.C.; Tulchinsky, D.; Williams, K.J. A Comparison of Front- and Backside-Illuminated High-Saturation Power Partially Depleted Absorber Photodetecters. *IEEE J. Quantum Electron.* **2004**, *40*, 1321–1325.
75. Effenberger, F.J.; Joshi, A.M. Ultrafast, Dual-Depletion Region, InGaAs/InP p-i-n Detector. *J. Light. Technol.* **1996**, *14*, 1859–1864. [CrossRef]
76. Williams, K.J. Comparisons between Dual-Depletion-Region and Uni-Travelling-Carrier p-i-n Photodetectors. *IEE Proc. Optoelectron.* **2002**, *149*, 131–137. [CrossRef]
77. Jun, D.-H.; Jang, J.-H.; Adesida, I.; Song, J.-I. Improved Efficiency-Bandwidth Product of Modified Uni-Traveling Carrier Photodiode Structures Using an Undoped Photo-Absorption Layer. *Jpn. J. Appl. Phys.* **2006**, *45*, 3475–3478. [CrossRef]
78. Li, Z.; Pan, H.; Chen, H.; Beling, A.; Campbell, J.C. High-Saturation-Current Modified Uni-Traveling-Carrier Photodiode with Cliff Layer. *IEEE J. Quantum Electron.* **2010**, *46*, 626–632. [CrossRef]
79. Shi, J.; Wu, Y.; Wu, C.; Chiu, P.; Hong, C. High-Speed, High-Responsivity, and High-Power Performance of Near-Ballistic Uni-Traveling-Carrier Photodiode at 1.55 Mm Wavelength. *IEEE Photonics Technol. Lett.* **2005**, *17*, 1929–1931. [CrossRef]
80. Shi, J.W.; Wu, Y.S.; Lin, Y.S. Near-Ballistic Uni-Traveling-Carrier Photodiode-Based V-Band Optoelectronic Mixers with Internal up-Conversion-Gain, Wide Modulation Bandwidth, and Very High Operation Current Performance. *IEEE Photonics Technol. Lett.* **2008**, *20*, 939–941. [CrossRef]
81. Tulchinsky, D.A.; Li, X.; Li, N.; Demiguel, S.; Campbell, J.C.; Williams, K.J. High-Saturation Current Wide-Bandwidth Photodetectors. *IEEE J. Sel. Top. Quantum Electron.* **2004**, *10*, 702–708. [CrossRef]
82. Ishibashi, T.; Furuta, T.; Fushimi, H.; Ito, H. Photoresponse Characteristics of Uni-Traveling-Carrier Photodiodes. In *Physics and Simulation of Optoelectronic Devices IX, Proceedings of the Symposium on Integrated Optics, San Jose, CA, USA, 20–26 January 2001*; Arakawa, Y., Blood, P., Osinski, M., Eds.; SPIE: Bellingham, WA, USA, 2001; Volume 4283, pp. 469–479.
83. Shimizu, N.; Watanabe, N.; Furuta, T.; Ishibashi, T. Electron Diffusivity in p -InGaAs Determined from the Pulse Response of InP/InGaAs Uni-Traveling-Carrier Photodiodes. *Appl. Phys. Lett.* **2000**, *76*, 1191–1193. [CrossRef]
84. Yasuoka, N.; Makiuchi, M.; Miyata, M.; Aoki, O.; Ekawa, M.; Okazaki, N.; Takechi, M.; Kuwatsuka, H.; Soda, H. High-Efficiency Pin Photo-Diodes with a Spot-Size Converter for 40 Gb/s Transmission Systems. In Proceedings of the 27th European Conference on Optical Communication (Cat. No.01TH8551), Amsterdam, The Netherlands, 30 September–4 October 2001; IEEE: New York, NY, USA, 2001; Volume 4, pp. 558–559.
85. Wang, G.; Tokumitsu, T.; Hanawa, I.; Sato, K.; Kobayashi, M. Analysis of High Speed P-I-N Photodiodes S-Parameters by a Novel Small-Signal Equivalent Circuit Model. *IEEE Microw. Wirel. Components Lett.* **2002**, *12*, 378–380. [CrossRef]
86. Wang, G.; Tokumitsu, T.; Hanawa, I.; Yoneda, Y.; Sato, K.; Kobayashi, M. A Time-Delay Equivalent-Circuit Model of Ultrafast p-i-n Photodiodes. *IEEE Trans. Microw. Theory Tech.* **2003**, *51*, 1227–1233. [CrossRef]
87. Demiguel, S.; Giraudet, L.; Joulaud, L.; Decobert, J.; Blache, F.; Coupeé, V.; Jorge, F.; Pagnod-Rossiaux, P.; Boucherez, E.; Achouche, M.; et al. Evanescently Coupled Photodiodes Integrating a Double-Stage Taper for 40-Gb/s Applications—Compared Performance with Side-Illuminated Photodiodes. *J. Light. Technol.* **2002**, *20*, 2004–2014. [CrossRef]
88. Lucovsky, G.; Schwarz, R.F.; Emmons, R.B. Transit-Time Considerations in p-i-n Diodes. *J. Appl. Phys.* **1964**, *35*, 622–628. [CrossRef]
89. Li, J.; Xiong, B.; Sun, C.; Miao, D.; Luo, Y. Analysis of Frequency Response of High Power MUTC Photodiodes Based on Photocurrent-Dependent Equivalent Circuit Model. *Opt. Express* **2015**, *23*, 21615–21623. [CrossRef] [PubMed]

90. Ito, H.; Nagatsuma, T.; Hirata, A.; Minotani, T.; Sasaki, A.; Hirota, Y.; Ishibashi, T. High-Power Photonic Millimetre Wave Generation at 100 GHz Using Matching-Circuit-Integrated Uni-Travelling-Carrier Photodiodes. *IEE Proc. Optoelectron.* **2003**, *150*, 138–142. [CrossRef]
91. Stöhr, A.; Malcoci, A.; Sauerwald, A.; Mayorga, I.C.; Güsten, R.; Jäger, D.S. Ultra-Wide-Band Traveling-Wave Photodetectors for Photonic Local Oscillators. *J. Light. Technol.* **2003**, *21*, 3062–3070. [CrossRef]
92. Jou, J.-J.; Liu, C.-K.; Hsiao, C.-M.; Lin, H.-H.; Lee, H.-C. Time-Delay Circuit Model of High-Speed p-i-n Photodiodes. *IEEE Photonics Technol. Lett.* **2002**, *14*, 525–527. [CrossRef] [PubMed]
93. Wu, Y.-S.; Shi, J.-W.; Chiu, P.-H. Analytical Modeling of a High-Performance near-Ballistic Uni-Traveling-Carrier Photodiode at a 1550 Nm Wavelength. *IEEE Photonics Technol. Lett.* **2006**, *18*, 938–940. [CrossRef]
94. Wu, Y.-S.; Shi, J.-W. Dynamic Analysis of High-Power and High-Speed Near-Ballistic Unitraveling Carrier Photodiodes at W-Band. *IEEE Photonics Technol. Lett.* **2008**, *20*, 1160–1162. [CrossRef]
95. Shi, J.W.; Kuo, F.M.; Bowers, J.E. Design and Analysis of Ultra-High-Speed near-Ballistic Uni-Traveling- Carrier Photodiodes under a 50-Ω Load for High-Power Performance. *IEEE Photonics Technol. Lett.* **2012**, *24*, 533–535. [CrossRef]
96. Meng, Q.Q.; Wang, H.; Gao, B.; Liu, C.Y.; Ang, K.S.; Guo, X.; Gao, J. Equivalent Circuit Model for InP-Based Uni-Traveling-Carrier Photodiodes with Dipole-Doped Structure. In Proceedings of the Asia Communications and Photonics Conference 2014, Shanghai, China, 11–14 November 2014; OSA: Washington, DC, USA, 2014; p. ATh3A.22.
97. Wun, J.M.; Lai, C.H.; Chen, N.W.; Bowers, J.E.; Shi, J.W. Flip-Chip Bonding Packaged THz Photodiode with Broadband High-Power Performance. *IEEE Photonics Technol. Lett.* **2014**, *26*, 2462–2464. [CrossRef]
98. Li, K.; Xie, X.; Li, Q.; Shen, Y.; Woodsen, M.E.; Yang, Z.; Beling, A.; Campbell, J.C. High-Power Photodiode Integrated with Coplanar Patch Antenna for 60-GHz Applications. *IEEE Photonics Technol. Lett.* **2015**, *27*, 650–653. [CrossRef]
99. Natrella, M.; Liu, C.-P.; Graham, C.; van Dijk, F.; Liu, H.; Renaud, C.C.; Seeds, A.J.; Fresnel, A.A.; Cedex, R.D.F.-P.; van Dijk, F.; et al. Accurate Equivalent Circuit Model for Millimetre-Wave UTC Photodiodes. *Opt. Express* **2016**, *24*, 4698–4713. [CrossRef]
100. Li, J.; Xiong, B.; Luo, Y.; Sun, C.; Wang, J.; Hao, Z.; Han, Y.; Wang, L.; Li, H. Ultrafast Dual-Drifting Layer Uni-Traveling Carrier Photodiode with High Saturation Current. *Opt. Express* **2016**, *24*, 8420–8428. [CrossRef] [PubMed]
101. Li, Q.; Li, K.; Fu, Y.; Xie, X.; Yang, Z.; Beling, A.; Campbell, J.C. High-Power Flip-Chip Bonded Photodiode with 110 GHz Bandwidth. *J. Light. Technol.* **2016**, *34*, 2139–2144. [CrossRef]
102. Zhou, G.; Runge, P. Nonlinearities of High-Speed p-i-n Photodiodes and MUTC Photodiodes. *IEEE Trans. Microw. Theory Tech.* **2017**, *65*, 2063–2072. [CrossRef]
103. Chen, Y.; Xie, Z.; Huang, J.; Deng, Z.; Chen, B. High-Speed Uni-Traveling Carrier Photodiode for 2 Mm Wavelength Application. *Optica* **2019**, *6*, 884–889. [CrossRef]
104. Naseem; Ahmad, Z.; Chao, R.-L.; Chang, H.-S.; Ni, C.-J.; Chen, H.-S.; Huang, J.J.-S.; Chou, E.; Jan, Y.-H.; Shi, J.-W. Enhancement in Speed and Responsivity of Uni-Traveling Carrier Photodiodes with $GaAs_{0.5}Sb_{0.5}/In_{0.53}Ga_{0.47}As$ Type-II Hybrid Absorbers. *Opt. Express* **2019**, *27*, 15495–15504. [CrossRef] [PubMed]
105. Mukherjee, C.; Natrella, M.; Seddon, J.; Graham, C.; Mounaix, P.; Renaud, C.C.; Maneux, C. Efficient Compact Modelling of UTC-Photodiode towards Terahertz Communication System Design. *Solid-State Electron.* **2020**, *170*, 107836. [CrossRef]
106. Xu, J. Design and Investigation of High Speed and High Power InGaAs/InP One-Sided Junction Photodiodes. Ph.D. Thesis, Concordia University, Montreal, QC, Canada, 2020.
107. Han, Y.; Xiong, B.; Sun, C.; Hao, Z.; Wang, J.; Han, Y.; Wang, L.; Li, H.; Yu, J.; Luo, Y. Distributed Parameter Circuit Model for Wideband Photodiodes with Inductive Coplanar Waveguide Electrodes. *Chin. Opt. Lett.* **2020**, *18*, 061301. [CrossRef]
108. Yang, J.; Ren, J.; Tang, X.; Feng, Z. An Equivalent Circuit Model for Uni-Traveling-Carrier Photodiode. In Proceedings of the 2021 4th International Conference on Electronics and Electrical Engineering Technology, Nanjing, China, 3–5 December 2021; ACM: New York, NY, USA, 2021; pp. 170–176.
109. Tian, Y.; Xiong, B.; Sun, C.; Hao, Z.; Wang, J.; Wang, L.; Han, Y.; Li, H.; Luo, Y. MUTC-PD with Enhanced D-Band Response by Shunt Capacitors on Submount. In Proceedings of the Asia Communications and Photonics Conference 2021, Shanghai, China, 24–27 October 2021; Optica Publishing Group: Washington, DC, USA, 2021; p. T4A.217.
110. Konstantinou, D.; Caillaud, C.; Rommel, S.; Johannsen, U.; Tafur Monroy, I. Investigation of De-Embedding Techniques Applied on Uni-Traveling Carrier Photodiodes. *Int. J. Microw. Wirel. Technol.* **2021**, *13*, 569–581. [CrossRef]
111. Chao, E.; Xiong, B.; Sun, C.; Hao, Z.; Wang, J.; Wang, L.; Han, Y.; Li, H.; Yu, J.; Luo, Y. D-Band MUTC Photodiodes With Flat Frequency Response. *IEEE J. Sel. Top. Quantum Electron.* **2022**, *28*, 3802208. [CrossRef]
112. Li, T.; Duan, X.; Li, G.; Li, K.; Yuan, W.; Liu, K.; Huang, Y. Modified Photodiode Equivalent Circuit Model Considering Coplanar Waveguide Electrodes. *Semicond. Sci. Technol.* **2023**, *38*, 085011. [CrossRef]
113. Tian, Y.; Xiong, B.; Sun, C.; Hao, Z.; Wang, J.; Wang, L.; Han, Y.; Li, H.; Gan, L.; Luo, Y. Ultrafast MUTC Photodiodes over 200 GHz with High Saturation Power. *Opt. Express* **2023**, *31*, 23790–23800. [CrossRef] [PubMed]
114. Wei, C.; Xie, X.; Wang, Z.; Chen, Y.; Zeng, Z.; Zou, X.; Pan, W.; Yan, L. 150 GHz High-Power Photodiode by Flip-Chip Bonding. *J. Light. Technol.* **2023**, *41*, 7238–7244. [CrossRef]
115. Morgan, J.S.; Tabatabaei, F.; Fatema, T.; Tang, C.W.; Sun, K.; Lau, K.M.; Beling, A. Bias-Insensitive GaAsSb/InP CC-MUTC Photodiodes for MmWave Generation up to 325 GHz. *J. Light. Technol.* **2023**, *41*, 7092–7097. [CrossRef]
116. Chaudhary, R.; Arabhavi, A.M.; Kulmer, L.; Hamzeloui, S.; Leich, M.; Ostinelli, O.; Leuthold, J.; Bolognesi, C.R. Wideband Type-II GaInAsSb/InP Uni-Traveling Carrier Photodiodes for Near 300 Gbps Communications. *J. Light. Technol.* **2023**, *14*, 1–6. [CrossRef]

117. Li, L.; Wang, L.; Chen, B. High-Speed Waveguide Modified Uni-Traveling Carrier Photodiodes with 130 GHz Bandwidth. In Proceedings of the 2023 Opto-Electronics and Communications Conference (OECC), Shanghai, China, 2–6 July 2023; IEEE: New York, NY, USA, 2023; pp. 1–3.
118. Li, L.; Wang, L.; Chen, B. High-Speed Evanescently-Coupled Waveguide MUTC Photodiodes with Bandwidth Over 220 GHz. In Proceedings of the 2023 International Topical Meeting on Microwave Photonics (MWP), Nanjing, China, 15–18 October 2023; IEEE: New York, NY, USA, 2023; pp. 1–4.
119. Li, L.; Wang, L.; Long, T.; Zhang, Z.; Lu, J.; Chen, B. Ultra-Fast Waveguide MUTC Photodiodes over 220 GHz. *arXiv* **2024**, arXiv:2402.07491. [CrossRef]
120. Natrella, M.; Liu, C.-P.; Graham, C.; van Dijk, F.; Liu, H.; Renaud, C.C.; Seeds, A.J. Modelling and Measurement of the Absolute Level of Power Radiated by Antenna Integrated THz UTC Photodiodes. *Opt. Express* **2016**, *24*, 11793–11807. [CrossRef]
121. Morgan, J.S.; Sun, K.; Li, Q.; Estrella, S.; Woodson, M.; Hay, K.; Mashanovitch, M.; Beling, A. High-Power Flip-Chip Bonded Modified Uni-Traveling Carrier Photodiodes with −2.6 DBm RF Output Power at 160 GHz. In Proceedings of the 2018 IEEE Photonics Conference (IPC), Reston, VA, USA, 30 September–4 October 2018; IEEE: New York, NY, USA, 2018; pp. 1–2.
122. Xie, X.; Li, K.; Shen, Y.; Li, Q.; Zang, J.; Beling, A.; Campbell, J.C. Photonic Generation of High-Power Pulsed Microwave Signals. *J. Light. Technol.* **2015**, *33*, 3808–3814. [CrossRef]
123. Nagatsuma, T.; Hirata, A.; Royter, Y.; Shinagawa, M.; Furuta, T.; Ishibashi, T.; Ito, H. A 120-GHz Integrated Photonic Transmitter. In Proceedings of the International Topical Meeting on Microwave Photonics MWP 2000 (Cat. No.00EX430), Oxford, UK, 11–13 September 2000; IEEE: New York, NY, USA, 2000; pp. 225–228.
124. Hirata, A.; Ishii, H.; Nagatsuma, T. Design and Characterization of a 120-GHz Millimeter-Wave Antenna for Integrated Photonic Transmitters. *IEEE Trans. Microw. Theory Tech.* **2001**, *49*, 2157–2162. [CrossRef]
125. Beling, A.; Xie, X.; Campbell, J.C. High-Power, High-Linearity Photodiodes. *Optica* **2016**, *3*, 328–338. [CrossRef]
126. Seeds, A.J. Microwave Photonics. *IEEE Trans. Microw. Theory Tech.* **2002**, *50*, 877–887. [CrossRef]
127. Seeds, A.J.; Williams, K.J. Microwave Photonics. *J. Light. Technol.* **2006**, *24*, 4628–4641. [CrossRef]
128. Cox, C.H.; Ackerman, E.I.; Betts, G.E.; Prince, J.L. Limits on the Performance of RF-over-Fiber Links and Their Impact on Device Design. *IEEE Trans. Microw. Theory Tech.* **2006**, *54*, 906–920. [CrossRef]
129. Chtioui, M.; Enard, A.; Carpentier, D.; Lelarge, F.; Rousseau, B.; Achouche, M.; Marceaux, A.; Renoult, A.; Feuillet, C.; Queguiner, M.; et al. High Power UTC Photodiodes Design and Application for Analog Fiber Optic Links. In Proceedings of the 2009 International Topical Meeting on Microwave Photonics, Valencia, Spain, 14–16 October 2009; pp. 1–4.
130. Xie, X.; Zhou, Q.; Li, K.; Beling, A.; Campbell, J. 1.8 Watt RF Power and 60% Power Conversion Efficiency Based on Photodiode Flip-Chip-Bonded on Diamond. In Proceedings of the CLEO: 2014 Postdeadline Paper Digest, San Jose, CA, USA, 8–13 June 2014; OSA: Washington, DC, USA, 2014; Volume 2014-January, p. JTh5B.9.
131. Zhou, Q.; Cross, A.S.; Fu, Y.; Beling, A.; Campbell, J.C. Development of Narrowband Modified Uni-Travelling-Carrier Photodiodes with High Power Efficiency. In Proceedings of the 2013 IEEE Avionics, Fiber-Optics and Photonics Technology Conference (AVFOP), San Diego, CA, USA, 1–3 October 2013; IEEE: New York, NY, USA, 2013; Volume 5, pp. 65–66.
132. Yi, L.; Iwamoto, K.; Yamamoto, T.; Ayano, F.; Li, Y.; Rolland, A.; Kuse, N.; Fermann, M.; Nagatsuma, T. 300-GHz-Band Wireless Communication Using a Low Phase Noise Photonic Source. In Proceedings of the 2019 49th European Microwave Conference (EuMC), Paris, France, 1–3 October 2019; IEEE: New York, NY, USA, 2019; Volume 12, pp. 816–819.
133. Yi, L.; Iwamoto, K.; Yamamoto, T.; Ayano, F.; Rolland, A.; Kuse, N.; Fermann, M.; Li, Y.; Nagatsuma, T. 300-GHz-Band Wireless Communication Using a Low Phase Noise Photonic Source. *Int. J. Microw. Wirel. Technol.* **2020**, *12*, 551–558. [CrossRef]
134. Ohara, T.; Ishibashi, T.; Kawamoto, Y.; Tojo, M.; Maekawa, K.; Nagatsuma, T. 2-MW-Output Power Uni-Traveling-Carrier Photodiodes in 300-GHz-Band. In Proceedings of the 2023 Asia-Pacific Microwave Conference (APMC), Taipei, Taiwan, 5–8 December 2023; IEEE: New York, NY, USA, 2023; pp. 676–678.
135. Nagatsuma, T.; Ohtake, H.; Kato, K.; Yumoto, J.; Ito, H. Photonics-Empowered Terahertz Wireless Communications. In Proceedings of the 2023 24th International Conference on Applied Electromagnetics and Communications (ICECOM), Dubrovnik, Croatia, 27–29 September 2023; IEEE: New York, NY, USA, 2023; pp. 1–3.
136. Tulchinsky, D.A.; Boos, J.B.; Park, D.; Goetz, P.G.; Rabinovich, W.S.; Williams, K.J. High-Current Photodetectors as Efficient, Linear, and High-Power RF Output Stages. *J. Light. Technol.* **2008**, *26*, 408–416. [CrossRef]
137. Sakai, K.; Nagatsuka, T.; Itakura, S.; Otsuka, H.; Hirano, Y. Backside-Illuminated High-Current Photodiode for Analog Optical Links. In Proceedings of the LEOS 2008—21st Annual Meeting of the IEEE Lasers and Electro-Optics Society, Newport Beach, CA, USA, 9–13 November 2008; IEEE: New York, NY, USA, 2008; pp. 288–289.
138. Itakura, S.; Sakai, K.; Nagatsuka, T.; Ishimura, E.; Nakaji, M.; Aoyagi, T. High-Current Backside-Illuminated InGaAs/InP p-i-n Potodiode. In Proceedings of the International Topical Meeting on Microwave Photonics, Valencia, Spain, 14–16 October 2009; pp. 1–4.
139. Yi, Y.; Umezawa, T.; Kanno, A.; Kawanishi, T. 50 GHz High Photocurrent PIN-PD and Its Thermal Effect. In Proceedings of the 2022 27th OptoElectronics and Communications Conference (OECC) and 2022 International Conference on Photonics in Switching and Computing (PSC), Toyama, Japan, 3–6 July 2022; IEEE: New York, NY, USA, 2022; pp. 1–3.
140. Fei, J.; Huang, Y.; Fang, W.; Liu, T.; Duan, X.; Liu, K.; Ren, X. High-Power Symmetric-Connected Uni-Traveling-Carrier Photodiode Array Integrated with Sub-Wavelength Gratings Based Beam-Splitter. *Opt. Express* **2017**, *25*, 21726–21734. [CrossRef] [PubMed]

141. Wang, X.; Huang, Y.; Tan, S.; Du, J.; Yang, M.; Liu, K.; Duan, X.; Ren, X. Bandwidth Optimization and Fabrication of High-Power MUTC-PD. *IEEE J. Quantum Electron.* **2024**, *60*, 4000206. [CrossRef]
142. Lin, X.; Natrella, M.; Seddon, J.; Graham, C.; Renaud, C.C.; Tang, M.; Wu, J.; Liu, H.; Seeds, A.J. High Performance Waveguide Uni-Travelling Carrier Photodiode Grown by Solid Source Molecular Beam Epitaxy. *Opt. Express* **2019**, *27*, 37065–37086. [CrossRef] [PubMed]
143. Rymanov, V.; Stöhr, A.; Dülme, S.; Tekin, T. Triple Transit Region Photodiodes (TTR-PDs) Providing High Millimeter Wave Output Power. *Opt. Express* **2014**, *22*, 7550–7558. [CrossRef] [PubMed]
144. Shi, J.W.; Kuo, F.M.; Wu, C.J.; Chang, C.L.; Liu, C.Y.; Chen, C.Y.; Chyi, J.I. Extremely High Saturation Current-Bandwidth Product Performance of a near-Ballistic Uni-Traveling-Carrier Photodiode with a Flip-Chip Bonding Structure. *IEEE J. Quantum Electron.* **2010**, *46*, 80–86. [CrossRef]
145. Wun, J.; Zeng, Y.; Shi, J. GaAs0.5Sb0.5/InP UTC-PD with Graded-Bandgap Collector for Zero-Bias Operation at Sub-THz Regime. In Proceedings of the Optical Fiber Communication Conference, Anaheim, CA, USA, 20–24 March 2016; OSA: Washington, DC, USA, 2016; p. Tu2D.4.
146. Shi, J.-W.; Kuo, F.-M.; Rodwell, M.; Bowers, J.E. Ultra-High Speed (270 GHz) near-Ballistic Uni-Traveling-Carrier Photodiode with Very-High Saturation Current (17 MA) under a 50 Ω Load. In Proceedings of the IEEE Photonic Society 24th Annual Meeting, Arlington, VA, USA, 9–13 October 2011; IEEE: New York, NY, USA, 2011; Volume 2, pp. 21–22.
147. Wun, J.M.; Liu, H.Y.; Zeng, Y.L.; Pan, C.L.; Huang, C.B.; Shi, J.W. High-Power THz-Wave Generation by Using Ultra-Fast (315 GHz) Uni-Traveling Carrier Photodiode with Novel Collector Design and Photonic Femtosecond Pulse Generator. In Proceedings of the Optical Fiber Communication Conference 2015, Los Angeles, CA, USA, 22–26 March 2015; pp. 1–3.
148. Wun, J.M.; Liu, H.Y.; Zeng, Y.L.; Da Yang, S.; Pan, C.L.; Huang, C.B.; Shi, J.W. Photonic High-Power Continuous Wave THz-Wave Generation by Using Flip-Chip Packaged Uni-Traveling Carrier Photodiodes and a Femtosecond Optical Pulse Generator. *J. Light. Technol.* **2016**, *34*, 1387–1397. [CrossRef]
149. Latzel, P.; Pavanello, F.; Billet, M.; Bretin, S.; Beck, A.; Vanwolleghem, M.; Coinon, C.; Wallart, X.; Peytavit, E.; Ducournau, G.; et al. Generation of MW Level in the 300-GHz Band Using Resonant-Cavity-Enhanced Unitraveling Carrier Photodiodes. *IEEE Trans. Terahertz Sci. Technol.* **2017**, *7*, 800–807. [CrossRef]
150. Song, H.J.; Ajito, K.; Muramoto, Y.; Wakatsuki, A.; Nagatsuma, T.; Kukutsu, N. Uni-Travelling-Carrier Photodiode Module Generating 300 GHz Power Greater than 1 MW. *IEEE Microw. Wirel. Compon. Lett.* **2012**, *22*, 363–365. [CrossRef]
151. Wun, J.M.; Wang, Y.W.; Shi, J.W. Ultrafast Uni-Traveling Carrier Photodiodes with $GaAs_{0.5}Sb_{0.5}/In_{0.53}Ga_{0.47}As$ Type-II Hybrid Absorbers for High-Power Operation at THz Frequencies. *IEEE J. Sel. Top. Quantum Electron.* **2018**, *24*, 8500207. [CrossRef]
152. Wakatsuki, A.; Furuta, T.; Muramoto, Y.; Yoshimatsu, T.; Ito, H. High-Power and Broadband Sub-Terahertz Wave Generation Using a J-Band Photomixer Module with Rectangular-Waveguide Output Port. In Proceedings of the 2008 33rd International Conference on Infrared, Millimeter and Terahertz Waves, Pasadena, CA, USA, 15–19 September 2008; IEEE: New York, NY, USA, 2008; pp. 1–2.
153. Ünlü, M.S.; Strite, S. Resonant Cavity Enhanced Photonic Devices. *J. Appl. Phys.* **1995**, *78*, 607–639. [CrossRef]
154. Ren, X.; Campbell, J.C. Theory and Simulations of Tunable Two-Mirror and Three-Mirror Resonant-Cavity Photodetectors with a Built-in Liquid-Crystal Layer. *IEEE J. Quantum Electron.* **1996**, *32*, 1903–1914.
155. Liu, K.; Huang, Y.; Ren, X. Theory and Experiments of a Three-Cavity Wavelength-Selective Photodetector. *Appl. Opt.* **2000**, *39*, 4263–4269. [CrossRef] [PubMed]
156. Duan, X.; Huang, Y.; Huang, H.; Ren, X.; Wang, Q.; Shang, Y.; Ye, X.; Cai, S. Monolithically Integrated Photodetector Array with a Multistep Cavity for Multiwavelength Receiving Applications. *J. Light. Technol.* **2009**, *27*, 4697–4702. [CrossRef]
157. Duan, X.; Huang, Y.; Ren, X.; Wang, W.; Huang, H.; Wang, Q.; Cai, S. Long Wavelength Multiple Resonant Cavities RCE Photodetectors on GaAs Substrates. *IEEE Trans. Electron Devices* **2011**, *58*, 3948–3953. [CrossRef]
158. Zhang, L.; Cao, Q.; Zuo, Y.; Xue, C.; Cheng, B.; Wang, Q. Wavelength-Tunable Si-Based InGaAs Resonant Cavity Enhanced Photodetectors Using Sol-Gel Wafer Bonding Technology. *IEEE Photonics Technol. Lett.* **2011**, *23*, 881–883. [CrossRef]
159. Fan, X.; Huang, Y.; Ren, X.; Duan, X.; Hu, F.; Wang, Q.; Cai, S.; Zhang, X. Hybrid Integrated Photodetector with Flat-Top Steep-Edge Spectral Response. *Appl. Opt.* **2012**, *51*, 5767–5772. [CrossRef]
160. Fan, X.; Huang, Y.; Ren, X.; Duan, X.; Zhang, X.; Hu, F.; Liu, Q.; Wang, Q.; Cai, S.; Zhang, X. A Novel Hybrid Integrated Photodetector Based on a Cone Absorption Cavity. *J. Light. Technol.* **2013**, *31*, 1234–1239. [CrossRef]
161. Duan, X.; Huang, Y.; Shang, Y.; Wang, J.; Ren, X. High-Efficiency Dual-Absorption InGaAs/InP Photodetector Incorporating GaAs/AlGaAs Bragg Reflectors. *Opt. Lett.* **2014**, *39*, 2447–2450. [CrossRef]
162. Liu, K.; Dong, X.; Huang, Y.; Duan, X.; Wang, Q.; Ren, X. Integrated Transceiving Chip Based on RCLED and PIN-PD for Plastic Optical Fiber Link. *Opt. Laser Technol.* **2023**, *158*, 108822. [CrossRef]
163. Umezawa, T.; Takamizawa, S.; Matsumoto, A.; Akahane, K.; Yamamoto, N.; Kanno, A.; Kawanishi, T. Resonant Cavity 4-λ Integrated 4×4 PD-Array for High Optical Alignment Robustness WDM-FSO Communications. *J. Light. Technol.* **2023**, *41*, 2465–2473. [CrossRef]
164. Chaudhary, R.; Arabhavi, A.M.; Ostinelli, O.; Leich, M.; Bolognesi, C.R. Type-II GaInAsSb Based Uni-Traveling Carrier Photodiodes with AlGaInAs/InP Bragg Reflectors for Improved 1.55 Mm Responsivity. In Proceedings of the 6th International Conference on Optics, Photonics and Lasers (OPAL'2023), Funchal, Portugal, 17–19 May 2023; pp. 80–82.

165. Umezawa, T.; Matsumoto, A.; Kanno, A.; Akahane, K.; Yamamoto, N. Application of Resonant-Cavity 4x4 Arrayed Photodetector for SDM-WDM-FSO Beam Detection. *J. Light. Technol.* **2024**, *42*, 1328–1334. [CrossRef]
166. Duan, X.; Huang, Y.; Ren, X.; Shang, Y.; Fan, X.; Hu, F. High-Efficiency InGaAs/InP Photodetector Incorporating SOI-Based Concentric Circular Subwavelength Gratings. *IEEE Photonics Technol. Lett.* **2012**, *24*, 863–865. [CrossRef]
167. Hu, J.; Huang, Y.; Ren, X.; Duan, X.; Li, Y.; Luo, Y. Realization of Quantum Efficiency Enhanced PIN Photodetector by Assembling Resonant Waveguide Grating. *Chin. Opt. Lett.* **2014**, *12*, 072301–072303.
168. Hu, J.H.; Huang, Y.Q.; Duan, X.F.; Wang, Q.; Zhang, X.; Wang, J.; Ren, X.M. Enhanced Absorption of Graphene Strips with a Multilayer Subwavelength Grating Structure. *Appl. Phys. Lett.* **2014**, *105*, 221113. [CrossRef]
169. Duan, X.; Wang, J.; Huang, Y.; Liu, K.; Shang, Y.; Zhou, G.; Ren, X. Mushroom-Mesa Photodetectors Using Subwavelength Gratings as Focusing Reflectors. *IEEE Photonics Technol. Lett.* **2016**, *28*, 2273–2276. [CrossRef]
170. Duan, X.; Zhang, M.; Huang, Y.; Liu, K.; Shang, Y.; Ren, X. Polarization-Independent Focusing Reflectors Using Two-Dimensional SWG. *IEEE Photonics Technol. Lett.* **2017**, *29*, 209–212. [CrossRef]
171. Chen, Q.; Fang, W.; Huang, Y.; Duan, X.; Liu, K.; Sharawi, M.S.; Ren, X. High Focusing-Reflection Subwavelength Gratings with Uni-Traveling-Carrier Photodetector for High Responsivity. In Proceedings of the Asia Communications and Photonics Conference (ACPC), Chengdu, China, 2–5 November 2019; p. S3D.4.
172. Liu, T.; Huang, Y.; Fei, J.; Wu, G.; Ma, X.; Duan, X.; Liu, K.; Ren, X. Research on Photodiode Integrated with Wide Spectrum Focusing Reflector Using Nonperiodic Subwavelength Gratings. *Chin. Opt. Lett.* **2018**, *16*, 051301.
173. Duan, X.; Chen, H.; Huang, Y.; Liu, K.; Cai, S.; Ren, X. Polarization-Independent High-Speed Photodetector Based on a Two-Dimensional Focusing Grating. *Appl. Phys. Express* **2017**, *11*, 012201. [CrossRef]
174. Wang, H.; Niu, H.; Jiang, C.; Fang, W.; Fan, X.; Zhang, X.; Bai, C. Symmetric Photodetector Integrated with Multilayer Dielectric Resonator Cavity for 400 Gb/s Optical Communication System. *Results Opt.* **2022**, *9*, 100324. [CrossRef]
175. Wang, J.; Liu, K.; Dong, X.; Duan, X.; Huang, Y.; Ren, X. Design of Uni-Traveling-Carrier Photodiode with Nanoscale Optical Microstructures. *Opt. Commun.* **2024**, *556*, 130305. [CrossRef]
176. Fattal, D.; Li, J.; Peng, Z.; Fiorentino, M.; Beausoleil, R.G. Flat Dielectric Grating Reflectors with Focusing Abilities. *Nat. Photonics* **2010**, *4*, 466–470. [CrossRef]
177. Yang, Y.; Huang, Y.; Ren, X.; Ye, X.; Duan, X.; Huang, H.; Wang, Q. Design Net-Grid Subwavelength Gratings for High Quantum Efficiency Photodetectors. *Adv. Mater. Res.* **2010**, *93–94*, 43–48. [CrossRef]
178. Li, G.; Duan, X.; Yuan, W.; Huang, Y.; Liu, K.; Ren, X. Quasi-Resonant Cavity Enhanced Photodetector with a Subwavelength Grating. *Chinese Opt. Lett.* **2022**, *20*, 031301. [CrossRef]
179. Liu, X.Q.; Yang, S.N.; Yu, L.; Chen, Q.D.; Zhang, Y.L.; Sun, H.B. Rapid Engraving of Artificial Compound Eyes from Curved Sapphire Substrate. *Adv. Funct. Mater.* **2019**, *29*, 1900037. [CrossRef]
180. Gao, Y.; Cansizoglu, H.; Polat, K.G.; Ghandiparsi, S.; Kaya, A.; Mamtaz, H.H.; Mayet, A.S.; Wang, Y.; Zhang, X.; Yamada, T.; et al. Photon-Trapping Microstructures Enable High-Speed High-Efficiency Silicon Photodiodes. *Nat. Photonics* **2017**, *11*, 301–308. [CrossRef]
181. Cansizoglu, H.; Bartolo-Perez, C.; Gao, Y.; Ponizovskaya Devine, E.; Ghandiparsi, S.; Polat, K.G.; Mamtaz, H.H.; Yamada, T.; Elrefaie, A.F.; Wang, S.-Y.; et al. Surface-Illuminated Photon-Trapping High-Speed Ge-on-Si Photodiodes with Improved Efficiency up to 1700 Nm. *Photonics Res.* **2018**, *6*, 734–742. [CrossRef]
182. Cansizoglu, H.; Wang, S.-Y.; Islam, M.S.; Mayet, A.S.; Ghandiparsi, S.; Gao, Y.; Bartolo-Perez, C.; Mamtaz, H.H.; Ponizovskaya Devine, E.; Yamada, T.; et al. Dramatically Enhanced Efficiency in Ultra-Fast Silicon MSM Photodiodes Via Light Trapping Structures. *IEEE Photonics Technol. Lett.* **2019**, *31*, 1619–1622. [CrossRef]
183. Gou, J.; Cansizoglu, H.; Bartolo-Perez, C.; Ghandiparsi, S.; Mayet, A.S.; Rabiee-Golgir, H.; Gao, Y.; Wang, J.; Yamada, T.; Devine, E.P.; et al. Rigorous Coupled-Wave Analysis of Absorption Enhancement in Vertically Illuminated Silicon Photodiodes with Photon-Trapping Hole Arrays. *Nanophotonics* **2019**, *8*, 1747–1756. [CrossRef]
184. Niu, H.; Huang, Y.; Yang, Y.; Liu, K.; Duan, X.; Wu, G.; Liu, T.; Wei, Q.; Ren, X. High Bandwidth-Efficiency Product MPIN Photodiode With Parallel-Connected Microstructure. *IEEE J. Quantum Electron.* **2020**, *56*, 4400305. [CrossRef]
185. Qarony, W.; Mayet, A.S.; Ponizovskaya-Devine, E.; Ghandiparsi, S.; Bartolo-Perez, C.; Ahamed, A.; Rawat, A.; Mamtaz, H.H.; Yamada, T.; Wang, S.-Y.; et al. Achieving Higher Photoabsorption than Group III-V Semiconductors in Ultrafast Thin Silicon Photodetectors with Integrated Photon-Trapping Surface Structures. *Adv. Photonics Nexus* **2023**, *2*, 056001. [CrossRef]
186. Shi, T.; Xiong, B.; Sun, C.; Luo, Y. Back-to-Back UTC-PDs with High Responsivity, High Saturation Current and Wide Bandwidth. *IEEE Photonics Technol. Lett.* **2013**, *25*, 136–139. [CrossRef]
187. Xie, X.; Zhou, Q.; Norberg, E.; Jacob-Mitos, M.; Chen, Y.; Ramaswamy, A.; Fish, G.; Bowers, J.E.; Campbell, J.; Beling, A. Heterogeneously Integrated Waveguide-Coupled Photodiodes on SOI with 12 DBm Output Power at 40 GHz. In Proceedings of the Optical Fiber Communication Conference Post Deadline Papers, Los Angeles, CA, USA, 22–26 March 2015; OSA: Washington, DC, USA, 2015; p. Th5B.7.
188. Liu, T.; Li, D.; Zhang, Y.; Yang, M.; Wang, Y.; Yang, T.; Shi, M.; Huang, Y.; Shi, Y.; Cao, Y.; et al. Design of Monolithic Distributed Bragg Reflector-Integrated Photodiode Using a Tapered Waveguide with InP and Polymer Cladding Layer. *Opt. Laser Technol.* **2021**, *144*, 107395. [CrossRef]

189. Liu, Y.; Jiang, K.; Jiang, Z.; Zhang, B.; Luo, D.; Liu, Y.; Qu, L.; Liu, W.; Wang, L. High Responsivity Evanescently Coupled Waveguide Photodiode Using Spot-Size Converter and Distributed Bragg Reflector at 1.55 Mm Wavelength. *Infrared Phys. Technol.* **2023**, *130*, 104619. [CrossRef]
190. Ghosh, S.; Lin, K.-C.; Tsai, C.-H.; Lee, K.H.; Chen, Q.; Son, B.; Mukhopadhyay, B.; Tan, C.S.; Chang, G.-E. Resonant-Cavity-Enhanced Responsivity in Germanium-on-Insulator Photodetectors. *Opt. Express* **2020**, *28*, 23739–23747. [CrossRef] [PubMed]
191. Jutzi, M.; Berroth, M.; Wöhl, G.; Oehme, M.; Kasper, E. Ge-on-Si Vertical Incidence Photodiodes with 39-GHz Bandwidth. *IEEE Photonics Technol. Lett.* **2005**, *17*, 1510–1512. [CrossRef]
192. D'Costa, V.R.; Cook, C.S.; Birdwell, A.G.; Littler, C.L.; Canonico, M.; Zollner, S.; Kouvetakis, J.; Menéndez, J. Optical Critical Points of Thin-Film $Ge_{(1-y)}Sn_{(y)}$ Alloys: A Comparative $Ge_{(1-y)}Sn_{(y)}/Ge_{(1-x)}Si_{(x)}$ Study. *Phys. Rev. B Condens. Matter Mater. Phys.* **2006**, *73*, 125207. [CrossRef]
193. Dong, Y.; Wang, W.; Lei, D.; Gong, X.; Zhou, Q.; Lee, S.Y.; Loke, W.K.; Yoon, S.-F.; Tok, E.S.; Liang, G.; et al. Suppression of Dark Current in Germanium-Tin on Silicon p-i-n Photodiode by a Silicon Surface Passivation Technique. *Opt. Express* **2015**, *23*, 18611–18619. [CrossRef] [PubMed]
194. Huang, B.-J.; Lin, J.-H.; Cheng, H.H.; Chang, G.-E. GeSn Resonant-Cavity-Enhanced Photodetectors on Silicon-on-Insulator Platforms. *Opt. Lett.* **2018**, *43*, 1215–1218. [CrossRef]
195. Chen, Q.; Wu, S.; Zhang, L.; Burt, D.; Zhou, H.; Nam, D.; Fan, W.; Tan, C.S. GeSn-on-Insulator Dual-Waveband Resonant-Cavity-Enhanced Photodetectors at the 2 Mm and 1.55 Mm Optical Communication Bands. *Opt. Lett.* **2021**, *46*, 3809–3812. [CrossRef]
196. Ito, H.; Hirota, Y.; Hirata, A.; Nagatsuma, T.; Ishibashi, T. 11 DBm Photonic Millimetre-Wave Generation at 100 GHz Using Uni-Travelling-Carrier Photodiodes. *Electron. Lett.* **2001**, *37*, 1225. [CrossRef]
197. Ito, H.; Ito, T.; Muramoto, Y.; Furuta, T.; Ishibashi, T. Rectangular Waveguide Output Unitraveling-Carrier Photodiode Module for High-Power Photonic Millimeter-Wave Generation in the F-Band. *J. Light. Technol.* **2003**, *21*, 3456–3462. [CrossRef]
198. Ito, H.; Furuta, T.; Nakajima, F.; Ito, T.; Muramoto, Y.; Ishibashi, T. Photonic Generation of Millimetre- and Submillimetre-Waves Using Rectangular-Waveguide-Output Uni-Travelling-Carrier Photodiode Module. *Electron. Lett.* **2004**, *40*, 387–388. [CrossRef]
199. Ito, H.; Furuta, T.; Muramoto, Y.; Ito, T.; Ishibashi, T. Photonic Millimetre- and Sub-Millimetre-Wave Generation Using J-Band Rectangular-Waveguide-Output Uni-Travelling-Carrier Photodiode Module. *Electron. Lett.* **2006**, *42*, 1424–1425. [CrossRef]
200. Furuta, T.; Ito, T.; Muramoto, Y.; Ito, H.; Tokumitsu, M.; Ishibashi, T. D-Band Rectangular-Waveguide-Output Uni-Travelling-Carrier Photodiode Module. *Electron. Lett.* **2005**, *41d*, 715–716. [CrossRef]
201. Kurokawa, T.; Ishibashi, T.; Shimizu, M.; Kato, K.; Nagatsuma, T. Over 300 GHz Bandwidth UTC-PD Module with 600 GHz Band Rectangular-Waveguide Output. *Electron. Lett.* **2018**, *54*, 705–706. [CrossRef]
202. Shen, X.; Costanzo, R.; Morgan, J.; Sun, K.; Woodson, M.; Estrella, S.; Beling, A.; Bowers, S.M. High-Power W-Band to G-Band Photonically-Driven Electromagnetic Emitter with 8.8 DBm EIRP. In Proceedings of the 2019 International Topical Meeting on Microwave Photonics (MWP), Ottawa, ON, Canada, 7–10 October 2019; IEEE: New York, NY, USA, 2019; pp. 1–4.
203. Nagatsuma, T. Photonic Measurement Technologies for High-Speed Electronics. *Meas. Sci. Technol.* **2002**, *13*, 1655–1663. [CrossRef]
204. Ito, H.; Nakajima, F.; Furuta, T.; Yoshino, K.; Hirota, Y.; Ishibashi, T. Photonic Terahertz-Wave Generation Using Antenna-Integrated Uni-Travelling-Carrier Photodiode. *Electron. Lett.* **2003**, *39*, 1828–1829. [CrossRef]
205. Ito, H.; Nagatsuma, T. High-Speed and High-Output-Power Unitraveling-Carrier Photodiodes. In *Active and Passive Optical Components for WDM Communications III, Proceedings of the ITCom, Orlando, FL, USA, 9–10 September 2003*; Dutta, A.K., Awwal, A.A.S., Dutta, N.K., Fujiura, K., Eds.; SPIE: Bellingham, WA, USA, 2003; Volume 5246, pp. 465–479.
206. Nellen, S.; Qian, T.; Schwanke, G.; Lauck, S.; de Felipe, D.; Kleinert, M.; Deumer, M.; Liebermeister, L.; Baier, M.; Globisch, B.; et al. Photonic-Enabled Beam Steering at 300 GHz Using a Photodiode-Based Antenna Array and a Polymer-Based Optical Phased Array. *Opt. Express* **2022**, *30*, 44701–44716. [CrossRef]
207. Renaud, C.C.; Robertson, M.; Rogers, D.; Firth, R.; Cannard, P.J.; Moore, R.; Seeds, A.J. A High Responsivity, Broadband Waveguide Uni-Travelling Carrier Photodiode. *Millim. Wave Terahertz Photonics* **2006**, *6194*, 61940C.
208. Hirata, A.; Nagatsuma, T.; Yano, R.; Ito, H.; Furuta, T.; Hirota, Y.; Ishibashi, T.; Matsuo, H.; Ueda, A.; Noguchi, T.; et al. Output Power Measurement of Photonic Millimetre-Wave and Sub-Millimetre-Wave Emitter at 100-800 GHz. *Electron. Lett.* **2002**, *38*, 798–800. [CrossRef]
209. Nakajima, F.; Furuta, T.; Ito, H. High-Power Continuous-Terahertz-Wave Generation Using Resonant-Antenna-Integrated Uni-Travelling-Carrier Photodiode. *Electron. Lett.* **2004**, *40*, 1297–1298. [CrossRef]
210. Monnai, Y.; Lu, X.; Sengupta, K. Terahertz Beam Steering: From Fundamentals to Applications. *J. Infrared Millim. Terahertz Waves* **2023**, *44*, 169–211. [CrossRef]
211. Che, M.; Kondo, K.; Doi, R.; Kato, K. Adaptive THz Beam Steering at UTC-PD Array by Genetic Algorithm. In Proceedings of the 2023 48th International Conference on Infrared, Millimeter, and Terahertz Waves (IRMMW-THz), Montreal, QC, Canada, 17–22 September 2023; IEEE: New York, NY, USA, 2023; pp. 1–2.
212. Che, M.; Matsuo, Y.; Kanaya, H.; Ito, H.; Ishibashi, T.; Kato, K. Optoelectronic THz-Wave Beam Steering by Arrayed Photomixers with Integrated Antennas. *IEEE Photonics Technol. Lett.* **2020**, *32*, 979–982. [CrossRef]
213. Che, M.; Kanaya, H.; Kato, K. Optically Controlled THz Power Tuning Based on Interference at Transmission Line. *Opt. Express* **2021**, *29*, 20034–20044. [CrossRef]
214. Siegel, P.H. Terahertz Technology. *IEEE Trans. Microw. Theory Tech.* **2002**, *50*, 910–928. [CrossRef]

215. Seddon, J.P.; Natrella, M.; Lin, X.; Graham, C.; Renaud, C.C.; Seeds, A.J. Photodiodes for Terahertz Applications. *IEEE J. Sel. Top. Quantum Electron.* **2022**, *28*, 3801612. [CrossRef]
216. Sengupta, K.; Nagatsuma, T.; Mittleman, D.M. Terahertz Integrated Electronic and Hybrid Electronic–Photonic Systems. *Nat. Electron.* **2018**, *1*, 622–635. [CrossRef]
217. Eichler, D.T. THz Generation and Analysiswith Electronic and Photonic. *Microw. J.* **2023**, *66*, 24–36.
218. Tonouchi, M. Cutting-Edge Terahertz Technology. *Nat. Photonics* **2007**, *1*, 97–105. [CrossRef]
219. Siegel, P.H. Terahertz Technology in Biology and Medicine. *IEEE Trans. Microw. Theory Tech.* **2004**, *52*, 2438–2447. [CrossRef]
220. Shur, M. Terahertz Technology: Devices and Applications. In Proceedings of the Proceedings of 35th European Solid-State Device Research Conference, Grenoble, France, 12–16 September 2005; ESSDERC 2005. IEEE: New York, NY, USA, 1974; Volume 83, pp. 13–21.
221. Giboney, K.S. Traveling-Wave Photodetector Theory. *IEEE Trans. Microw. Theory Tech.* **1997**, *45*, 1310–1319. [CrossRef]
222. Rouvalis, E.; Renaud, C.C.; Moodie, D.G.; Robertson, M.J.; Seeds, A.J. Continuous Wave Terahertz Generation from Ultra-Fast InP-Based Photodiodes. *IEEE Trans. Microw. Theory Tech.* **2012**, *60*, 509–517. [CrossRef]
223. Shi, J.-W.; Kuo, F.-M.; Chiueh, T.; Teng, H.-F.; Tsai, H.J.; Chen, N.-W.; Wu, M. Photonic Generation of Millimeter-Wave White-Light at W-Band Using a Very Broadband and High-Power Photonic Emitter. *IEEE Photonics Technol. Lett.* **2010**, *22*, 847–849. [CrossRef]
224. Ito, H.; Furuta, T.; Hirota, Y.; Ishibashi, T.; Hirata, A.; Nagatsuma, T.; Matsuo, H.; Noguchi, T.; Ishiguro, M. Photonic Millimetre-Wave Emission at 300 GHz Using an Antenna-Integrated Uni-Travelling-Carrier Photodiode. *Electron. Lett.* **2002**, *38*, 989–990. [CrossRef]
225. Wun, J.-M.; Chen, N.-W.; Shi, J.-W. THz Photonic Transmitters with Type-II Hybrid Absorber UTC-PDs and Dual-Ridged Horn Antennas for High-Power and Extremely Wide Fractional Bandwidth Performances. In Proceedings of the 2018 IEEE Photonics Conference (IPC), Reston, VA, USA, 30 September–4 October 2018; IEEE: New York, NY, USA, 2018; pp. 1–2.
226. Beck, A.; Ducournau, G.; Zaknoune, M.; Peytavit, E.; Akalin, T.; Lampin, J.F.; Mollot, F.; Hindle, F.; Yang, C.; Mouret, G. High-Efficiency Uni-Travelling-Carrier Photomixer at 1550 Nm and Spectroscopy Application up to 1.4 THz. *Electron. Lett.* **2008**, *44*, 1320–1321. [CrossRef]
227. Ito, H.; Ishibashi, T. Photonic Terahertz-Wave Generation Using Slot-Antenna-Integrated Uni-Traveling-Carrier Photodiodes. *IEEE J. Sel. Top. Quantum Electron.* **2017**, *23*, 3800907. [CrossRef]

applied sciences

MDPI

Article

Water Pipeline Leakage Detection Based on Coherent φ-OTDR and Deep Learning Technology

Shuo Zhang [1], Zijian Xiong [1], Boyuan Ji [1], Nan Li [1,2], Zhangwei Yu [3], Shengnan Wu [2,*] and Sailing He [1,2,4,*]

1 National Engineering Research Center for Optical Instruments, College of Optical Science and Engineering, Zhejiang University, Hangzhou 310058, China
2 Ningbo Research Institute, Zhejiang University, Ningbo 315100, China
3 Key Laboratory of Optical Information Detection and Display Technology of Zhejiang, Zhejiang Normal University, Jinhua 321004, China
4 Department of Electromagnetic Engineering, School of Electrical Engineering, Royal Institute of Technology, SE-100 44 Stockholm, Sweden
* Correspondence: wushengnan@zju.edu.cn (S.W.); sailing@kth.se (S.H.)

Abstract: Leakage in water supply pipelines remains a significant challenge. It leads to resource and economic waste. Researchers have developed several leak detection methods, including the use of embedded sensors and pressure prediction. The former approach involves pre-installing detectors inside pipelines to detect leaks. This method allows for the precise localization of leak points. The stability is compromised because of the wireless signal strength. The latter approach, which relies on pressure measurements to predict leak events, does not achieve precise leak point localization. To address these challenges, in this paper, a coherent optical time-domain reflectometry (φ-OTDR) system is employed to capture vibration signal phase information. Subsequently, two pre-trained neural network models based on CNN and Resnet18 are responsible for processing this information to accurately identify vibration events. In an experimental setup simulating water pipelines, phase information from both leaking and non-leaking pipe segments is collected. Using this dataset, classical CNN and ResNet18 models are trained, achieving accuracy rates of 99.7% and 99.5%, respectively. The multi-leakage point experiment results indicate that the Resnet18 model has better generalization compared to the CNN model. The proposed solution enables long-distance water-pipeline precise leak point localization and accurate vibration event identification.

Keywords: leakage detection; φ-OTDR; deep leaning; Mels spectrograms

1. Introduction

Water is an indispensable and important substance for the survival of human beings and all living things and is an irreplaceable and extremely valuable natural resource for industrial and agricultural production, economic development, and environmental improvement. However, many countries in the world are facing the problem of water shortage. Take China as an example. Although China is a water-rich nation, it also suffers from severe water shortages on a per capita basis. According to statistics, China's per capita water resources comprise less than 2200 cubic meters, less than one-third of the world's per capita water resources [1]. Among the 660 cities in China, approximately 400 cities are short of water. With the development and construction of the country, the water usage rates of various cities are increasing year by year, and the lengths of urban water supply pipelines are also increasing. Consequently, society places higher demands on the quality of urban water supply networks, which is crucial for ensuring stable development [2–4]. However, in terms of the construction of China's overall municipal water supply network, many places still have the problem of pipeline leakage, which causes residents to be forced to cut off water and even causes accidents such as road collapse, resulting in the wastage of water resources and economic damage. The primary problem for reducing the leakage rate in a city concerns how to accurately know that the water supply pipeline has leaked, and it is necessary to accurately locate the location of the leakage point so as to be able to quickly repair and reduce the wastage of resources. Therefore, it is necessary to develop a system that can monitor the precise location of leaks in real time when leaks occur in the water supply network.

Mainstream pipeline leak detection technologies include the magnetic flux leakage (MFL) detection method, pipeline robot inspection method, water pressure prediction method, and Distributed Acoustic Sensing (DAS) method. The MFL detection method is a commonly used method for pipeline inspection [5]. The principle behind MFL detection involves fully magnetizing the pipeline, causing magnetic field lines to deform at the locations of defects. These deformed lines flow out from a defect and then return to the interior of the pipeline. Sensors carried by the inspection system detect and store this information, allowing the accurate localization of pipeline leaks. While this approach offers high precision in leak localization, it is limited to ferromagnetic pipelines. The pipeline robot inspection method uses robots within the pipelines to monitor their condition. Simultaneously, a wireless sensor network is installed on the ground to receive data [6,7]. However, this approach faces several challenges. Firstly, in practical scenarios, pipes are often buried deep underground, hindering wireless signal transmission. Additionally, there is a risk of robots becoming stuck within the pipes, leading to blockages. The water pressure prediction method uses water pressure values to predict water leakage. Mashford et al. employed Support Vector Machines (SVMs) to classify data from the EPANET hydraulic modeling system, resulting in improved leak prediction [8]. Leu et al. further enhanced leak prediction accuracy through Bayesian learning processes [9]. The proposed solution suffers from a limitation: it cannot provide precise information about the exact locations of leaks. Distributed Acoustic Sensing (DAS) tackles water leakage detection by capturing vibration signals associated with leak locations. DAS is widely applicable to both ferromagnetic and non-ferromagnetic pipelines, offering the precise localization of vibration signals. In 2015, Huijuan Wu et al. successfully detected water pipe leaks by deploying communication optical fibers along the inner walls of pipelines, achieving leak detection when under conditions where the aperture exceeded 4 mm and internal pressure exceeded 0.2 MPa [10]. In 2022, Shichong Fu et al. utilized DAS systems to achieve meter-level spatial resolution for pinpointing gas pipeline leak points. Their study also explored how leak hole size and leak direction impact the spectral characteristics of leak signals [11]. However, none of the aforementioned works in the literature delved into the discussion of leak event recognition accuracy.

In practice, while achieving leak localization is feasible, false positives and false negatives remain significant challenges for DAS systems. Initially, researchers relied on threshold settings for accurate vibration event identification. In 2014, Zhu et al. set a level threshold to determine the event category [12]. In 2018, Jiang et al. used the amplitude of the vibration signal as the basis for judging whether there was a vibration event by setting a threshold and achieved a recognition accuracy of 95.57% for a specific "instantaneous destructive disturbance" in a laboratory environment [13]. However, pattern recognition is only performed through thresholds, and its recognition accuracy and stability are poor in practical application scenarios. It is difficult to ensure the stability of the signal-to-noise ratio of the received signal due to the attenuation of the signal in the process of long-distance transmission, coupled with the influence of coherent fading and polarization fading, which leads to the inaccuracy of the frequency characteristics demodulated. The disturbance signals on the optical fiber are intricately intertwined, and even the vibration signals that need to be detected are dynamically changing, which further leads to a reduction in the recognition accuracy of the system.

Pattern recognition through machine learning is a popular solution. By analyzing existing data and selecting appropriate algorithms, a model can be built and continuously optimized, and finally, various events can be identified quickly and accurately. The concept of machine learning was first proposed by the American computer scientist Arthur Samuel in 1959 as a branch of artificial intelligence (AI) and computer science that focuses on the use of data and algorithms to enable AI to mimic human learning methods and gradually improve its recognition accuracy [14]. In 1998, LeCun et al. proposed a model architecture for convolutional neural networks (CNNs) and applied them to the task of handwriting number recognition [15]. The LeNet-5 model proposed by LeCun achieved excellent performance on the MNIST dataset, marking the birth of deep learning and achieving more complex feature extraction. In 2015, Kaiming He et al. proposed the ResNet18 neural network, which solves the gradient vanishing problem that may occur when the number of layers in traditional CNN networks increases [16]. The emergence of these models provides a strong guarantee for the accurate realization of pattern recognition.

In this paper, a coherent φ-OTDR system is employed to capture vibration signal phase information and a pre-trained neural network model is responsible for processing this information to accurately identify leakage events and locations. The proposed method is evaluated in a simulated water pipelines setup. Phase information from both leaking and non-leaking pipe segments is collected to create separate datasets for training different models: CNN and ResNet18. During practical testing, the test accuracy values of the CNN model and ResNet18 model achieve 99.7% and 99.5%, respectively. What is more, in the multi-leakage point detection experiment, the ResNet18 model performs better, which indicates that Resnet18 has better generalization ability compared to the CNN model.

2. Sensing Principle of φ-OTDR

Optical pulses propagate through optical fibers, and due to the uneven refractive index distribution of fibers, Rayleigh scattering occurs [17]. Some of the scattered light will be back-guided towards the input end of the fiber and will be detected with a delay proportional to the distance of the diffusing point, which is called optical time-domain reflection.

There is a relationship between the length d of the optical fiber link through which the incident light passes and the time delay τ between the incident light and the scattered light:

$$d = \frac{c}{2n}\tau \tag{1}$$

Here, c represents the speed of light in a vacuum and n denotes the effective refractive index of the fundamental mode of SMF.

The system's positioning accuracy, denoted as z, primarily depends on the pulse width, represented by T; then, the spatial resolution z can be written as follows:

$$z = \frac{Tc}{2n} \tag{2}$$

When the refractive index of fiber section at the corresponding location undergoes slight changes, an initial determination of the signal position can be achieved. However, within the optical fiber, the power of Rayleigh scattering light is weak, making it challenging to work over long-distance fiber links. To enhance the sensitivity of scattered light detection, coherent detection is commonly employed. The optical time-domain reflectometer (OTDR) technology based on coherent detection is referred to as a Coherent Optical Time-Domain Reflectometer (COTDR) [18,19]. The schematic diagram of a COTDR is shown in Figure 1.

The laser beam generated by the narrow-linewidth laser (NLL) is divided into two paths through a fiber coupler. The upper path serves as the probing light while the other acts as the local oscillator light. The probing light passes through an acousto-optic modulator (AOM), where it is modulated into optical pulses. These pulses are injected into the optical fiber through a circulator, resulting in Rayleigh scattering along the fiber under test (FUT). The scattered light then enters the coupler and interferes with the local oscillator light. The interference results in an optical intensity proportional to the local oscillator's light amplitude. Coherent detection effectively amplifies the scattered light.

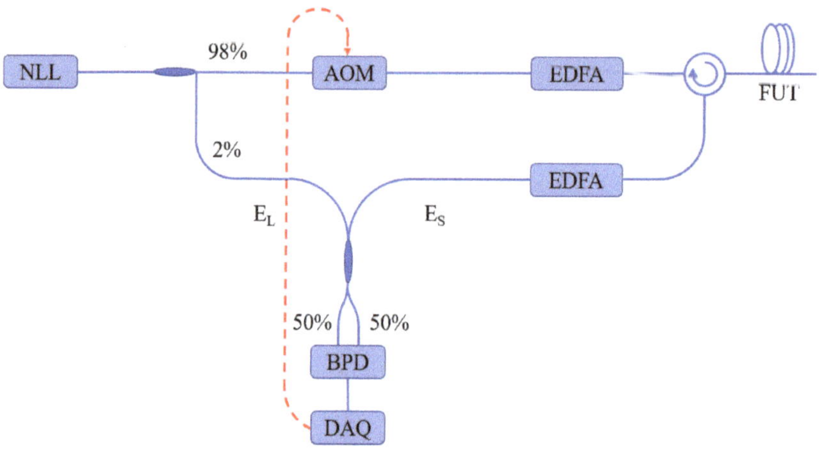

Figure 1. Framework diagram of COTDR. Abbreviations—NLL: narrow-linewidth laser; AOM: acousto-optic modulator; EDFA: erbium-doped fiber amplifier; BPD: balanced photodetector; DAQ: data acquisition card.

In Figure 1, the local oscillator is denoted as E_L and the scattered light is denoted as E_S, and according to Maxwell's equations, we can derive the following:

$$E_L = E_0{}^{j\omega_0 t} \tag{3}$$

Here, E_0 represents the amplitude of the local oscillator, ω_0 represents the angular frequency of the local oscillator, and t represents the detection time.

$$E_s = rE_1 e^{-\alpha z} e^{j(\omega_s t + 2\beta z + \varphi_s)} \tag{4}$$

Here, r represents the Rayleigh scattering coefficient, E_1 represents the incident light amplitude, α represents the fiber attenuation coefficient, z represents the location of the Rayleigh scattering point, ω_s represents the angular frequency of the detection light, β represents the propagation constant of the propagating mode at the operating wavelength, t represents the detection time, and φ_s represents the Rayleigh scattering phase shift.

The intensity of the light after the interference between the scattered light and the local oscillator is denoted as I. The light output of the light source is linearly polarized; in the case of considering that the scattered light is also ideally linearly polarized, the expression for I can be obtained according to the interference principle:

$$I = E_0{}^2 + r^2 E_1{}^2 e^{-2\alpha z} \pm 2rE_0 E_1 e^{-\alpha z} \gamma \cos\theta \cos(\Delta\omega t + \varphi) \tag{5}$$

Here, the symbols represent the following parameters: γ denotes the spectral correlation factor, θ represents the polarization angle, $\Delta\omega$ signifies the angular frequency shift introduced by the modulator ($\Delta\omega = \omega_s - \omega_0$), and φ represents the system phase shift ($\varphi = 2\beta z + \varphi_s$, where $2\beta z$ corresponds to the transmission phase shift and φ_s represents the Rayleigh scattering phase shift).

When a water pipe experiences a leak, it generates acoustic waves at a certain frequency near the leakage point, which will introduce phase shift η_s when applied to the sensing fiber. The schematic diagram of an optical pulse passing through external disturbances is shown in Figure 2. Here, point A represents the position before the external disturbance test point and point B represents the position after the test point. The phase of the optical pulse at point A is denoted as η_0. After encountering the external disturbance, it acquires an additional phase shift η_s, resulting in a phase of $\eta_0 + \beta\Delta z + \eta_s$ at point B. Upon the return of scattered light, it again encounters external disturbances, introducing an additional phase shift η_s, resulting in a phase of $\eta_0 + 2\beta\Delta z + 2\eta_s$ at point A. It should be noted that Δz is the distance between points A and B. Leveraging this, demodulating phase information enables the extraction of disturbance frequency information, facilitating subsequent event type recognition using deep learning techniques.

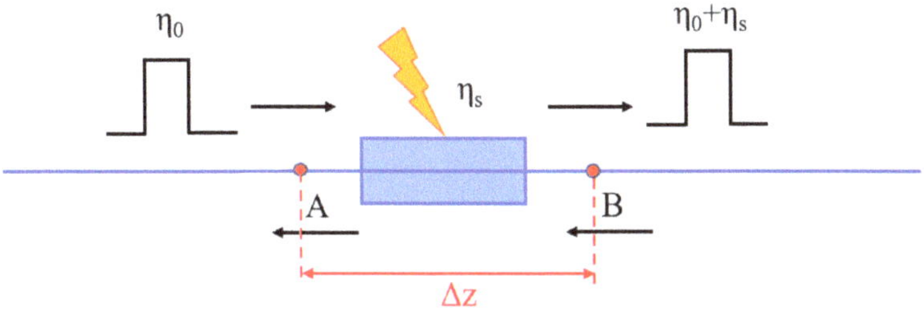

Figure 2. Schematic diagram of the optical signal at the disturbance location.

3. Data Collection and Preprocessing

3.1. Experimental Setup

The schematic of the φ-OTDR system in this paper is illustrated in Figure 1. A narrow-linewidth laser (UNFSRL-1550-20-PMF-FC/APC-M4, Tanguang Technology Co., Ltd., Wuhan, China) with a line width of 1 kHz emits continuous light, which is split into two beams after passing through a coupler. Among these beams, the probing light is modulated by an acousto-optic modulator (G-1550-80-L-D-T-AA-G3-T-L, Tanguang Technology Co., Ltd., Wuhan, China) to generate pulsed signals with a pulse width of 100 ns and a repetition frequency of 10 kHz. Additionally, this modulation introduces an 80 MHz frequency shift. The modulated probing light then enters the tested optical fiber (G657A1, Guangzhou Pingtong Optical Cable Co., Ltd., Guangzhou, China) via an erbium-doped fiber amplifier (PMEDFA-PA35-09-1-1-M1, Tanguang Technology Co., Ltd., Wuhan, China) following amplification. The scattered signal from the probing light, further amplified by the erbium-doped fiber amplifier, combines with the local oscillator light emitted by the narrow-linewidth laser and enters the coupler. Subsequently, the signal is directed to a balanced detector (MBD-350M-A, Tanguang Technology Co., Ltd., Wuhan, China) with a bandwidth of 350 MHz. Finally, the received signal is sampled by a data acquisition card (PCIE-6920-DAQ-250 MSps, Guangyi Intelligent Technology Co., Ltd, Guilin, China) operating at a sampling rate of 250 MHz.

3.2. Sensing Characteristic

In our study, the accuracy of vibration signal restoration was validated using a Piezoelectric Transducer (PZT) as the vibration source. Multiple sinusoidal signals with peak voltages of 600 mV and frequencies of 100 Hz, 150 Hz, 200 Hz, and 250 Hz were applied to the PZT. By utilizing a φ-OTDR system, the phase evolution chart of the demodulated signal within a time window of 0.1 s was obtained. This approach successfully reconstructed the original vibration signal. The corresponding power spectral density (PSD) is displayed beneath the respective time domain chart in this paper. From Figure 3, it is evident that the peak frequencies in the four sets of PSD charts align precisely with the vibration signal frequencies, thereby confirming the accuracy of vibration signal restoration. The signal-to-noise ratios for the datasets were 40.0 dB, 42.0 dB, 45.0 dB, and 51.9 dB.

A sinusoidal signal with a frequency of 100 Hz and a peak voltage from 100 mV to 600 mV in 100 mV increments was applied to the PZT, and the corresponding differential phase response peak was recorded. The obtained voltage–phase response curve was linearly fitted to obtain its fitting curve. To demonstrate the reproducibility of the experiment, the above operations were repeated at 150 Hz, 200 Hz, and 250 Hz, and three new fitting curves were obtained; these are summarized and displayed in Figure 4. The slopes of the four curves were almost identical, being 4.56 rad/V, 4.56 rad/V, 4.52 rad/V, and 4.64 rad/V, and the correlation coefficients were 0.99913, 0.99896, 0.99927, and 0.99924, respectively. This confirms the stability and reproducibility of the system's demodulation.

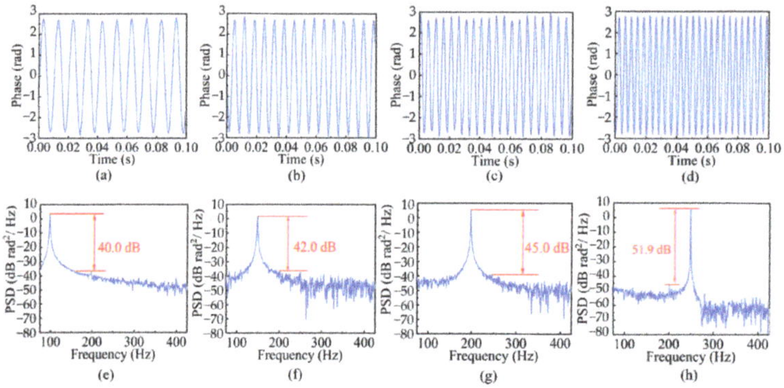

Figure 3. Phase demodulation results of vibration signals at different frequencies: (**a**) 100 Hz; (**b**) 150 Hz; (**c**) 200 Hz; (**d**) 250 Hz. Phase demodulation PSD plots of vibration signals at different frequencies: (**e**) 100 Hz; (**f**) 150 Hz; (**g**) 200 Hz; (**h**) 250 Hz.

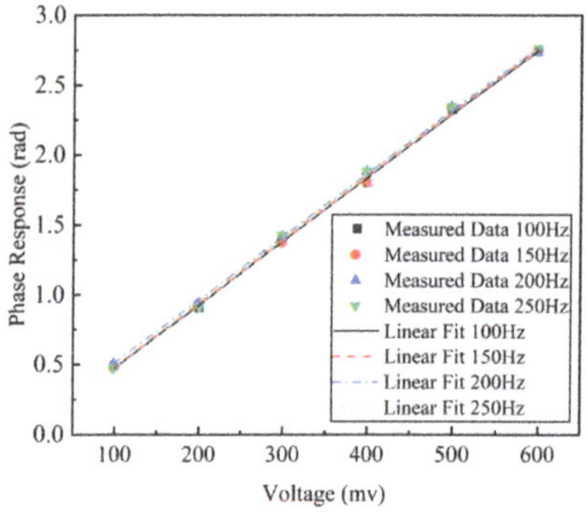

Figure 4. Phase response results at different voltages and frequencies.

3.3. Dataset Production

The data collection platform in this paper simulates a water pipeline leakage environment. The sensing fiber cable is tightly installed to the water pipeline using nylon ribbon, and an opening is created on the water pipe to simulate a leak. To simulate long-distance detection scenarios, the fiber length is 4 km, and the pulse width is set to 100 ns, corresponding to a spatial resolution of 10 m. Figure 5a is the picture simulating the water pipeline and Figure 5b is the picture of a leakage point, which is a fixed-size hole with a diameter of about 5 mm that is located at the back of the water pipe with a distance of a few centimeters from the sensing cable.

Data from operation both with and without leakage events in the water pipe were collected on the aforementioned experimental platform. The data were then divided into training and testing sets. The sample quantities for the two distinct event categories are shown in Table 1, with each individual sample containing 10,240 data points. Figure 6 illustrates five sets of the original phase signals for the water pipe under both non-leakage and leakage conditions. Notably, even during non-leakage periods, the water pipe exhibits low-amplitude, low-frequency vibrations, which are caused by water flow impacting the pipe walls and other environmental disturbances. In addition, since some leakage signals are quite similar to the case of non-leakage, it is difficult to distinguish leakage events through original phase signals.

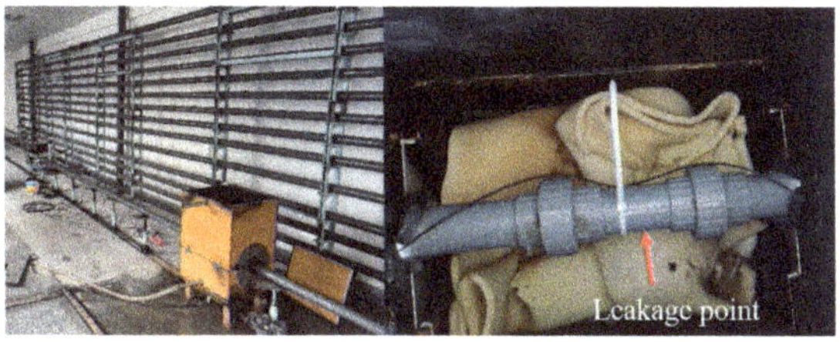

(a) (b)

Figure 5. (**a**) Simulating water pipeline; (**b**) leakage point.

Table 1. Number of samples per category.

Dataset	No Leakage	Leakage
Training set	13,699	1951
Test set	1574	120

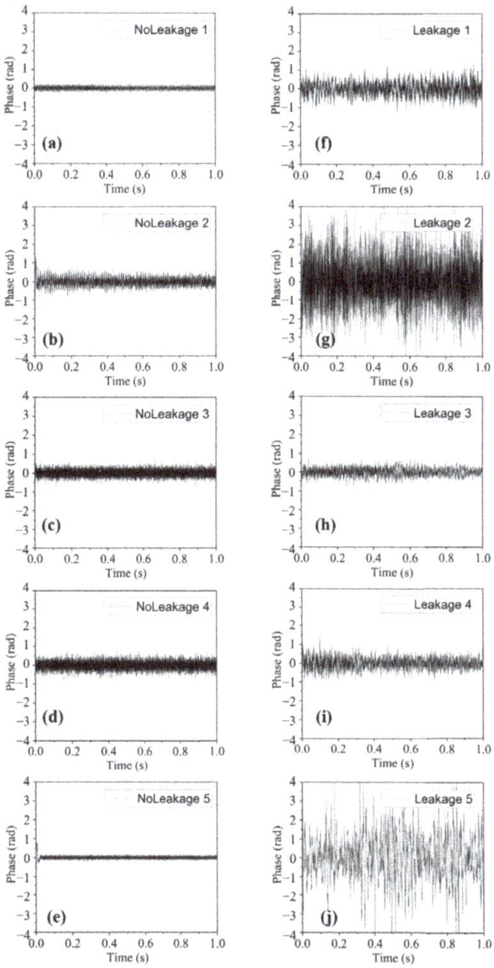

Figure 6. Phase signals of different events: (**a–e**) no leakage; (**f–j**) leakage.

To extract as much effective information as possible from the collected data, original phase data were converted into Mel spectrograms. Mel spectrograms are commonly used in audio signal processing as feature representations. They decompose audio signals into energy distributions across different frequencies, providing richer information. Specifically, Mel spectrograms map linear spectra (typically obtained through Fast Fourier Transform) onto the Mel scale, which better aligns with human auditory perception.

The data were sampled at a rate of 10 KHz, with an FFT window size of 1024 and a frame shift size of 256 and Mel filter number of 40. Figure 7 provides five sets of Mel spectrograms corresponding to both non-leakage and leakage conditions in the water pipe.

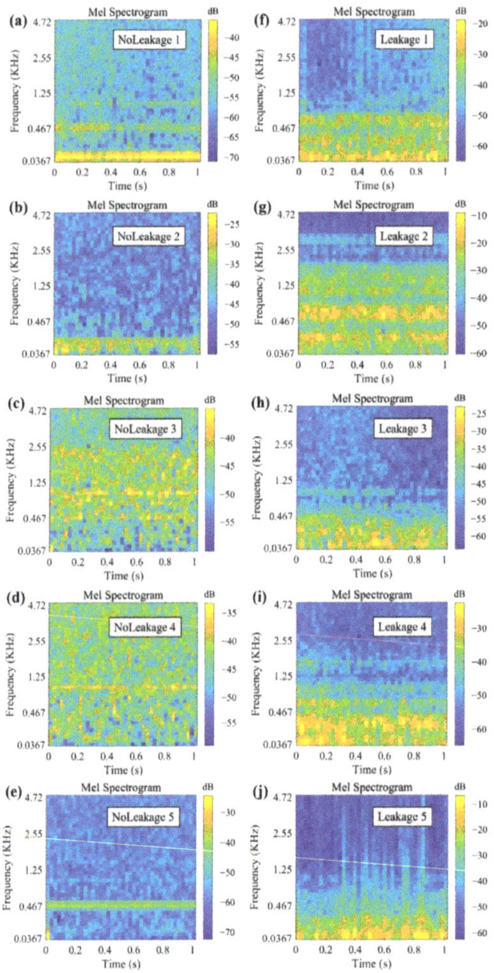

Figure 7. Mel spectrum diagrams of different events: (**a**–**e**) no leakage; (**f**–**j**) leakage.

4. Deep Learning Training

4.1. CNN with Mel Spectrograms as Input

When dealing with vibration audio signals, more complex feature representations are often required. Simple one-dimensional data input may not suffice, leading to suboptimal model performance. Therefore, as described in this section, Mel spectrograms were transformed from the one-dimensional phase data, then the Mel spectrograms that served as the input for the CNN model.

After generating Mel spectrograms, original phase data were transferred into two-dimensional images, where the horizontal axes represented time and the vertical axes represented energy at different frequencies. This allowed us to input them into the CNN model, capturing audio signal features more effectively. The model was built and trained using TensorFlow 2.6 GPU and Python 3.8. When using Mel spectrograms as input data, the corresponding model employs two-dimensional convolutional layers with a kernel size of 3×3 and a pooling layer window size of 2×2. Other basic settings remain unchanged, and the neural network architecture has been appropriately optimized, as illustrated in the diagram shown in Figure 8:

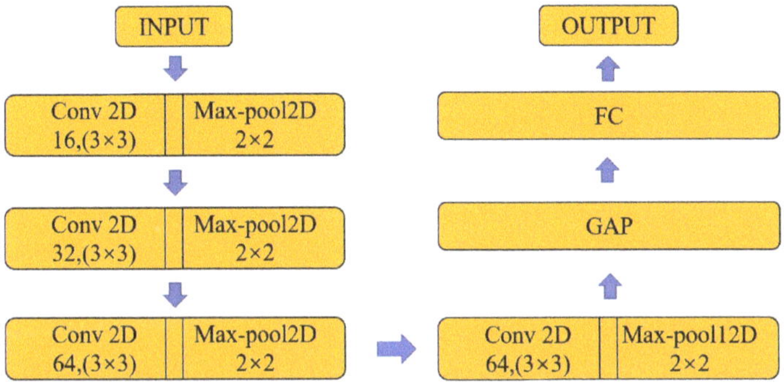

Figure 8. Structure of the CNN model. Abbreviations—FC: fully connected layer; GAP: global average pooling.

In the process of model training, we continued to utilize the Adam optimizer, with batch size, learning rate, and training epoch values set at 32, 0.001, and 200. The corresponding training curve is depicted in Figure 9a. Using the model, we identified the test dataset and obtained a confusion matrix as shown in Figure 9b. The achieved accuracy was 99.7%.

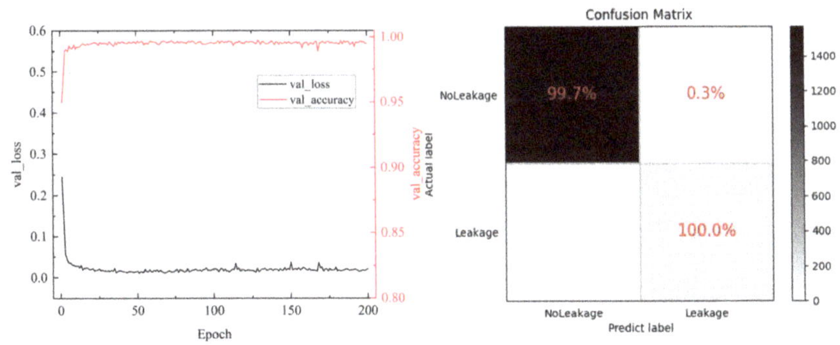

Figure 9. Training results of the CNN model: (**a**) training curves; (**b**) confusion matrix.

4.2. ResNet18 with Mel Spectrograms as Input

In traditional neural networks, increasing the depth of the network is a common strategy to enhance recognition accuracy by extracting higher-level features from images. However, when the network depth reaches a certain point, the issue of gradient vanishing arises. Deep networks may perform worse than shallow ones due to this phenomenon. To address this challenge, in 2015, He et al. proposed the ResNet18 neural network, which introduced residual blocks with shortcut connections to mitigate gradient vanishing. In conventional deep neural networks, during backpropagation, gradients are multiplied layer by layer with weight matrices, leading to gradual reduction and the eventual disappearance of gradients. In the ResNet model, the shortcut connections directly transmit gradients from the input to subsequent layers, effectively preventing gradient vanishing. A simple residual block structure is illustrated in Figure 10, where input data x_i is transformed via weighted operations into f(x) and the final output x_{i+1} is obtained by adding x and f(x). The specific model architecture is shown in Figure 11.

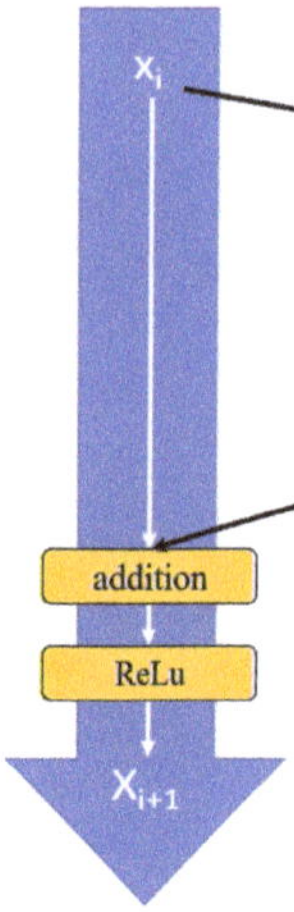

Figure 10. Residual block structure diagram.

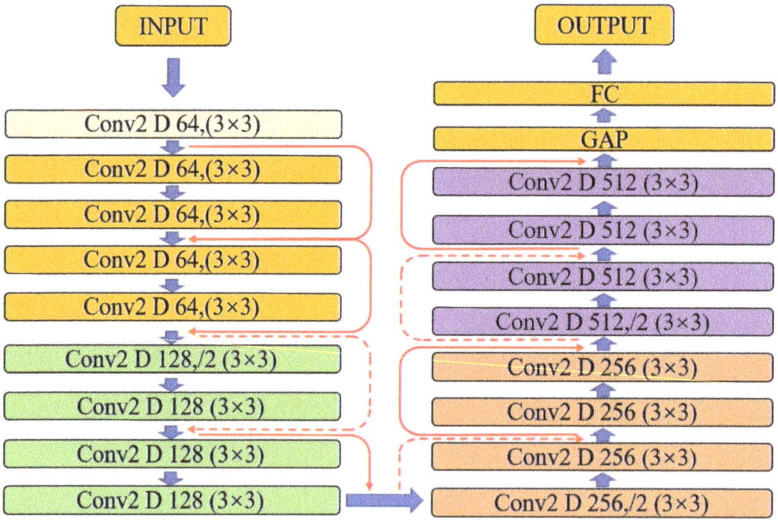

Figure 11. Structure of the ResNet18 model.

Specifically, ResNet18 consists of multiple residual blocks, each containing two 3×3 convolutional layers and a skip connection that adds the original input to the output. These skip connections facilitate the easier training of deep networks. The ResNet18 architecture comprises eighteen layers, including sixteen convolutional layers and two fully connected layers. Convolutional layers are organized into several residual blocks, each incorporating multiple convolutional layers, batch normalization, and ReLU activation. Compared to deeper variants like ResNet50 and ResNet101, ResNet18 strikes a balance between training speed and network depth, with lower complexity and computational resource requirements. In our study, we constructed and trained the ResNet18 model using PyTorch-GPU 0.0.1 and Python 3.8. We adjusted the output dimension of the fully connected layer to 2 for binary classification. Input data consisted of Mel spectrograms derived from raw data. During training, we employed the Adam optimizer with a learning rate of 0.01 and conducted 200 epochs. The corresponding training curve is depicted in Figure 12a. Using the model, we identified the test dataset and obtained a confusion matrix as shown in Figure 12b. The achieved accuracy was 99.5%.

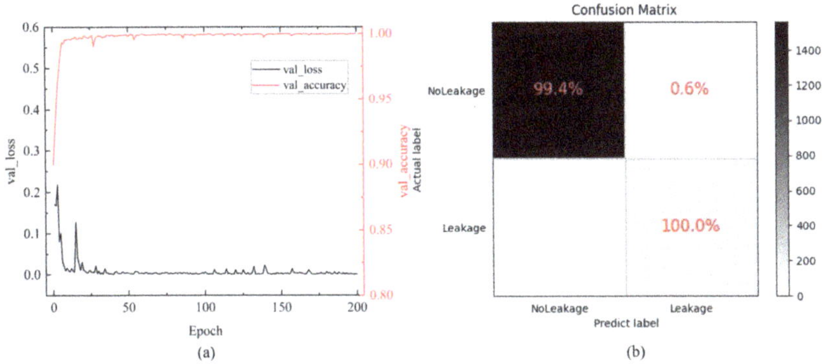

Figure 12. Training results of the ResNet18 model: (**a**) training curves; (**b**) confusion matrix.

4.3. Multi-Leakage Point Detection

In order to validate and compare the two models described above, we further performed experiments with multiple leak points. In this experiment, three leakage points were set at about 2070 m, 2100 m, and 2150 m from the start of the fiber. The two leak points at 2070 m and 2150 m were simulated by ball valves, and the maximum opening diameter was 20 mm. The leak point at 2100 m was a fixed-size hole with a diameter of about 5 mm. Then, using the phase-sensitive OTDR system described above to collect the experimental data, a total of 100 groups of raw phase data were collected, and each group contained 767 phase data, that is, corresponding to the phase data of all sampling points on the entire FUT. Then, these 100 groups of data were input to the previously trained CNN and ResNet18 models for prediction and to perform statistical analyses of the prediction results, as shown in Figure 13a,b. Experimental results showed that Resnet18 could effectively locate the locations of three leakage points while CNN could only identify leakage events at 2100 m, and the performance of the ResNet18 model at other non-leakage points was also significantly better than that of the CNN model.

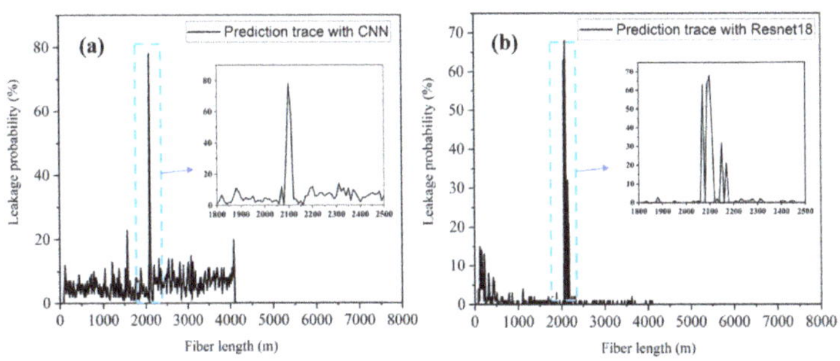

Figure 13. Multi-leakage points: predicted results of the (**a**) CNN model and (**b**) ResNet18 model.

In summary, the CNN and ResNet18 models trained on Mel spectrograms, which contain more informative features, demonstrated good validation accuracy in the classification. However, although, in the previous training and testing phases, the CNN model and the ResNet18 model had similar accuracy, in the multi-leakage point detection experiment, the ResNet18 model performed better, which indicates that Resnet18 has better generalization ability compared to the CNN model when processing new input data. ResNet18 is also the most complex in terms of architecture, computational requirements, and training time.

5. Conclusions

In response to the issue of water pipeline leaks in long-distance transportation, we analyzed, in this paper, the fundamental principles of the phase-sensitive optical time-domain reflectometry system for locating and deciphering disturbance signals. Furthermore, the φ-OTDR system was combined with a deep learning technique to enhance the detection rate of leakage events. By utilizing Mel spectrograms as inputs to a convolutional neural network (CNN) and ResNet18, recognition accuracies of 99.7% and 99.5% were achieved on the validation dataset. In addition, in the multi-leakage point detection experiment, the ResNet18 neural network had better generalization ability compared to the CNN model and successfully located three pre-set leakage points. This has demonstrated that it has good application prospects in the field of monitoring the leakage of long-distance water pipelines.

Author Contributions: Conceptualization, S.W. and S.H.; methodology, S.W. and S.H.; software, S.Z., B.J. and S.W.; validation, S.W. and S.H.; formal analysis, S.Z., Z.Y. and S.W.; investigation, S.Z., Z.X., N.L. and S.W.; resources, S.H.; data curation, S.Z., Z.X. and N.L.; writing—original draft preparation, S.Z.; writing—review and editing, S.W. and S.H.; visualization, S.Z.; supervision, S.W. and S.H.; project administration, S.H.; funding acquisition, S.H. All authors have read and agreed to the published version of the manuscript.

Funding: This work was supported partially by the Ningbo Science and Technology Project (2021Z030, 2023Z179), the National Key Research and Development Program of China (2022YFC3601002), the Key Research and Development Program of Zhejiang Province (2021C03178, 2022C03051, 2023C03135), and the Scientific Research Foundation for Talent Introduction of Zhejiang University's Ningbo Campus (20201203Z0180).

Institutional Review Board Statement: Not applicable.

Informed Consent Statement: Not applicable.

Data Availability Statement: Data underlying the results presented in this paper are not publicly available at this time but may be obtained from the authors upon reasonable request.

Acknowledgments: The authors are grateful to Julian Evans of Zhejiang University and Wei Chen of Ningbo Water Meter (Group) Company Ltd. for valuable discussions and also grateful to Ningbo Water Meter (Group) Company Ltd. for providing the water leakage test environment.

Conflicts of Interest: The authors declare no conflicts of interest.

References

1. Zhuo, L.; Feng, B.; Wu, P. Water Footprint Study Review for Understanding and Resolving Water Issues in China. *Water* **2020**, *12*, 2988. [CrossRef]
2. Momeni, A.; Piratla, K.R. A Proof-of-Concept Study for Hydraulic Model-Based Leakage Detection in Water Pipelines Using Pressure Monitoring Data. *Front. Water* **2021**, *3*, 648622. [CrossRef]
3. Korlapati, N.V.S.; Khan, F.; Noor, Q.; Mirza, S.; Vaddiraju, S. Review and Analysis of Pipeline Leak Detection Methods. *J. Pipeline Sci. Eng.* **2022**, *2*, 100074. [CrossRef]
4. Sousa, D.P.; Du, R.; Mairton Barros da Silva, J., Jr.; Cavalcante, C.C.; Fischione, C. Leakage Detection in Water Distribution Networks Using Machine-Learning Strategies. *Water Supply* **2023**, *23*, 1115–1126. [CrossRef]
5. Peng, L.; Huang, S.; Wang, S.; Zhao, W. A Simplified Lift-Off Correction for Three Components of the Magnetic Flux Leakage Signal for Defect Detection. *IEEE Trans. Instrum. Meas.* **2021**, *70*, 6005109. [CrossRef]
6. Wu, D.; Chatzigeorgiou, D.; Youcef-Toumi, K.; Mekid, S.; Ben-Mansour, R. Channel-Aware Relay Node Placement in Wireless Sensor Networks for Pipeline Inspection. *IEEE Trans. Wirel. Commun.* **2014**, *13*, 3510–3523. [CrossRef]

7. Wu, D.; Chatzigeorgiou, D.; Youcef-Toumi, K.; Ben-Mansour, R. Node Localization in Robotic Sensor Networks for Pipeline Inspection. *IEEE Trans. Ind. Inform.* **2016**, *12*, 809–819. [CrossRef]
8. Mashford, J.; De Silva, D.; Marney, D.; Burn, S. An Approach to Leak Detection in Pipe Networks Using Analysis of Monitored Pressure Values by Support Vector Machine. In Proceedings of the 2009 Third International Conference on Network and System Security, Gold Coast, QLD, Australia, 19–21 October 2009; pp. 534–539.
9. Leu, S.-S.; Nha, B. Leak Prediction Model for Water Distribution Networks Created Using a Bayesian Network Learning Approach. *Water Resour. Manag.* **2016**, *30*, 2719–2733. [CrossRef]
10. Wu, H.; Sun, Z.; Qian, Y.; Zhang, T.; Rao, Y. A Hydrostatic Leak Test for Water Pipeline by Using Distributed Optical Fiber Vibration Sensing System. In Proceedings of the Fifth Asia-Pacific Optical Sensors Conference, Jeju, Republic of Korea, 20–22 May 2015; Lee, B., Lee, S.-B., Rao, Y., Eds.; SPIE: Bellingham, WA, USA, 2015; p. 965543.
11. Fu, S.; Zhang, D.; Peng, Y.; Shi, B.; Yedili, N.; Ma, Z. A Simulation of Gas Pipeline Leakage Monitoring Based on Distributed Acoustic Sensing. *Meas. Sci. Technol.* **2022**, *33*, 095108. [CrossRef]
12. Zhu, H.; Pan, C.; Sun, X. Vibration Pattern Recognition and Classification in OTDR Based Distributed Optical-Fiber Vibration Sensing System. *Proc. SPIE-Int. Soc. Opt. Eng.* **2014**, *9062*, 906205. [CrossRef]
13. Jiang, F.; Li, H.; Zhang, Z.; Zhang, Y.; Zhang, X. Localization and Discrimination of the Perturbation Signals in Fiber Distributed Acoustic Sensing Systems Using Spatial Average Kurtosis. *Sensors* **2018**, *18*, 2839. [CrossRef] [PubMed]
14. Samuel, A.L. Some Studies in Machine Learning Using the Game of Checkers. *IBM J. Res. Dev.* **1959**, *3*, 210–229. [CrossRef]
15. Lecun, Y.; Bottou, L.; Bengio, Y.; Haffner, P. Gradient-Based Learning Applied to Document Recognition. *Proc. IEEE* **1998**, *86*, 2278–2324. [CrossRef]
16. He, K.; Zhang, X.; Ren, S.; Sun, J. Deep Residual Learning for Image Recognition. In Proceedings of the 2016 IEEE Conference on Computer Vision and Pattern Recognition (CVPR), Las Vegas, NV, USA, 27–30 June 2016; pp. 770–778.
17. Jia, H.; Liang, S.; Lou, S.; Sheng, X. A k-Nearest Neighbor Algorithm-Based Near Category Support Vector Machine Method for Event Identification of φ-OTDR. *IEEE Sens. J.* **2019**, *19*, 3683–3689. [CrossRef]
18. Zhang, M.; Li, Y.; Chen, J.; Song, Y.; Zhang, J.; Wang, M. Event Detection Method Comparison for Distributed Acoustic Sensors Using φ-OTDR. *Opt. Fiber Technol.* **2019**, *52*, 101980. [CrossRef]
19. Pan, Z.; Liang, K.; Ye, Q.; Cai, H.; Qu, R.; Fang, Z. Phase-Sensitive OTDR System Based on Digital Coherent Detection. In Proceedings of the 2011 Asia Communications and Photonics Conference and Exhibition (ACP), Shanghai, China, 13–16 November 2011; pp. 1–6.

applied sciences

MDPI

Article

Optical Cable Lifespan Prediction Method Based on Autoformer

Mengchao Niu, Yuan Li * and Jiaye Zhu

School of Computer Science, Central China Normal University, Wuhan 430079, China; mengchaon@163.com (M.N.); kennaedcampe@gmail.com (J.Z.)
* Correspondence: yuanli@mail.ccnu.edu.cn

Abstract: We proposed a novel method for predicting the service life of optical cables based on the Autoformer model combined with the calculation method. Leveraging historical weather data from Guangzhou and employing specific cable length calculation techniques, our study comprehensively considers factors impacting cable lifespan. Moreover, through comparative analysis with alternative deep learning models and parameter assessments, our method validates the superiority and stability of the Autoformer model in predicting cable lifespan, which can offer a more reliable approach for ensuring cable technology reliability and the management of associated industries.

Keywords: fiber optic cable lifespan; prediction; deep learning models; Autoformer

1. Introduction

As optical communication networks continue to evolve and advance, fiber optic communication has become the principal mode of transmission within modern communication systems, assuming a pivotal role. Many communication lines that were laid early on are now nearing or have reached their predetermined lifespan. Therefore, accurately predicting the service life of OPGW (optical ground wire) cables to facilitate their timely replacement or extend the operation of aging lines is crucial for the sustained stability of communication networks.

Numerous studies have probed into unraveling the longevity of these cables. Grunvalds R conducted measurements on the polarization mode dispersion of OPGW to gauge the infrastructure's expected lifespan [1]. Burdin VA introduced an approach to forecast the remaining lifespan of field-aged optical cables using test results obtained from samples extracted from a cable line [2]. They also developed a model to estimate the service life of optical cables within an operational cable line [3]. Lastly, they provided a formula to calculate the remaining lifespan of optical fibers in cables based on actual fiber strength estimations derived from selected cable samples on the production line [4]. Nizhgorodov A O introduced a method to forecast the service life of optical fibers, providing simple approximate formulas to determine the communication cable lines' lifetime with a certain reliability probability [5].

Although these methods lay the groundwork for predicting the lifespan of optical cables, they also present noticeable deficiencies. They tend to concentrate on the impact of solitary factors without adequately accounting for the array of complex scenarios and environmental conditions that optical cables face in actual operation. Moreover, the experimental conditions employed by these methods are typically idealized, leading to overly optimistic projections that struggle to apply to real-world contexts. Consequently, lifespan estimations for optical cables derived from these approaches often lack practical accuracy and comprehensiveness.

To break through the limitations of conventional optical cable lifespan prediction methods and enhance the accuracy and reliability of forecasts, it is particularly crucial to explore new predictive technologies. Deep learning, a powerful class of machine learning algorithms known for its ability to process large datasets, has shown promising results in various fields and is being actively explored for applications in optical cable lifespan prediction, particularly for OPGW cables. For instance, Weixing Han and his team have developed an innovative digital twin method for cable temperature prediction based on RF-GPR (random forest–Gaussian process regression), which combines finite element analysis with artificial intelligence techniques [6]. Furthermore, they have proposed a data-driven cable insulation defect model based on a convolutional neural network approach to address the problem that the electric field is not easily measured under the operating condition of transmission cables with insulation defects [7]. Inspired by these approaches, this paper introduces deep learning into the realm of optical cable lifespan prediction. Autoformer, a novel sequence model, demonstrates proven success in fields including natural language processing and image recognition. Adopting an approach that diverges from traditional sequence analysis, it combines time series decomposition with an auto-correlation mechanism, allowing for a more accurate grasp of data trends and dependencies [8,9]. Considering its high compatibility with the characteristics of optical cable data, this study applies the Autoformer model to the task of predicting the remaining service life of optical cables. Optical cable data typically exhibit strong time-dependencies, reflecting the gradual degradation process influenced by factors such as temperature, wind, and load. The Autoformer model, combining time series decomposition with an auto-correlation mechanism, is particularly well suited for capturing these long-range dependencies. The model's ability to effectively analyze temporal patterns and extract long-term correlations makes it a promising approach for accurately predicting the remaining service life of optical cables.

This paper examines how fiber length, stress, and specific loads under varied weather conditions (e.g., temperature, lightning strikes, icing, wind vibrations) affect optical cable lifespans. Using the Autoformer model, it forecasts trends in excess fiber length alterations, confirming the model's superiority over others through simulation validation. This method not only offers a more accurate and reliable means of predicting cable lifespans but also presents a new technical direction for the optical cable industry by leveraging deep learning and time series analysis for lifespan prediction, moving away from traditional methods that rely on limited factors and idealized conditions. This shift towards data-driven approaches can provide vital support for ensuring the stability and reliability of communication networks.

While this study focuses on the impact of temperature, wind, and residual length on optical cable lifespan, other factors such as humidity, UV exposure, and installation quality can also play a significant role. We acknowledge the importance of these additional factors and plan to investigate their influence on cable lifespan in our future research.

2. Factors Impacting OPGW Lifespan

2.1. Environmental Impacts on OPGW Lifespan

The persistent operation of fiber optic cables post installation is contingent on various factors, with temperature fluctuations emerging as a crucial determinant of the cable's longevity. Environments experiencing considerable temperature shifts can induce thermal expansion and contraction within the cable, leading to cyclic stress and subsequent material aging and degradation. Elevated temperatures accelerate material aging, thereby increasing the risk of optical attenuation and signal loss, while lower temperatures render the cable more brittle and prone to breakage. Prolonged temperature fluctuations may alter the cable's internal microstructures, potentially causing micro-cracks or damage, ultimately compromising the cable's performance and lifespan [10,11].

In addition to temperature, wind and ice loads surpassing the cable's weight significantly impact the longevity of OPGWs. Concurrently, the high-frequency, low-amplitude vibrations induced by wind forces influence both the cable's fatigue performance and the transmission characteristics of the optical fibers. Furthermore, events like short circuits and lightning currents can generate excessive heat, potentially exceeding the temperature threshold of the optical fiber coating and causing material damage [12].

2.2. Effect of Residual Length on OPGW Lifespan

Throughout the installation and operational phases, faults in optical fiber length significantly impact the cable's lifespan and can be categorized into three stages [13]. During the initial 1–3 years of deployment, wind-induced vibrations and thermal stresses arising from temperature fluctuations can induce minor variations in the residual length of OPGW. This effect tends to balance the residual fiber length, line pitch, and cable characteristics. Hence, if during this phase, the OPGW cable experiences increased optical fiber loss or experiences a sudden surge, it is likely attributed to insufficient fiber optic residual length. Between 3 and 15 years of the cable's lifespan, it undergoes expansion due to the creep characteristics of aluminum-clad steel raw materials. This expansion diminishes the available margin for the optical fiber, which remains relatively constant in length compared to the cable's growth. The decrease in the residual length of the optical fiber results in increased optical fiber loss, particularly when the remaining length of the optical fiber decreases significantly. At the end of the 15–25 year period, an inspection will reveal that the fiber optic paste will gradually denature over time. This begins with the formation of small particles, followed by evaporation, decomposition, drying and solidification. Notably, significant denaturation usually starts around 18 years. The direct interaction of the optical fiber with the optical fiber paste initiates a notable elevation in its acid value, triggering an augmented precipitation of hydrogen. In addition, paste oxidation affects the stability of the optical unit, reducing the fiber's ability to adapt to various stresses or strains, weakening the buffering effect, and ultimately shortening the life of the cable.

2.3. Intrinsic Influences on OPGW Lifespan

The lifespan of a fiber optic cable is largely dependent on the integrity of its core material, the fiber itself. Micro-cracks represent a critical factor influencing the overall lifespan of these fibers. These tiny fissures, stemming from manufacturing imperfections or extended exposure to environmental elements over prolonged use, possess the potential to gradually expand. As these fissures grow, they compromise the mechanical robustness of the optical fiber, eventually leading to its structural failure. Furthermore, sustained stress due to various adverse conditions also contributes to the gradual deterioration of optical fiber cable lifespan.

3. Related Work

3.1. Definition

This study is dedicated to predicting the remaining service life of optical cables. During operation, optical cables are subject to various weather-related factors, including strong winds, high temperatures, low temperatures, and cumulative impacts, all of which can cause the cables to expand or contract. These environmental changes consequently affect the residual length of the internal fibers and thereby the lifespan of the cables. It forecasts future cable length trends based on historical changes and associated characteristics, considering the key factors influencing cable life. When the cable length reaches a critical threshold, it is determined that the cable has reached the end of its service life. The prediction of optical cable lifespan can be seen as a form of time series forecasting. This involves using a historical data series containing C features, $X_h = \{X_1^t, \cdots, X_C^t\}_{t=1}^{T}$, along with a known future data series following T timesteps, $X_f = \{X_1^t, \cdots, X_{C-1}^t\}_{t=T+1}^{T+L}$, to predict the cable length for the upcoming L timesteps, represented as $Y = \{y_T, \cdots, y_{T+L}\}$. The calculation formula is shown in Equation (1), where $F()$ represents the fitted model, and θ denotes the model's learnable parameters, which are optimized during the training process to minimize the prediction error.

$$Y = F\left(X_h, X_f, \theta\right), \tag{1}$$

3.2. Optical Cable Length Calculation Method

Throughout the service life of optical cables, changes in environmental temperature and varying loads prompt stress, deformation, and fluctuations in cable length [14,15]. This situation arises from two principal causes:

1. Temperature-Induced Length Alterations: Changes in ambient temperature prompt thermal expansion or contraction in the optical cable materials, thereby inducing fluctuations in their length.
2. Load-Related Deformation: Variations in stress or load can lead to deformation in the optical cable. While the optical fibers themselves primarily exhibit elastic behavior, other components may experience different types of deformation, consequently leading to corresponding alterations in length.

Assuming the optical cable transitions from state m (temperature t_m, loading g_m, and stress σ_m) to state n (temperature t_n, loading g_n, and stress σ_n), the length changes due to thermal variations and stress variations can be expressed by Equations (2) and (3), respectively, where α and E represent the thermal expansion coefficient and elastic modulus of the optical cable sheath, respectively.

$$L_t = [1 + \alpha(t_n - t_m)]L_m, \tag{2}$$

$$L_n = \left[1 + \frac{1}{E}(\sigma_n - \sigma_m)\right]L_t, \tag{3}$$

Substituting (2) into (3) yields the following:

$$L_n = L_m[1 + \alpha(t_n - t_m)]\left[1 + \frac{1}{E}(\sigma_n - \sigma_m)\right], \tag{4}$$

Further expanding (4) results in the term $\left[\frac{\alpha}{E}(t_n - t_n)(\sigma_n - \sigma_m)\right]$, which due to its minute magnitude, can be disregarded within an acceptable margin of error, simplifying (4) to

$$L_n = L_m\left[1 + \alpha(t_n - t_m) + \frac{1}{E}(\sigma_n - \sigma_m)\right], \tag{5}$$

It is important to note that this calculation simplifies the thermal expansion of the optical cable by using a single value (α) for the thermal expansion coefficient. In reality, different components of the cable, such as the optical fibers, buffer tubes, and sheath, may have different thermal expansion coefficients. However, for the scope of this study, we simplified this aspect by using the thermal expansion coefficient of the sheath as a representative value. This simplification is common in preliminary cable length change calculations as the sheath's properties often dominate the overall thermal behavior. Future work could explore incorporating the individual thermal expansion coefficients of different cable components for a more precise calculation.

3.3. Optical Cable Stress Calculation Method

The term "optical cable stress" typically refers to external forces acting upon a cable, primarily originating from its installation and operating environment, significantly impacting its performance and lifespan.

In accordance with the cable length formula, the correlation between cable length and stress under specific meteorological conditions is articulated as follows:

$$L = l + \frac{g^2 l^3}{24\sigma_0^2}, \tag{6}$$

through point-to-point similarities, which enhances the model's efficiency and precision during learning. In time series data analysis, sub-sequences at the same position across different periods may exhibit similar trends. The auto-correlation mechanism introduced by Autoformer further strengthens the model's ability to capture the characteristics of sequential data, thereby improving the accuracy and stability of predictions.

The structure of the Autoformer model is illustrated in Figure 1. It primarily consists of three modules: the auto-correlation mechanism, the series decomposition module, and the feed forward network layer [8].

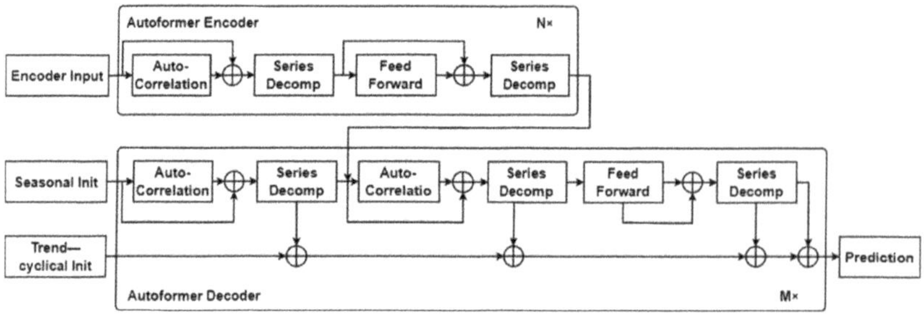

Figure 1. Autoformer architecture. The encoder eliminates the long-term trend–cyclical part by series decomposition blocks and focuses on seasonal patterns modeling. The decoder accumulates the trend part extracted from hidden variables progressively. The past seasonal information from encoder is utilized by the encoder–decoder auto-correlation.

The encoder primarily focuses on the periodic elements, using the preceding I time steps to compose the input time series. Governed by autocorrelation mechanisms, the encoder performs an initial sequence decomposition, dividing the original input time series into trend and periodic elements sent to the decoder. As a result, the decoder's input consists of both trend and periodic elements, signifying long-term trends and periodicity. The specific calculation formula is outlined as follows:

$$x_t = AvgPool(Padding(x)), \tag{16}$$

where x denotes the input time series, and x undergoes a sliding average operation (AvgPool) to derive the trend component x_t [17]. The AvgPool operation utilizes a sliding window of 60 days, meaning each data point in x_t represents the trend value, calculated from the preceding 60 days of the input time series x. Additionally, the Padding operation preserves the sequence length [18].

$$x_s = x - x_t, \tag{17}$$

where x_s represents the periodic component, obtained by subtracting the trend component derived from (16) from the original input time series.

The decoder, fed with seasonal and trend components, engages in a layer-by-layer sequence dissection governed by the autocorrelation mechanism. Moreover, each decoding layer manipulates complete time series data, not discrete points, ensuring step-by-step predictions to enhance forecast reliability and accuracy.

The decoder operates through two key components: firstly, it iteratively extracts trend-related information by synthesizing inputs from the encoder's trend output. Then, for processing the encoder's seasonal output, it applies a stacked autocorrelation mechanism to detect patterns and integrate related sub-processes. This sequence aims to progressively refine the model's ability to predict by understanding and leveraging the temporal structure inherent within the data. Considering a decoder with N encoding layers, let us take the first decoding layer as an example to illustrate its operation. Given that the decoding layer manages operations entailing both the input time series and output derived from the encoding layer, the depiction of the initial decoding layer is articulated by $x_d^l = Decoder\left(x_{de}^{l-1}, x_{en}^N\right)$. "Decoder" represents the decoding operations, inclusive of

$$s_{de}^{l,1}, \tau_{de}^{l,1} = SD\left(AC\left(x_{de}^{l-1}\right) + x_{de}^{l-1}\right), \tag{18}$$

where AC embodies the auto-correlation mechanism processing; SD represents sequence decomposition; x_{de}^{l-1} stands for the input time series of the initial decoding layer; and $S_d^{l,1}$, $\tau_d^{l,1}$ reflects the extracted trend information within the decoding layer.

Autoformer capitalizes on a sequence-based periodic autocorrelation mechanism. This method entails convolutions implemented in local regions at diverse positions within the input vector. It selects data within a specified time delay to acquire similar subprocesses, thereby achieving the autocorrelation of discrete-time processes. The formula for computing autocorrelation coefficients is delineated as

$$R_{xx}(\tau) = \lim_{L \to \infty} \frac{1}{L} \sum_{t=1}^{L-1} x_t x_{t-\tau}, \tag{19}$$

where $R_{xx}(\tau)$ represents the autocorrelation coefficient, measuring the similarity between x_t and its lagged value $x_{t-\tau}$ at time t, and L denotes the length of the time series. This formula gauges the similarity between the time series post shifting and the initial time series, where a substantial resemblance suggests potential periodic patterns within those shifts. Autoformer optimizes its computation efficiency by utilizing fast Fourier transform for computing autocorrelation coefficients [19]. The procedure is elucidated as

$$S_{xx}(f) = F(x_t)F^*(x_t) = \int_{-\infty}^{+\infty} x_t{}^{-2\pi t f i} dt \sqrt{\int_{-\infty}^{+\infty} x_t{}^{-2\pi t f i} t}, \tag{20}$$

where x_t signifies the original input time series, F represents the Fourier transform and F^* stands for the conjugate operation. The symbol f denotes frequency, multiplied by 2π to yield angular frequency. F and F^* integrate with the trend component outcomes, enabling the transformation of time series data into the frequency domain denoted by S_{xx}.

$$R_{xx}(\tau) = F^{-1}(S_{xx}(f)) \int_{-\infty}^{-\infty} S_{xx}(f) e^{2\pi \tau i} f, \tag{21}$$

where F^{-1} symbolizes the inverse Fourier transform. Applying the inverse Fourier transform once to the result obtained from (20) yields autocorrelation coefficients, simplifying the autocorrelation solving complexity to $O(LlogL)$.

4.2. Prediction Process Based on the Autoformer Model

This paper employs the Autoformer model for predicting the lifespan of optical cables. The proposed forecasting approach is primarily divided into two segments: data preprocessing and model construction, with the specific workflow illustrated in Figure 2.

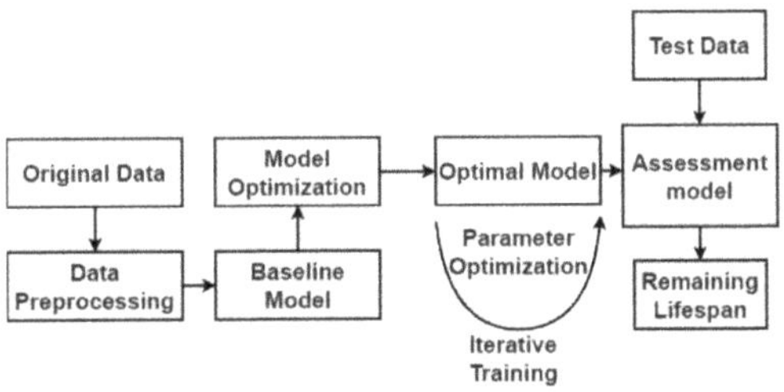

Figure 2. Prediction process.

1. Data Preprocessing: This study utilizes measured meteorological data from January 2011 to December 2023 in Guangzhou, including daily maximum and minimum temperatures and wind speed. Subsequently, leveraging the daily average temperature and wind speed, the corresponding optical cable lengths are calculated using a dedicated formula. For anomalous values, the Lagrange interpolation method is employed for estimation and imputation.
2. Model Construction: The model is constructed following the architectural schematic of Autoformer, with input comprising the processed data and labels representing the residual lifespan at that time step. The output is a time series for the remaining lifespan prediction.
3. Model Training and Testing: The model is trained on the training dataset and then evaluated on a separate test dataset.

5. Analysis of Cable Life Prediction Model Based on Autoformer

5.1. Comparison of Cable Life Prediction Model Results

Predicting the lifespan of fiber optic cables demands a rigorous model assessment. This paper utilizes root mean square error (RMSE), mean absolute error (MAE), and mean square error (MSE) as pivotal metrics to gauge the accuracy of predictive models against actual values.

In this paper, the optical cable information dataset was divided into training, validation, and test sets in a 4:1:1 ratio to ensure comprehensive coverage across different data distributions for model training, validation, and evaluation. During the experimental process, the model was configured to utilize a time window of 60 days as input, employing data from the past 60 days as input to capture the historical change trends in cable length. The encoder and decoder layers were set to two and one layers, respectively, balancing model complexity with the effective learning of data features and dependencies. The d_model was set to 4, correlating with the count of autocorrelation coefficients, to capture the temporal relationships in sequence data while ensuring computational efficiency. The learning rate was set to 0.0001, with 15 training epochs and a batch size of 16, to ensure stable model convergence during training and minimizing the occurrence of overfitting. To ensure the fairness and validity of the experiment, the configurations for the comparison models were kept identical to those of the proposed model. The comparative analysis results, as depicted in Table 1, reveal that the model proposed in this paper outperforms the comparative models across three evaluative metrics. Compared to the Bi-LSTM + Attention model, the Autoformer model achieved reductions in the RMSE, MAE, and MSE metrics by 35.46%, 49.63%, and 58.34% respectively.

Table 1. Performance metrics for different models.

Model	RMSE	MAE	MSE
LSTM	1.3184	1.0828	1.7383
Bi-LSTM	1.2361	0.9582	1.5278
Bi-LSTM + Attention	0.9131	0.7321	0.8338
Autoformer	0.5893	0.3687	0.3473

Figure 3 illustrates a comparison of forecast results for changes in optical cable lengths in 2023 within the test set between the Autoformer model and comparison models. As can be observed in Figure 3, each model is proficient in accurately predicting future information, with a minor deviation from the actual data. This indicates that the models are proficient in mining time series sequences, effectively and precisely forecasting future trends in cable length variations. Notably, the Autoformer model proposed in this paper achieves a better fit in its predictions compared to actual values, displaying heightened predictive accuracy. Furthermore, the Autoformer model's smooth and coherent prediction line overall suggests its stability and consistency in analyzing and projecting optical cable lifespan, particularly manifesting this trend when the data span a broader time range.

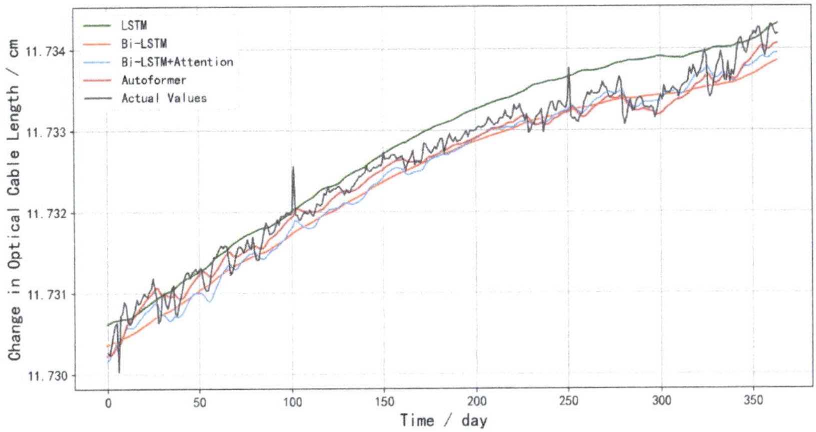

Figure 3. Comparison of predicted optical cable length changes in 2023: LSTM, Bi-LSTM, Bi-LSTM + Attention, and Autoformer Models. Actual values are shown in gray. The Autoformer model's predictions are shown in red. Green, orange, and blue lines represent the predictions of the LSTM, Bi-LSTM, and Bi-LSTM + Attention models, respectively.

It is important to note that the chosen values for the time window length and d_model are based on our understanding of the problem and the experimental results presented in this study. These values may not be optimal for all scenarios, and further investigation and fine-tuning may be required for specific applications.

5.2. The Impact of Iteration Count on the Performance of the Autoformer Model

The iteration count reflects the decrease in the loss function of deep learning models and their convergence status. Utilizing MSE as the loss function, as demonstrated in Figure 4, the predictive performance of the Autoformer model markedly improves, with an increase in iteration count during parameter adjustment experiments. When the iteration count reaches 4, the model has already converged optimally. However, further training beyond this point not only fails to enhance the model's predictive accuracy but also significantly diminishes it. This phenomenon suggests that the model might be overfitting the training data, losing its ability to generalize to new data. This attests to the Autoformer model's robust parallel computing capacity in feature extraction, facilitating faster convergence. Considering the training time and computational costs, this study sets the iteration count for the model at 4.

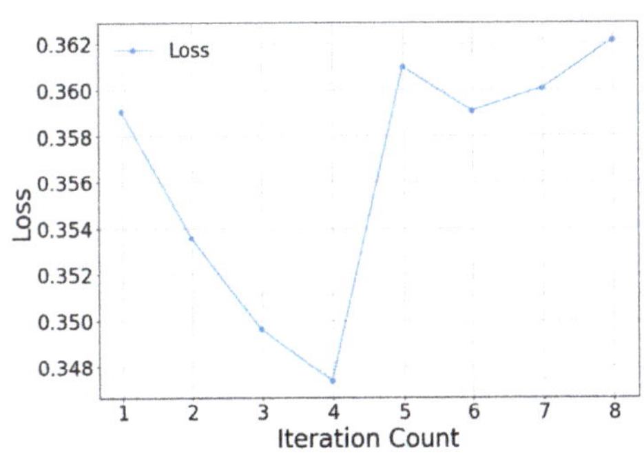

Figure 4. Variation in loss with iteration count.

5.3. The Impact of Learning Rate on the Performance of the Autoformer Model

The learning rate is a crucial factor in deep learning, directly impacting the convergence and speed of the optimization process. A learning rate that is too small can result in a slow convergence process, thereby extending the training time. Conversely, an excessively large learning rate may hinder convergence. As depicted in Table 2, the predictive efficacy of the Autoformer model, as measured by MSE, exhibits a non-linear relationship with the learning rate: while decreasing the learning rate initially improves model performance by allowing for more precise convergence, learning rates below 2.5×10^{-5} lead to prolonged training times and ultimately overfitting, as evidenced by the increasing MSE observed on the validation set when the learning rate falls below 2.5×10^{-5}. Therefore, this study adopts the optimal learning rate of 2.5×10^{-5}, which balances model accuracy and training efficiency.

Table 2. Evaluation metrics for different learning rates.

Learning Rate	RMSE	MAE	MSE
1×10^{-3}	0.6091	0.3711	0.3535
5×10^{-5}	0.6088	0.3707	0.3496
2.5×10^{-5}	0.5893	0.3687	0.3473
1.25×10^{-5}	0.6141	0.3771	0.3609
6.25×10^{-6}	0.6186	0.3827	0.3591
3.125×10^{-6}	0.6121	0.3746	0.3601

5.4. The Impact of Time Window Length on the Performance of the Autoformer Model

The time window length is a crucial parameter that can significantly affect the performance of time series forecasting models. To delve into the effect of the time window on model efficacy, this experiment employed models trained over various window lengths for comparative analysis.

Figure 5 illustrates the impact of time window length on the Autoformer model's performance, as measured by RMSE, MAE, and MSE. Initially, increasing the time window length leads to improved performance, reaching an optimal point at a length of 60 days. However, further extending the time window results in performance degradation. This phenomenon can be attributed to two primary factors. Firstly, a wider time window may cause the model to overemphasize earlier data points, potentially neglecting the significance of more recent information. Secondly, larger time windows require processing a greater volume of data, which can increase the risk of overfitting. This is evidenced by the rising validation error observed with longer time windows, indicating a reduced ability to generalize to unseen data. Therefore, selecting an appropriate time window length is crucial for balancing model complexity and generalization capability.

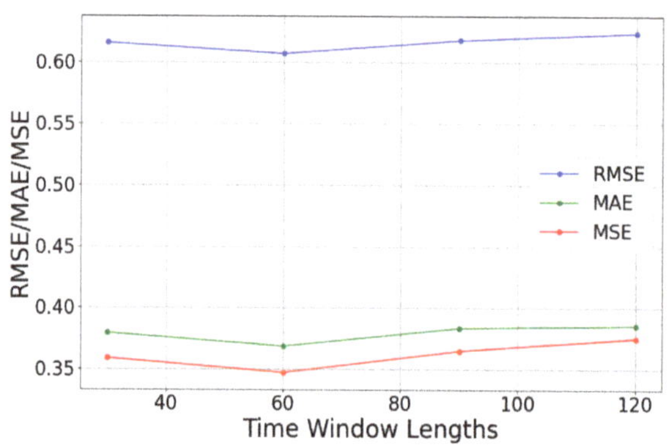

Figure 5. Impact of time window lengths on Autoformer model performance.

Figure 6 presents the results of a similar experiment conducted with the Bi-LSTM + Attention model. As with the Autoformer model, when the input window is set to 60, the model achieves peak performance. However, deviating from this value leads to a decline in predictive accuracy, suggesting limitations in the model's ability to capture long-sequence temporal features.

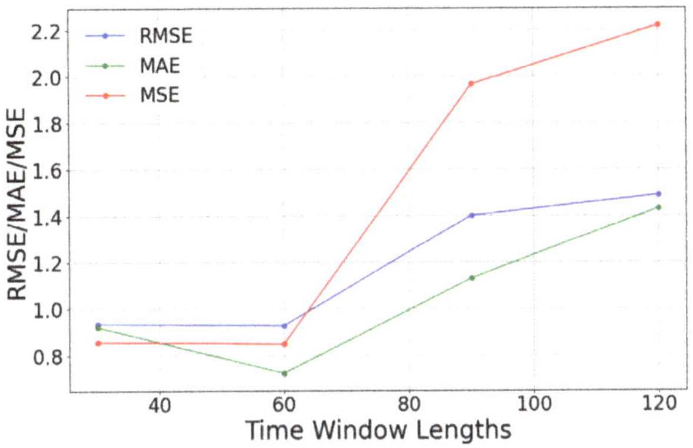

Figure 6. Impact of time window length on Bi-LSTM + Attention model performance.

A comparison of the results depicted in Figures 5 and 6 reveals that the Autoformer model demonstrates greater robustness to variations in time window length compared to the Bi-LSTM + Attention model. Both models exhibit performance degradation with time windows exceeding 90 days, but the decline in RMSE is significantly more pronounced for the Bi-LSTM + Attention model. This suggests that the Autoformer model is more adept at capturing long-range dependencies within time series data, a capability not as pronounced in the Bi-LSTM + Attention model. These findings provide further evidence supporting the suitability of the Autoformer model for optical cable lifespan prediction tasks.

6. Conclusions

Based on the results of our numerical experiments, this paper introduces a novel approach for predicting the remaining lifespan of optical cables utilizing the Autoformer model. The proposed model leverages deep learning and computational methodologies, utilizing temperature fluctuations, wind loads, and the cable's residual length to predict the cable's lifespan. Experimental results demonstrate that the Autoformer model outperforms other deep learning models, indicating its potential for accurate and reliable lifespan prediction. This research contributes a more effective method to the field of optical cable lifespan prediction, providing network operators and communication services with more accurate residual life predictions. This facilitates timely maintenance scheduling, reduces potential network failure risks, and ensures service continuity and stability. Furthermore, this work also offers enterprises more precise predictions of a cable's remaining life, aiding in planning equipment maintenance and production scheduling, thereby enhancing production efficiency and equipment utilization. These contributions are significant not only in the academic sphere but also have a profound impact on industrial practices and commercial applications.

Author Contributions: Conceptualization, M.N. and Y.L.; methodology, M.N. and Y.L.; software, M.N.; validation, M.N. and J.Z.; formal analysis, M.N.; investigation, M.N.; resources, Y.L.; data curation, M.N. and J.Z.; writing—original draft preparation, M.N.; writing—review and editing, M.N. and Y.L.; visualization, M.N.; supervision, Y.L.; project administration, Y.L.; funding acquisition, Y.L. All authors have read and agreed to the published version of the manuscript.

Funding: This research was funded by Wuhan National Laboratory for Optoelectronics and China Southern Power Grid Company Co., Ltd.

Institutional Review Board Statement: Not applicable.

Informed Consent Statement: Not applicable.

Data Availability Statement: The data presented in this study are available on request from the corresponding author due to the sensitive nature of the data and legal restrictions preventing public disclosure.

Acknowledgments: The authors are grateful to those who provided support for this paper. Weihua Lian, representing China Southern Power Grid Company Co., Ltd., provided the data for this paper. Lei Dong, representing Wuhan WUTOS Co., Ltd., contributed to the selection of detection parameters for OPGW optical cables, the provision of practical engineering parameters, and suggestions for model optimization. Hualong Xiao, from Central China Normal University, made significant contributions to this work through the computation of new neural network models and other related tasks.

Conflicts of Interest: The authors declare that this study received funding from Wuhan National Laboratory for Optoelectronics and China Southern Power Grid Company Co., Ltd. The funder was involved in the data collection of this study. However, the funder was not involved in the study design, analysis, interpretation of data, the writing of this article, or the decision to submit it for publication.

References

1. Grunvalds, R.; Ciekurs, A.; Porins, J.; Supe, A. Evaluation of Fibre Lifetime in Optical Ground Wire Transmission Lines. *Latv. J. Phys. Tech. Sci.* **2017**, *54*, 40–49. [CrossRef]
2. Burdin, V.; Andreev, V.; Bourdine, A.; Dashkov, M.; Nizhgorodov, A. Reliability and lifetime estimations for field-aged optical cable. In Proceedings of the Internet of Things, Smart Spaces, and Next Generation Networks and Systems: 20th International Conference, NEW2AN 2020, and 13th Conference, ruSMART 2020, St. Petersburg, Russia, 26–28 August 2020.
3. Burdin, V.A.; Nizhgorodov, A.O. Lifetime prediction algorithm for an optical cable of cable link under exploitation. In Proceedings of the Optical Technologies for Telecommunications, Bellingham, WA, USA, 22 May 2020.
4. Burdin, V.A.; Nikulina, T.G.; Praporshchikov, D.E. Method for predicting the lifetime of an optical cable after the maintenance cycle. In Proceedings of the Optical Technologies for Telecommunications, Bellingham, WA, USA, 25 July 2022.
5. Nizhgorodov, A.O. Simple Approximate Formula for Forecast of The Lifetime of the Optical Fibers of a Cable Communication Line. *Infokommunikacionnye Tehnol.* **2021**, *19*, 29–34.
6. Han, W.; Hao, C.; Kong, D.; Yang, G. Cable Temperature Prediction Based on RF-GPR for Digital Twin Applications. *Appl. Sci.* **2023**, *13*, 7700. [CrossRef]
7. Han, W.; Yang, G.; Hao, C.; Kong, D.; Dong, Y. A Data-Driven Model of Cable Insulation Defect Based on Convolutional Neural Networks. *Appl. Sci.* **2022**, *12*, 8374. [CrossRef]
8. Wu, H.; Xu, J.; Wang, J.; Long, M. Autoformer: Decomposition transformers with auto-correlation for long-term series forecasting. *Adv. Neural Inf. Process. Syst.* **2021**, *34*, 22419–22430.
9. Zhou, S.; Zhang, X.; Liu, J.; Zhang, Y.; Wei, P.; Wang, Y.; Zhang, J. Long-Term Prediction of Particulate Matter2.5 Concentration with Modal Autoformer Based on Fusion Modal Decomposition Algorithm. *Atmosphere* **2024**, *15*, 4. [CrossRef]
10. Borzycki, K. Influence of temperature and aging on polarization mode dispersion of tight-buffered optical fibers and cables. *J. Telecommun. Inf. Technol.* **2005**, *30*, 96–104. [CrossRef]
11. Günday, A.; Karlık, S.E. Optical fiber distributed sensing of temperature, thermal strain and thermo-mechanical force formations on OPGW cables under wind effects. In Proceedings of the 2013 8th International Conference on Electrical and Electronics Engineering (ELECO), Piscataway, NJ, USA, 28 November 2013.
12. Malyszko, O.; Zenczak, M. Current-carrying capacity of overhead power transmission lines in different weather conditions. In Proceedings of the XV International Symposium on Theoretical Engineering, Berlin, Germany, 22 June 2009.
13. Burdin, V.A. The method for a measurement of the excess fiber length on the cable delivery length by using the polarization reflectometry. In Proceedings of the Optical Technologies for Telecommunications, Bellingham, WA, USA, 6 April 2017.
14. Setayeshgar, A. Principles of Mechanical Design in Overhead Transmission Lines. Master's Thesis, Aalto University, Espoo, Finland, 2016.
15. Li, Y.-Q.; Zhao, H.-W.; Yue, Z.-X.; Li, Y.W.; Zhang, Y.; Zhao, D.C. Real-Time Intelligent Prediction Method of Cable's Fundamental Frequency for Intelligent Maintenance of Cable-Stayed Bridges. *Sustainability* **2023**, *15*, 4086. [CrossRef]
16. Ashfaq, A.; Chen, Y.; Yao, K.; Sun, G.; Yu, J. Microbending loss caused by stress in optical fiber composite low voltage cable. In Proceedings of the 12th International Conference on the Properties and Applications of Dielectric Materials (ICPADM), Xi'an, China, 20 May 2018.

17. Qin, J.; Gao, C.; Wang, D. LCformer: Linear Convolutional Decomposed Transformer for Long-Term Series Forecasting. In Proceedings of the International Conference on Neural Information Processing, Changsha, China, 20 November 2023.
18. Alrasheedi, F.; Zhong, X.; Huang, P.C. Padding module: Learning the padding in deep neural networks. *IEEE Access* **2023**, *11*, 7348–7357. [CrossRef]
19. Henry, M. An ultra-precise Fast Fourier Transform. *Meas. Sens.* **2024**, *32*, 101039. [CrossRef]

applied sciences

MDPI

Article

A Multi-Format, Multi-Wavelength Erbium-Doped Fiber Ring Laser Using a Tunable Delay Line Interferometer

Cheng-Kai Yao, Amare Mulatie Dehnaw and Peng-Chun Peng *

Department of Electro-Optical Engineering, National Taipei University of Technology, Taipei 10608, Taiwan; t109658093@ntut.org.tw (C.-K.Y.); mulatieamare7@gmail.com (A.M.D.)
* Correspondence: pcpeng@ntut.edu.tw; Tel.: +886-2-2771-2171 (ext. 4671)

Abstract: This work demonstrates the use of an erbium-doped fiber amplifier (EDFA), a tunable bandpass filter (TBF), and a tunable delay line interferometer (TDLI) to form a ring laser that produces multi-format, multi-wavelength laser beams. The TDLI serves as the core of the proposed laser generation system. TDLI harnesses the weak Fabry–Pérot (FP) interferences generated by its built-in 50/50 beamsplitter (BS) with unalterable filtering characteristics and the interferences with free spectral range (FSR) adjustable from each of its two outputs with nearly complementary phases to superpose and generate a variable interference standing wave. The interferometric standing wave and weak FP interferences are used to form a spatial-hole burning to promote the excitation of multi-format and multi-wavelength lasers. The proposed system enables dual-wavelength spacing ranging from 0.3 nm to 3.35 nm, with a switchable wavelength position at approximately 1527 nm to 1535 nm, providing flexible tunability.

Keywords: multi-wavelength lasers; erbium-doped fiber amplifier; fiber ring laser; tunable delay line interferometer; spatial hole-burning; Fabry–Pérot interference

1. Introduction

Fiber optic lasers represent a significant advancement in laser technology, offering a multitude of benefits when compared to traditional lasers. These advantages include notably increased efficiency, enabling faster data transmission speeds, and enhanced precision [1–5]. Moreover, fiber optic lasers possess the unique capability of transmitting data over considerably longer distances without experiencing any degradation in the signal quality. Their exceptional reliability and durability make them particularly well suited for diverse applications in telecommunications [6–10], sensing [11–13], medical procedures [14–16], manufacturing [1], and research [16], where consistent and robust performance is essential. Continuous-wave (CW) lasers and pulsed-wave lasers are two primary operating types. CW lasers emit a continuous, unmodulated beam of light and are used in activities such as light pumping, laser cooling, and continuous-wave spectroscopy [17–19]. On the other hand, pulsed lasers emit light in the form of pulses; the duration and repetition rate of the pulses can vary widely from nanoseconds to femtoseconds, and they are utilized in diverse applications, such as material processing, medical procedures, and other specialized uses [20]. Fiber optic lasers can be further subdivided into several categories, including mode-locked lasers, single longitudinal mode lasers, swept lasers, and multi-wavelength lasers, each with unique characteristics and applications. Mode-locked lasers produce very short pulses in the picosecond or femtosecond range by locking the phases of modes. They are useful for fast and accurate tasks such as time-resolved spectroscopy, medical imaging, and high-precision material processing [21–24]. Single longitudinal mode lasers emit light at a single frequency with narrow linewidth, high coherence, and stability, making them suitable for high-resolution spectroscopy, interferometry, and coherent communication systems [25,26]. Swept lasers (tunable lasers) can rapidly change their emitted wavelengths over a wide range, making them highly adaptable for various applications in fields such as spectroscopy, optical coherence tomography, and telecommunications [27,28]. Multi-wavelength lasers can emit multiple wavelengths of light, either simultaneously or sequentially, finding applications in areas such as wavelength-division multiplexing (WDM) in telecommunications, optical spectroscopy, and medical diagnostics [29].

The gain medium housed within the fiber laser's resonance cavity is essential for amplifying light and generating laser output through population inversion and stimulated emission. Commonly used gain mediums include erbium-doped fiber amplifiers (EDFAs) [8], semiconductor optical amplifiers (SOAs) [30], reflective semiconductor optical amplifiers (RSOAs) [31], Raman amplifiers [32], and fiber optic Brillouin amplifiers [33]. EDFAs use erbium-ion-doped optical fiber to provide a high gain of up to 20–30 dB with a low noise factor. They are suitable for long-distance communication, with a main amplification area aimed at the C-band (1530–1565 nm) and L-band (1565–1625 nm). SOAs use a semiconductor as the gain medium and can amplify a wide range of wavelengths. Still, they have a high noise figure, are sensitive to polarization, and may introduce nonlinear effects. RSOAs are similar to SOAs but have a reflective end, making them suitable for bidirectional communication and remote amplification. Raman amplifiers enhance light signals using the Raman scattering effect and can be used with various types of optical fibers and wavelengths, making them suitable for WDM (wavelength-division multiplexing) systems. Raman amplification can be distributed along the fiber to reduce noise and enhance signal quality over long distances. However, it is important to note that Raman amplifiers can be costly, structurally complex, and may introduce non-linear effects. Brillouin amplifiers are capable of providing gain levels of up to 30–40 dB at pump powers typically below 100 mW. However, they are only suitable for applications requiring narrow frequency amplification, are sensitive to temperature, and are complex to implement.

In recent years, there has been significant research into multi-wavelength laser devices that are constructed using the optical fiber ring architecture as the resonant cavity [34–60]. These devices have garnered considerable interest due to their more complex production techniques compared to single-wavelength lasers. They offer a wide range of practical applications, including optical communications, optical fiber sensing, optical signal processing, microwave photonics, high-resolution spectroscopy, wavelength-division multiplexing, time-division multiplexing, and mode division multiplexing systems [29,34]. One of the most preferred options for these devices is the use of erbium-doped fiber (EDF) as the gain medium, primarily because of its high gain and low noise properties [35–60]. Utilizing EDFAs with high pump conversion efficiency presents a challenge since an increase in temperature results in the deterioration of the homogeneous broadening effect [61]. This, in turn, leads to intense competition between different modes and makes it difficult to generate stable multi-wavelength lasers at room temperature. Researchers have made efforts to address this issue by cooling the EDFA using liquid nitrogen [62,63], and various alternatives have been explored, including the implementation of fiber Bragg grating systems [35–37], polarization hole burning [38], and interferometer filters [39–60]. Among these options, fiber ring architecture with the interferometer filter is particularly favored for shaping multi-wavelength lasers because of its ability to easily produce different numbers of wavelengths, as well as its adjustable wavelength spacing and positions. In the pursuit of improved multi-wavelength laser performance, it has been found that a single interferometer filter may not fully meet the requirements. This limitation arises from the potential for reduced effectiveness in suppressing mode competition due to the large FSR of the interferometric spectrum. As a solution, researchers have proposed the use of cascaded interferometer-based filter architectures [46–63]. This approach involves the superposition of two interferometric spectra, which can lead to the compression of the FSR or the creation of a vernier effect. These effects are beneficial, as they can more effectively suppress mode competition or mode hopping, enabling the generation of more wavelength outputs, narrower laser line width, or facilitating the conversion of wavelength spacing and position. However, the issues with the architecture of a single interferometer filter or a cascaded interferometer filter stem from irregularities in the wavelength distribution positions of constructive and destructive interferences. These irregularities can occur within the interferometric spectrum itself or when superimposing two interferometric spectra. As a result, inconsistencies in the spacings between multi-wavelength lasers or in the spacings between wavelength tuning positions can arise. Additionally, there are instances where the FSR of the interference is restrictively tuned or is non-tunable, leading to irregular spacing changes between multi-wavelength lasers or non-tunable spacing between wavelength-tuning positions.

This study aims to demonstrate a novel multi-wavelength laser system using EDFA within a fiber ring architecture. The objectives include enhancing the stability and tunability of laser wavelengths through the integration of a TDLI and a TBF, thereby addressing the challenges posed by homogeneous broadening in EDFA. The TDLI can offer two phase-complementary interferences, a strong extinction ratio, and high tunability in terms of the interference's FSR [64]. Two power-different complementary interferences can form a switchable standing wave interference by tuning FSR. This leads to a spatial hole-burning effect, causing the laser to be excited at the half-upper-waist of the lobe in the interferometric pattern. Moreover, the built-in BS of TDLI creates weak FP interferences, thus shaping a deeper layer of filtering to deepen the spatial hole-burning effect. Hence, by changing the FSR of the TDLI to create various interference superposition effects, it is feasible to generate multi-format, multi-wavelength lasers with switchable wavelength positions and spacings.

2. Experimental Setup and Procedures

The diagram in Figure 1 illustrates the setup of an erbium-doped fiber ring laser based on a TDLI filter (Kylia WT-MINT, Kylia, Paris, France) as the main protagonist. The beam is first emitted from the EDFA (EDPA-NE6000) and through the fiber connector at port 1, and then it enters the TDLI, where it travels along an interference path with a delay line route. After the beam passes through the TDLI, it is routed from the fiber connector at port 3 through the circulator to the TBF (BVF-200CL) to filter out the unwanted wavelengths. After passing through the TBF in the initial beam path, half of the energy of the light will enter the optical spectral analyzer (OSA-Anritsu MS9740A, Anritsu, Atsugi, Japan) through the 3dB coupler, and the other half of the energy of the light will re-enter the TDLI through the circulator and the fiber optic connector of port 3. On the other hand, this time, the interferometric path is traveled without a delay line path, and its phase is 180 degrees different from that of the interferometric path with a delay line path. The superposition of these two interferences passes through the fiber connector of port 2, and an isolator that prevents the output light at the input of the EDFA from interfering with the stability of the system is finally amplified by the EDFA.

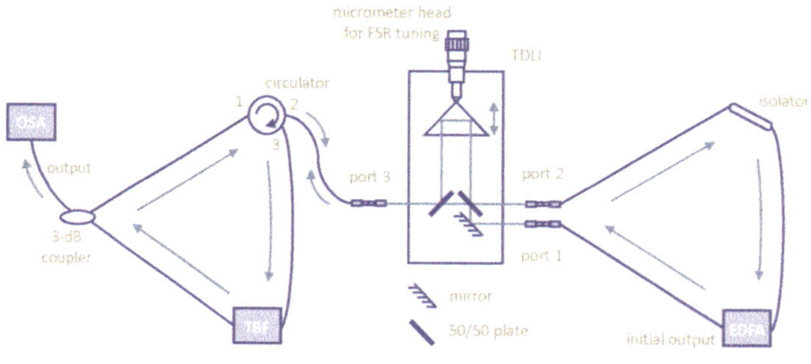

Figure 1. An experimental framework involves a multi-wavelength erbium-doped fiber ring laser with a delayed line interferometer (internal structure enlargement) as the primary component. (TDLI: tunable delay line interferometer; EDFA: erbium-doped fiber amplifier; TBF: tunable bandpass filter; OSA: optical spectral analyzer).

The internal TDLI consists of two plate beam splitters, a triangular prism with a displaceable micrometer scale, and a mirror. The incident light enters port 1 and is reflected from the mirror, which is placed at an angle of 45 degrees, to a BS, which is also placed at an angle of 45 degrees. Half of the light is diverged from the BS to the forward-placed triangular prism and reflected to another BS placed at a reverse 45-degree angle; the other half of the light is diverged directly from the BS to the BS placed at a reverse 45-degree angle. The two beams of light converge together again due to the difference in optical paths, resulting in interference. This is interference with a delayed path. If the incident light enters from port 2, it does not go through the path between the mirror and the BS, which is interference without a delayed path. The distance between the mirror and the BS is intentionally designed so that the two main output interferences of the TDLI are phase-complementary and can be utilized simultaneously.

It is worth mentioning that when the beam passes through the BS inside the TDLI, multiple FP interferences will occur due to the reflection inside the BS [65,66], which forms some irregular interfering pattern embedded in the main interfering pattern at the output of the TDLI, and the spectrum is shown in Figure 2a. Since the reflected light inside the BS is relatively weak, the energy of the main interference needs to be weakened to observe the spectrum. The interferometric pattern of different FSRs shown in Figure 2a is observed from the incident light of EDFA entering the TDLI at port 1 and then reflected from the fiber connector at port 3 to the fiber connector at port 2 for OSA observation. It can be seen that the wavelengths of the peaks and troughs of the interference spectrum caused by the internal reflections of the BS are fixed regardless of the FSR of the main interference, although the powers of the peaks and troughs are modulated by the peaks and troughs of the main interference and have a difference in power. The yellow and blue lines in Figure 2b show the main interference spectra generated by the port 3 to port 2 and port 3 to port 1 paths of the TDLI, respectively. It can be seen that the intrinsic interference spectrum generated by the BS is no longer visible because its lower power has been overshadowed by the main interference with higher power. Nevertheless, the intrinsic interference generated by the BS still has an effect on the laser wavelength output, as will be shown in the subsequent experimental results. For the two complementary interference spectra depicted in Figure 2b, it is observed that the yellow line exhibits a higher interference power compared to the blue line. Additionally, the difference in power between the interference peak of the yellow line and the interference trough of the blue line is greater than that between the interference peak of the blue line and the interference trough of the yellow line. Consequently, the fiber ring laser structure in question displays a multi-wavelength laser spectrum, as indicated by the dark line in Figure 2b. Each lobe of interference is characterized by a pair of head horns. In addition, the lasing of the laser wavelength will be at the half-upper-waist of the lobe of the two main interferences, where the two main interferences are superimposed and have the highest power. The concept of "spatial hole-burning" as an effect of lasers is derived from [67,68]. The operating principle of hole-burning in a laser involves selectively depleting certain frequency components within the gain profile of the amplifying medium, creating a "hole" in the spectral line. This process is achieved by preferentially exciting atoms or molecules in the upper energy level of the laser transition, causing a reduction in the population of those specific energy states. As a result, the gain at those frequencies decreases, leading to a spectral "hole" where the gain is lower than the surrounding frequencies. Hole burning can be induced by various mechanisms, such as optical pumping, where external light sources excite the atoms to specific energy levels or have inhomogeneous broadening effects in the medium. The creation of these holes in the gain profile can have significant implications for the operation and performance of the laser, affecting parameters such as linewidth, power dependence, and mode pulling. Therefore, in this scheme, the hole will be formed in the gain spectrum corresponding to the node of the interference pattern according to inhomogeneous broadening, resulting in a decrease in the lasing threshold at the half-upper-waist of the lobe of the interference spectrum.

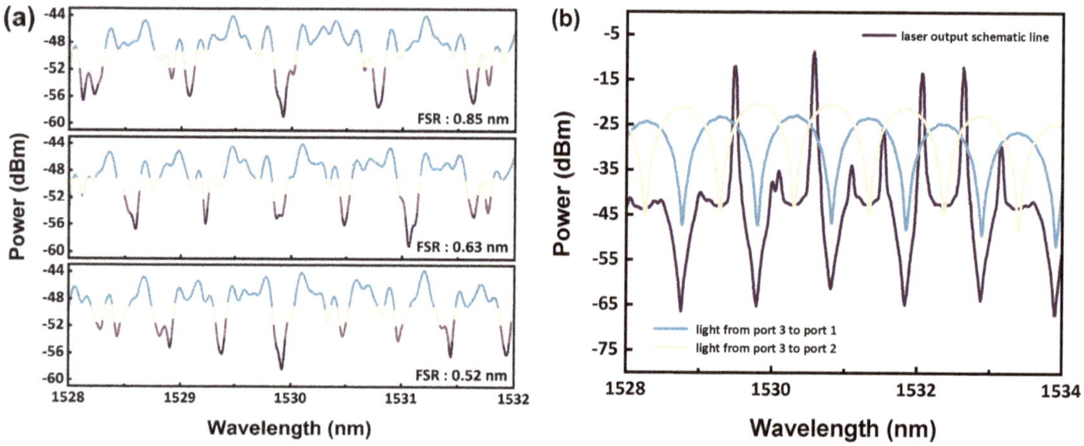

Figure 2. (**a**) The spectrum shows multiple FP interferences resulting from the combination of two BS and the primary interference caused by TDLI with different FSRs by virtue of reflections from the end face of the fiber head. The EDFA emits light from port 1 into the TDLI, which is then reflected by port 3 and the BS into port 2 for observation by the OSA. (**b**) Transmission spectra of two-phase complementary interferences formed by TDLI and spectral characteristics of a multi-wavelength laser capable of output in the Figure 1 framework. The EDFA emits light, which passes from ports 3 to 2 and 1, respectively, and then yellow and blue line interferograms can be obtained with the OSA, respectively.

3. Experimental Results

In this research, the FSR variation is achieved by adjusting the axial position of the prism through the movement of the micrometer head of the TDLI. The main function of the TBF is to filter out wavelengths other than 1527 nm to 1534 nm, and it is also used to ensure laser output by filtering a wider range of wavelengths when the laser wavelength spacing is small. Additionally, the power faults at the edges of the interferometric pattern indicate the areas where the TBF filters are applied. In addition, the OSA is set up with a spectral sampling range of 10 nm and 2001 points. Figure 3 shows various types of dual-wavelength lasers shaped by the Figure 1 architecture. Figure 3a shows the dual-wavelength laser with variable wavelength spacing by adjusting the interferometric pattern position and FSR size of the TDLI and filtering out the unnecessary wavelengths with the TBF. The wavelength spacing between the two wavelengths in the dual-wavelength laser varies from smallest to largest, which are 0.5 nm, 1.71 nm, 2.55 nm, and 3.35 nm. The wavelength positions of the dual-wavelength lasers are listed from smallest to largest wavelength spacing, which are 1530.15 nm and 1530.65 nm (light blue pattern); 1529.54 nm and 1531.25 nm (light green pattern); 1529.23 nm and 1531.78 nm (dark currant pattern); and 1528.72 nm and 1532.07 nm (yellow pattern), respectively. Moreover, the wavelength spacing variation of the dual-wavelength laser is adjusted according to the position of 1530.4 nm as the center, and the difference between the midpoint of the wavelength intervals of the dual-wavelength lasers from the smallest to the largest wavelength spacing and 1530.4 nm is 0 nm, 0.005 nm, 0.105 nm, and 0.005 nm, respectively. It can be noticed that no matter how the FSR varies, the wavelength position of the laser is always situated at the half-upper-waist of the lobe. It is also clear that the multiple interferences generated by the BS form a burr in the main interference pattern, which is caused by the amplification of the EDFA. It is also clear from the subfigure of Figure 3a that the wavelength positions of the dual-wavelength lasers with different spacings are highly overlapped with the peaks of the multiple interferences generated by the BS. This means that the sub-interferences generated by the BS deepen the spatial hole-burning effect, and the laser generation position is more likely to fall in the node of the half-upper-waist of the lobe, where it also overlaps with the peaks of the sub-interferences. As a result, the displacement between 1530.4 nm and the center point of the interval in a dual-wavelength laser with a wavelength spacing of 2.55 nm is larger. Figure 3b illustrates a dual-wavelength laser with staggered positions of several wavelengths. The laser configurations exhibit wavelength spacings of 0.8 nm, 1.15 nm, and 1.42 nm, with corresponding wavelength positions of 1529.85 nm and 1530.65 nm (light green pattern); 1530.12 nm and 1531.27 nm (dark currant pattern); and 1529.62 nm and 1531.04 nm (yellow pattern). It is noticeable that the midpoints of the wavelength intervals for the dual-wavelength lasers are distinct. This difference is attributed to the fact that the FSR of the TDLI can be adjusted arbitrarily to change the excitation points of the lasers, which should be close to the wave crests of the sub-interferences. The graph in Figure 3c depicts the regular tuning of the wavelength position of a dual-wavelength laser with almost all wavelength spacings of 1.42 nm. The shorter and longer wavelengths of the five dual-wavelength lasers are as follows: 1529.62 nm and 1531.04 nm (light blue pattern); 1529.92 nm and 1531.33 nm (light green pattern); 1530.22 nm and 1531.55 nm (dark currant pattern); 1530.42 nm and 1531.84 nm (light currant pattern); and 1530.72 nm and 1532.14 nm (yellow pattern), respectively. The difference in the positions of each pair of dual-wavelength lasers is approximately 0.3 nm. The position of the third pair of dual-wavelength lasers is slightly off due to the effects of sub-interference generated by the BS. It is worth noting that the dual-wavelength laser in the yellow spectrum of Figure 3b and the dual-wavelength laser in the light blue spectrum of Figure 3c have the same wavelength position and spacing, and their interference patterns are complementary. One is that the two lasers arise on the same lobe of the interference pattern, and the other is that the two lasers separately arise on the neighboring lobe of the interference pattern. This proves that the interference pattern position can be shifted; as long as the node of the interference pattern corresponds to the peak of the sub-interference, the same two dual-wavelength lasers can arise. Figure 3c illustrates a dual-wavelength laser featuring a narrow tuning range. To showcase a dual-wavelength laser with an extended tuning range, it is essential to fully utilize the TBF function to filter out unnecessary wavelengths and prevent mode competition. The outcomes, depicted in Figure 3d, demonstrate dual wavelengths and their respective spacings: 1527.71 nm and 1529.12 nm (light blue pattern), with a spacing of 1.41 nm; 1529.91 nm and 1531.32 nm (light green pattern), with a spacing of 1.41 nm; 1531.85 nm and 1533.29 nm (dark currant pattern), with a spacing of 1.44 nm; and 1533.9 nm and 1535.03 nm (yellow pattern), with a spacing of 1.13 nm. In the yellow part of the spectrum, the lasers have a smaller wavelength spacing because the power of EDFA is higher in the range of 1527 nm to 1534 nm. This makes it challenging for the laser to be stimulated at wavelengths beyond 1535 nm.

Figure 4 shows the case where the dual-wavelength laser position remains the same under different interferometric FSRs, and the laser spectrum at different times under this experimental scheme is verified to verify the stability of the system. The wavelengths of the dual-wavelength lasers in Figure 4a,b are both 1530.14 nm and 1531.56 nm, but the FSR of Figure 4a is 1.45 nm, while the FSR of Figure 4b is 0.9 nm. As mentioned in Figure 3, the same interference FSR, but with complementary phases, can produce the same dual-wavelength lasers. Here, the same dual-wavelength laser can be generated with different FSRs, and any point in the interference pattern cannot correspond to each other neatly. This serves as confirmation that the experimental setup is capable of generating multi-format lasers, indicating that the same laser wavelength can coexist with different interference states, a doing that has not been previously demonstrated in previous studies. Figure 5a shows the dual-wavelength laser case with the narrowest wavelength spacing, and Figure 5b shows the dual-wavelength laser case with several interfering lobes in the lasing wavelength gap. Additionally, the dual-wavelength laser states are recorded at different times to confirm their stability. Figure 5a shows a dual-wavelength laser with wavelengths of 1531.85 nm and 1532.15 nm at a wavelength spacing of 0.3 nm, which is achieved by filtering out unnecessary bands with TBF and setting the trough of the interferometric pattern between the two lasing wavelengths to maintain a balanced energy distribution. The dual-wavelength lasing involves wavelengths of 1529.65 nm and 1532.65 nm with a wavelength spacing of 3 nm, as shown in Figure 5b. There are eight nodes between the lasing wavelengths located at the half-upper-waist of the interference lobes. Despite this, it is noteworthy that the laser was not successfully excited at points on these eight nodes. The main reason for this is that the lasing wavelength aligns most accurately with the peak of the sub-interference, while the other eight nodes do not align as precisely with the peak of the sub-interference. Figure 6a–d illustrate the fluctuation of the wavelength position of the dual-wavelength laser in the cases of Figures 4a,b and 5a,b and the change in the power value corresponding to the peak of the lasing wavelength every minute, respectively. In Figure 6, it is evident that each wavelength position of different dual-wavelength laser cases fluctuates by less than 0.1 nm. The maximum change in the value corresponding to the peak of each lasing wavelength exceeds 3 dB due to the lack of temperature control for the EDFA in this scheme. The temperature variation of the EDFA impacts its output spectrum [69] and consequently affects the stability of the lasing wavelength output in this fiber ring laser setup. Figure 7 shows the case of triple-wavelength lasers; Figure 7a shows their spectra recorded every 10 min; and Figure 7b shows their wavelength position fluctuations and power variations at different times. The triple-wavelength lasers emit light at 1530.1 nm, 1531.22 nm, and 1532.34 nm, with a consistent 1.12 nm spacing between wavelengths. The fluctuation of each wavelength of triple-wavelength lasers is consistently less than 0.1 nm every 10 min, and the power corresponding to the peak of the lasing wavelength does not change by more than 2 dB.

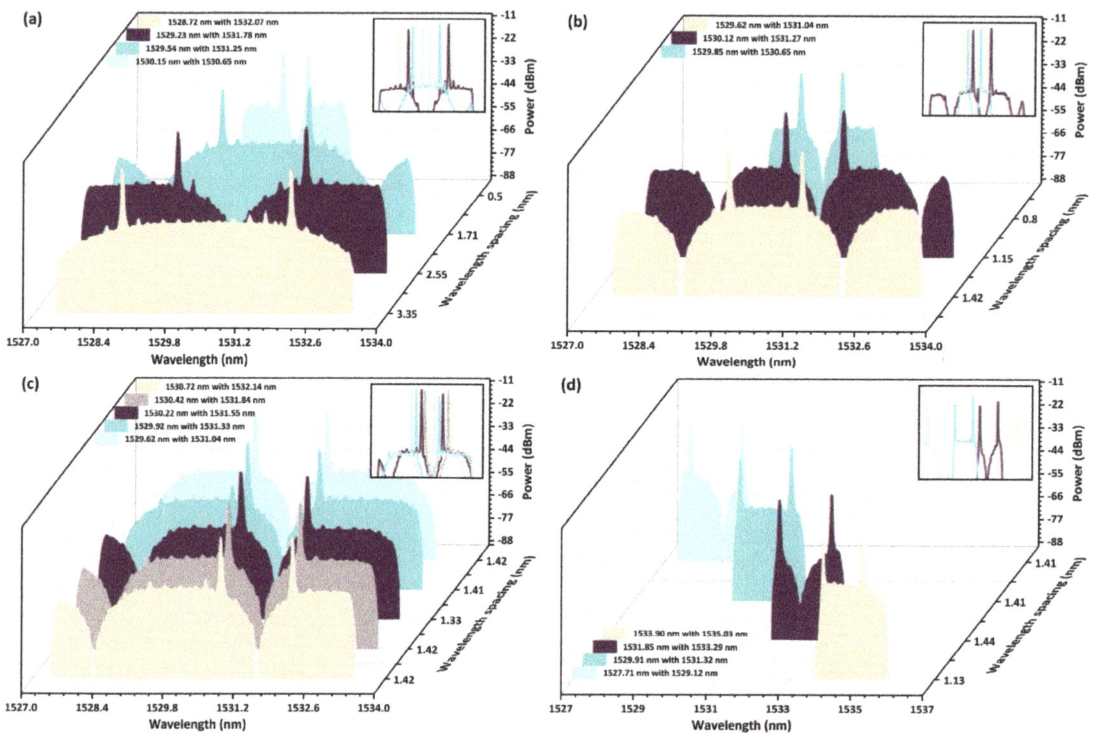

Figure 3. (**a**) Dual-wavelength laser with variable spacing and consistent center position of dual wavelengths. (**b**) Dual-wavelength laser with intertwined wavelength positions. (**c**,**d**) Dual-wav-length laser with regular positional tuning and similar wavelength spacing (the subplot is a superimposed drawing of the spectrum that does not correspond to the power scale of the vertical axis).

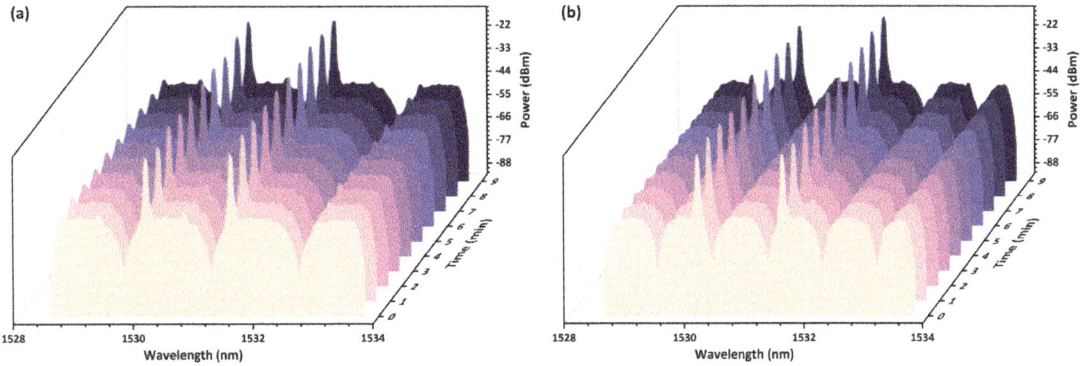

Figure 4. (**a**,**b**) The same dual-wavelength laser with different interferences and the spectral observation at different times.

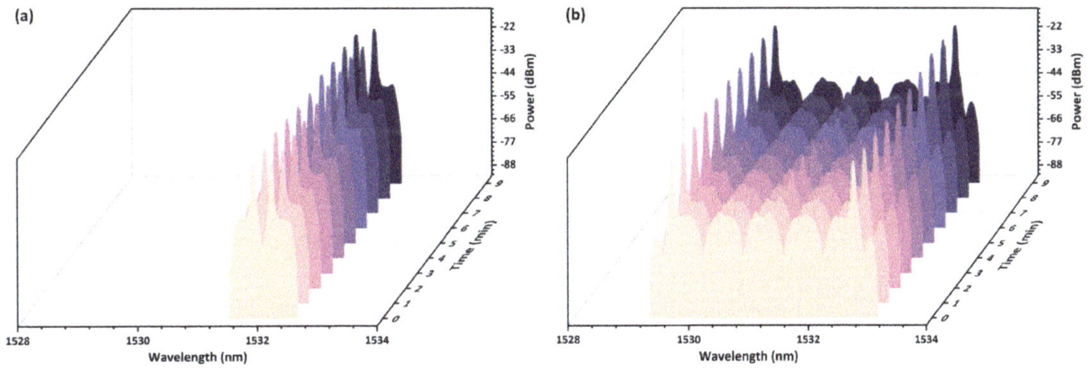

Figure 5. (**a**) Dual-wavelength laser with the narrowest wavelength spacing. (**b**) Dual-wavelength laser with several interfering lobes between the lasing wavelengths.

Maintaining a clear separation between telecommunication channels is essential to avoiding crosstalk. The wavelength separation of WDM in the C-band is usually 0.8 nm (100 GHz) or 0.4 nm (50 GHz). However, with the development of wavelength selective switch (WSS) components, it will be possible to support WDM with wavelength intervals less than 0.4 nm. WSS utilizes switch arrays that operate on wavelength-dispersed light to decouple or complex any single wavelength to a selected common or output port [70]. In a recent study [71], data communication in the C-band using two WDM channels at 0.3 nm intervals has been successfully demonstrated by WSS. Hence, this program can be effectively implemented in real-world scenarios. Furthermore, minimizing fiber loss, monitoring system temperature, monitoring laser output power, and adjusting the looped-in system's power will be essential for maintaining stability in laser power output during practical applications. Moreover, the operations of TDLI and TBF can cover the 1520–1570 nm wavelength range. Consequently, it is theoretically feasible to use other bands of EDFA to amplify and produce lasers within the respective bands.

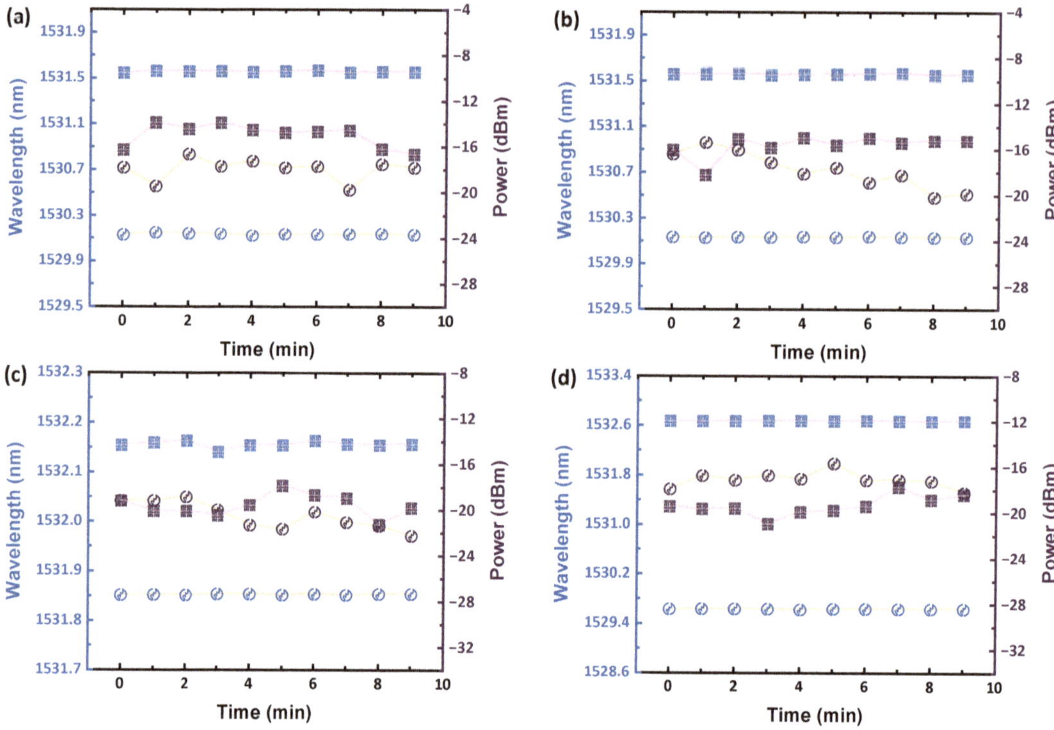

Figure 6. (**a**,**b**) The wavelength position and the power corresponding to the peak of the lasing wavelength per minute are obtained from Figure 4a,b, respectively. (**c**,**d**) The wavelength position and the power corresponding to the peak of the lasing wavelength per minute are obtained from Figure 5a,b, respectively. (The yellow lines with circle signs represent the shorter wavelength; the pink lines with square signs represent the longer wavelength; the blue signs correspond to wavelength positions; and the dark signs correspond to power values).

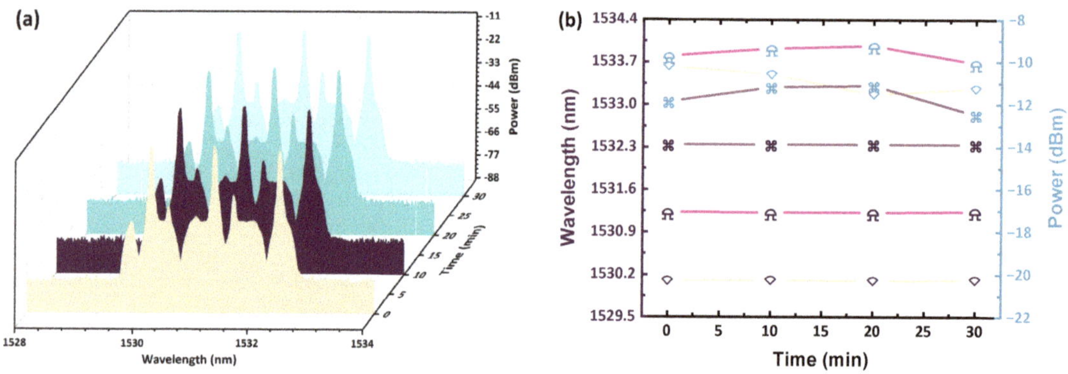

Figure 7. (**a**) A record of triple-wavelength lasers with the same wavelength spacing every ten minutes. (**b**) The wavelength position and the power corresponding to the peak of the lasing wavelength per ten minutes are obtained from (**a**). (The yellow line is at 1530.1 nm, the bright red line is at 1531.22 nm, and the coffee line is at 1532.34 nm; the dark signs correspond to wavelength positions, and the blue signs correspond to power values).

4. Conclusions

This study illustrates the utilization of erbium-doped fiber as a gain medium within a fiber ring architecture to establish a laser system with a principal lasing wavelength of approximately 1531 nm, coinciding with the amplification band of an EDFA. The TDLI serves a primary role in facilitating the generation of multi-format and multi-wavelength lasers through the superposition of two interferences with a 180-degree phase difference, thereby disrupting the homogeneous broadening effect of EDFA. Additionally, the TBF is deployed to eliminate extraneous spectral components, mitigating mode competition. The lasing wavelength assumes a position at the confluence of two complementary interference superposition patterns stemming from spatial hole-burning effects. Crucially, this lasing wavelength must correspond with the peaks of the FP interferences engendered by the BS within the TDLI. Hence, the presentation of highly adjustable FSR by TDLI allows interference standing wave patterns to be interchanged in real-time, supplementing the TBF to effectuate controlled variations in wavelength spacing and position in the multi-wavelength laser. As a consequence, this approach enables the deliberate orchestration of regular or irregular changes in the multi-wavelength laser, fostering the creation of multi-format, multi-wavelength lasers characterized by different interference states but a consistent wavelength position.

Author Contributions: Conceptualization, C.-K.Y. and P.-C.P.; methodology, C.-K.Y.; data curation, C.-K.Y.; model validation, C.-K.Y.; formal analysis, C.-K.Y., A.M.D. and P.-C.P.; investigation, C.-K.Y., A.M.D. and P.-C.P.; visualization: C.-K.Y.; writing—original draft preparation: C.-K.Y.; writing—review and editing: C.-K.Y., A.M.D. and P.-C.P. All authors have read and agreed to the published version of the manuscript.

Funding: This work was supported by the National Science and Technology Council, Taiwan, under Grant NSTC 112-2221-E-027-076-MY2.

Institutional Review Board Statement: Not applicable.

Informed Consent Statement: Not applicable.

Data Availability Statement: The original contributions presented in the study are included in the article, further inquiries can be directed to the corresponding author.

Conflicts of Interest: The authors declare no conflicts of interest.

References

1. Nemova, G. Brief Review of Recent Developments in Fiber Lasers. *Appl. Sci.* **2024**, *14*, 2323. [CrossRef]
2. Churkin, D.V.; Sugavanam, S.; Vatnik, I.D.; Wang, Z.; Podivilov, E.V.; Babin, S.A.; Rao, Y.J.; Turitsyn, S.K. Recent advances in fundamentals and applications of random fiber lasers. *Adv. Opt. Photon.* **2015**, *7*, 516–569. [CrossRef]
3. Shi, W.; Fang, Q.; Zhu, X.; Norwood, R.A.; Peyghambarian, N. Fiber lasers and their applications. *Appl. Opt.* **2014**, *53*, 6554–6568. [CrossRef]
4. Addanki, S.; Amiri, I.S.; Yupapin, P. Review of optical fibers-introduction and applications in fiber lasers. *Results Phys.* **2018**, *10*, 743–750. [CrossRef]
5. Kim, T.H. Analysis of optical communications, fiber optics, sensors and laser applications. *J. Mach. Comput.* **2023**, *3*, 115–125. [CrossRef]
6. Lu, H.H.; Huang, X.H.; Li, C.Y.; Xu, Y.Z.; Jin, J.L.; Chen, W.X.; Lin, C.H.; Wu, T.M. Two-way free-space optics-based interface between fibre and 5G communication employing polarisation-orthogonal modulation. *Commun. Eng.* **2023**, *2*, 89. [CrossRef]
7. Dehnaw, A.M.; Manie, Y.C.; Du, L.Y.; Yao, C.K.; Li, Y.L.; Hayle, S.T.; Peng, P.C. Bidirectional Free Space Optics Communication for Long-Distance Sensor System. *J. Lightw. Technol.* **2023**, *41*, 5870–5878. [CrossRef]
8. Peng, P.C.; Jhang, J.J.; Peng, Y.W.; Bitew, M.A.; Chi, Y.C.; Wu, W.C.; Wang, H.Y.; Lin, G.R.; Li, C.Y.; Lu, H.H. Tunable C-and L-band laser source based on colorless laser diode. *Laser Phys. Lett.* **2017**, *14*, 035806. [CrossRef]
9. Manie, Y.C.; Yao, C.K.; Yeh, T.Y.; Teng, Y.C.; Peng, P.C. Laser-based optical wireless communications for internet of things (IoT) application. *IEEE Internet Things J.* **2022**, *9*, 24466–24476. [CrossRef]
10. Dehnaw, A.M.; Shiu, R.K.; Chen, R.B.; Li, J.W.; Manie, Y.C.; Liang, H.C.; Peng, P.C. Self-start multi-wavelength laser source with tunable delay-line interferometer and optical fiber reflector for wireless communication system. *Appl. Sci.* **2021**, *11*, 9553. [CrossRef]
11. Murray, J.B.; Cerjan, A.; Redding, B. Distributed Brillouin fiber laser sensor. *Optica* **2022**, *9*, 80–87. [CrossRef]

12. Yao, C.K.; Manie, Y.C.; Chen, H.M.; Hsu, W.Y.; Lin, T.C.; Peng, P.C. Involvement of free-space optics in Raman distributed temperature sensing. *Opt. Lett.* **2023**, *48*, 6340–6343. [CrossRef]
13. Jati, M.P.; Luthfi, M.I.; Yao, C.K.; Dehnaw, A.M.; Manie, Y.C.; Peng, P.C. An Extremely Close Vibration Frequency Signal Recognition Using Deep Neural Networks. *Appl. Sci.* **2024**, *14*, 2855. [CrossRef]
14. Soares, L.; Perez-Herrera, R.A.; Novais, S.; Ferreira, A.; Frazão, O.; Silva, S. Paracetamol concentration-sensing scheme based on a linear cavity fiber laser configuration. *Opt. Fiber Technol.* **2023**, *80*, 103407. [CrossRef]
15. Soares, L.; Pérez-Herrera, R.A.; Novais, S.; Ferreira, A.; Silva, S.; Frazão, O. Linear Fiber Laser Configurations for Optical Concentration Sensing in Liquid Solutions. *Photonics* **2024**, *11*, 393. [CrossRef]
16. Nemova, G.; Jin, X.; Chen, L.R.; Firstov, S.V.; Sezerman, O. Modeling and experimental characterization of a dual-wavelength Bi-doped fiber laser with cascaded cavities. *J. Opt. Soc. Am. B* **2020**, *37*, 1453–1460. [CrossRef]
17. Nemova, G.; Kashyap, R. Yb3+-doped fiber laser with integrated optical cooler. In Proceedings of the Laser Technology for Defense and Security VI, Orlando, FL, USA, 5–6 April 2010; SPIE: Bellingham, WA, USA, 2015; Volume 7686, pp. 241–247.
18. Tokita, S.; Murakami, M.; Shimizu, S.; Hashida, M.; Sakabe, S. Liquid-cooled 24 W mid-infrared Er: ZBLAN fiber laser. *Opt. Lett.* **2009**, *34*, 3062–3064. [CrossRef]
19. Topper, B.; Peysokhan, M.; Albrecht, A.R.; Flores, A.S.; Kuhn, S.; Hässner, D.; Hein, S.; Hupel, C.; Nold, J.; Haarlammert, N.; et al. Laser cooling of a Yb doped silica fiber by 18 Kelvin from room temperature. *Opt. Lett.* **2021**, *46*, 5707–5710. [CrossRef]
20. Nishizawa, N. Ultrashort pulse fiber lasers and their applications. *Jpn. J. Appl. Phys.* **2014**, *53*, 090101. [CrossRef]
21. Zhang, L.; Tian, Z.; Chen, N.K.; Grattan, K.T.; Yao, Y.; Rahman, B.A.; Li, X.H.; Yao, C.K.; Han, H.L.; Chui, H.C.; et al. Pulse Dynamics of an All-Normal-Dispersion Ring Fiber Laser Under Four Different Pulse Regimes. *IEEE Access* **2020**, *8*, 115263–115272. [CrossRef]
22. Chen, N.K.; Lin, J.W.; Liu, F.Z.; Liaw, S.K. Wavelength-Tunable Er^{3+}-Doped fs Mode-Locked Fiber Laser Using Short-Pass Edge Filters. *IEEE Photon. Technol. Lett.* **2010**, *22*, 700–702. [CrossRef]
23. Chen, N.K.; Liu, F.Z.; Chuang, H.P.; Lai, Y.; Yang, S.D.; Lin, J.W.; Liaw, S.K.; Chang, Y.C.; Huang, C.B.; Chi, S. Highly efficient femtosecond pulse stretching by tailoring cavity dispersion in erbium fiber lasers with an intracavity short-pass edge filter. *Opt. Express* **2011**, *19*, 15879–15884. [CrossRef] [PubMed]
24. Han, H.; Chen, N.K.; Zhang, L.; Xie, Y.; Tian, Z.; Yao, Y.; Huang, Y.; Zhang, X. Output pulse characteristics of a Mamyshev fiber oscillator. *Photonics* **2021**, *8*, 590. [CrossRef]
25. Yeh, C.H.; Liu, L.H.; Ko, H.S.; Wang, B.Y.; Hsu, W.H.; Chow, C.W.; Chen, J.H. Quad-ring based erbium fiber laser for switchable and stable single-longitudinal-mode operation. *Opt. Fiber Technol.* **2021**, *61*, 102450. [CrossRef]
26. Li, G.H.; Lin, Y.Z.; Chang, H.W.; Yeh, C.H.; Chen, J.R.; Wang, B.Y.; Hsu, W.H.; Chow, C.W.; Liaw, S.K. Stabilized single-longitudinal-mode fiber laser with broadband and flat wavelength output. *Phys. Scr.* **2020**, *96*, 015503. [CrossRef]
27. Yamashita, S.; Takubo, Y. Wide and fast wavelength-swept fiber lasers based on dispersion tuning and their application to optical coherence tomography. *Photon. Sens.* **2013**, *3*, 320–331. [CrossRef]
28. Jiang, H.; Zhao, Z.; Jin, L.; Set, S.Y.; Yamashita, S. Self-wavelength-sweeping in a bi-directional thulium-doped fiber ring laser. *Appl. Phys. Express* **2019**, *12*, 042006. [CrossRef]
29. Chen, H.; Jiang, X.; Xu, S.; Zhang, H. Recent progress in multi-wavelength fiber lasers: Principles, status, and challenges. *Chin. Opt. Lett.* **2020**, *18*, 041405. [CrossRef]
30. Kbashi, H.J.; Sharma, V.; Sergeyev, S.V. Tunable Multiwavelength SOA-Based Fiber Laser. *Electronics* **2023**, *12*, 3277. [CrossRef]
31. Imbu, J.J.; Bitew, M.A.; Chang, C.H.; Tang, W.C.; Manie, Y.C.; Peng, P.C. Multi-wavelength generation? Based on RSOA for passive optical networks. In Proceedings of the 2017 IEEE International Conference on Consumer Electronics-Taiwan (ICCE-TW), Taipei, Taiwan, 12–14 June 2017.
32. Peng, P.C.; Tseng, H.Y.; Chi, S. Long-distance FBG sensor system using a linear-cavity fiber Raman laser scheme. *IEEE Photonics Technol. Lett.* **2004**, *16*, 575–577. [CrossRef]
33. Awsaj, M.K.; Al-Mashhadani, T.F.; Al-Mashhadani, M.K.S.; Hammudi, R.N.; Ali, A.Y.; Zan, M.S.D.; Arsad, N. Tunable 60 GHz Multiwavelength Brillouin Erbium Fiber Laser. *Appl. Sci.* **2023**, *13*, 3275. [CrossRef]
34. Wang, T.; Yang, A.; Shi, F.; Huang, Y.; Wen, J.; Zeng, X. High-order mode lasing in all-FMF laser cavities. *Photon. Res.* **2019**, *7*, 42–49. [CrossRef]
35. Radzi, N.M.; Latif, A.A.; Ismail, M.F.; Liew, J.Y.; Awang, N.A.; Lee, H.K.; Ahmad, F.; Norizan, S.F.; Ahmad, H. Tunable spacing dual-wavelength Q-switched fiber laser based on tunable FBG device. *Photonics* **2021**, *8*, 524. [CrossRef]
36. Gao, M.; Yin, B.; Lv, Y.; Sang, G.; Hou, B.; Li, H.; Wang, M.; Wu, S. Tunable and switchable dual-wavelength SLM Narrow-linewidth fiber laser with a PMFBG-FP filter cascaded by multi-ring cavity. *Photonics* **2022**, *9*, 756. [CrossRef]
37. He, W.; Li, D.; Zhu, L.; Dong, M.; Luo, F. Tunable multiwavelength erbium-doped fiber laser employing PM-FBG and Mach–Zehnder interferometer with optical fiber delay line. *IEEE Photon. J.* **2017**, *9*, 7202108. [CrossRef]
38. Zhou, X.; Li, Z.; Zhou, Y. Tunable and switchable multi-wavelength fiber laser based on polarization hole burning effect and Sagnac loop mirrors. *Opt. Quantum Electron.* **2020**, *52*, 451. [CrossRef]
39. Ahmad, H.; Jasim, A.A. Stable C-band fiber laser with switchable multi-wavelength output using coupled microfiber Mach-Zehnder interferometer. *Opt. Fiber Technol.* **2017**, *36*, 105–114. [CrossRef]

40. Martinez-Ramirez, L.G.; Alvarado, E.S.; Gallegos-Arellano, E.; Fernandez-Jaramillo, A.A.; Estudillo-Ayala, J.M.; Jauregui-Vazquez, D.; Rojas-Laguna, R.; Sierra-Hernandez, J.M. Select-cutoff Mach-Zehnder interferometer based on waist-enlarged technique and its multi-wavelength fiber laser application. *Infrared Phys. Technol.* **2023**, *128*, 104508. [CrossRef]
41. Lopez-Dieguez, Y.; Estudillo-Ayala, J.M.; Jauregui-Vazquez, D.; Herrera-Piad, L.A.; Sierra-Hernandez, J.M.; Garcia-Mina, D.F.; Gallegos-Arellano, E.; Hernandez-Garcia, J.C.; Rojas-Laguna, R. Erbium ring fiber laser cavity based on tip modal interferometer and its tunable multi-wavelength response for refractive index and temperature. *Appl. Sci.* **2018**, *8*, 1337. [CrossRef]
42. Tian, Z.; Chen, N.K.; Shum, P.P.; Yao, C.K.; Zhang, L.; Yao, Y.; Grattan, K.T.V.; Ren, S.; Wu, Q. Erbium-doped dual wavelength fiber laser interferometric proximity sensor with ±16 nm measurement accuracy. In Proceedings of the CLEO: QELS_Fundamental Science, San Jose, CA, USA, 9–14 May 2021; Optica Publishing Group: Washington, DC, USA, 2021.
43. Geng, X.; Jiang, Y.; Gu, H.; Luo, S.; Sun, M.; Li, L. Switchable multi-wavelength fiber lasers based on asymmetric biconical fiber tapers. *Opt. Commun.* **2023**, *548*, 129837. [CrossRef]
44. Torres-Torres, M.; Salceda-Delgado, G.; Rodríguez-Carreón, V.C.; Martinez-Rios, A.; Sierra-Hernandez, J.M.; Antonio-Lopez, J.E.; Amezcua-Correa, R.; Estudillo-Ayala, J. Reshaping the output of fiber lasers by using a variable intra-cavity filter based on a reconfigurable Fabry–Perot interferometer. *Laser Phys.* **2021**, *31*, 035102. [CrossRef]
45. Chang, Y.; Pei, L.; Ning, T.; Zheng, J. Switchable multi-wavelength fiber laser based on hybrid structure optical fiber filter. *Opt. Laser Technol.* **2020**, *124*, 105985. [CrossRef]
46. Tang, Z.; Lou, S.; Wang, X. Stable and widely tunable single-/dual-wavelength erbium-doped fiber laser by cascading a twin-core photonic crystal fiber based filter with Mach-Zehnder interferometer. *Opt. Laser Technol.* **2019**, *109*, 249–255. [CrossRef]
47. Filoteo-Razo, J.D.; Hernandez-Garcia, J.C.; Estudillo-Ayala, J.M.; Pottiez, O.; Jauregui-Vazquez, D.; Sierra-Hernandez, J.M.; Lauterio-Cruz, J.P.; Carrillo-Delgado, C.M.; Rojas-Laguna, R. Multi-wavelength Er–Yb-doped fibre ring laser using a double-pass Mach–Zehnder interferometer with a Sagnac interferometer. *Opt. Laser Technol.* **2021**, *139*, 106994. [CrossRef]
48. Chang, Y.; Pei, L.; Ning, T.; Zheng, J.; Li, J.; Xie, C. Switchable and tunable multi-wavelength fiber ring laser employing a cascaded fiber filter. *Opt. Fiber Technol.* **2020**, *58*, 102240. [CrossRef]
49. Huang, B.; Sheng, X.; Tang, Z.; Lou, S. High SMSR and widely tunable multi-wavelength erbium doped fiber laser based on cascaded filters. *Infrared Phys. Technol.* **2022**, *122*, 104082. [CrossRef]
50. Yao, C.K.; Lee, M.C.; Peng, P.C. EDFA-Based Multi-Wavelength Laser Using Tunable Filter and Delay-Interferometer for WDM Systems. In Proceedings of the 2022 IEEE International Conference on Consumer Electronics-Taiwan (ICCE-TW), Beitou, Taipei, Taiwan, 6–8 July 2022.
51. Fan, B.; Zhou, X.; Li, Z.; Zhou, Y.; Wang, T. Experimental research on an L-band multi-wavelength erbium-doped fiber laser based on a cascaded Sagnac loop and M–Z filters. *Laser Phys.* **2019**, *29*, 065102. [CrossRef]
52. He, W.; Yuan, H.; Lou, X.; Zhu, L.; Dong, M. Multi-wavelength switchable erbium-doped fiber laser based on a hybrid filter incorporating a bi-tapered Mach–Zehnder interferometer and Sagnac loop. *Phys. Scr.* **2019**, *94*, 125502. [CrossRef]
53. Zhou, X.; Ni, Y.; Wei, Y. L-Band multi-wavelength erbium-doped fibre laser based on non-linear optical loop mirror and dual-channel Mach–Zehnder interferometer. *J. Mod. Opt.* **2017**, *64*, 210–213. [CrossRef]
54. Herrera-Piad, L.A.; Marrujo-García, S.; Hernández-Romano, I.; May-Arrioja, D.A.; Minkovich, V.P.; Torres-Cisneros, M. Highly Stable Switchable Emissions of an Erbium-Doped Fiber Ring Laser Using Cascaded MZIs Based on CHCF. *Machines* **2022**, *10*, 962. [CrossRef]
55. Tian, X.; Zhang, L.; Yao, Y.; Xu, H.; Zhu, K.; Zhang, X.; Nie, Z. Interval Adjustable Dual-Wavelength Erbium-Doped Fiber Laser Based on Cascaded Two Mach-Zehnder Interferometers. *Photonics* **2023**, *10*, 1156. [CrossRef]
56. Zhang, L.; Tian, Z.; Chen, N.K.; Han, H.; Liu, C.N.; Grattan, K.T.; Rahman, B.M.A.; Zhou, H.; Liaw, S.K.; Bai, C. Room-temperature power-stabilized narrow-linewidth tunable erbium-doped fiber ring laser based on cascaded Mach-Zehnder interferometers with different free spectral range for strain sensing. *J. Light. Technol.* **2020**, *38*, 1966–1974. [CrossRef]
57. Peng, Y.; Peng, J.; Pan, H.; Zhang, A.; Liu, Z.; Chen, Z.; Zhang, B.; Liu, X.; Cao, C.; Du, P. Switchable multi-wavelength actively Q-switched erbium-doped fiber laser based on nonlinear polarization rotation and Sagnac filter. *Optik* **2023**, *284*, 170955. [CrossRef]
58. Al-Rubaiyee, H.A.; Al-Hayali, S.K.; Harun, S.W.; Al-Janabi, A.H. Vernier effect based on cascading two Mach–Zehnder interferometers for selectable comb filter and saturable absorber applications in erbium-doped fiber laser. *Opt. Fiber Technol.* **2024**, *84*, 103757. [CrossRef]
59. Al-Rubaiyee, H.A.; Al-Hayali, S.K.; Al-Janabi, A.H. Vernier effect based on hybrid fiber interferometers: A new tool for wavelength switchability and adjustable free spectral range fiber lasing. *Opt. Contin.* **2023**, *2*, 1203–1215. [CrossRef]
60. Yin, G.; Lou, S.; Wang, X.; Han, B. Tunable multi-wavelength erbium-doped fiber laser by cascading a standard Mach–Zehnder interferometer and a twin-core fiber-based filter. *Laser Phys. Lett.* **2013**, *10*, 125110. [CrossRef]
61. Jachpure, D.; Vijaya, R. Mixed line broadening in the saturable absorption of erbium-doped fiber. *Opt. Quantum Electron.* **2023**, *55*, 113. [CrossRef]
62. Yamashita, S.; Hotate, K.J.E.L. Multiwavelength erbium-doped fibre laser using intracavity etalon and cooled by liquid nitrogen. *Electron. Lett.* **1996**, *32*, 1298–1299. [CrossRef]
63. Park, N.; Wysocki, P.F. 24-line multiwavelength operation of erbium-doped fiber-ring laser. *IEEE Photon. Technol. Lett.* **1996**, *8*, 1459–1461. [CrossRef]

64. Dejband, E.; Yao, C.K.; Manie, Y.C.; Huang, P.Y.; Lee, H.K.; Tan, T.H.; Peng, P.C. Utilizing a Tunable Delay Line Interferometer to Improve the Sensing Accuracy of an FBG Sensor System. *Photonics* **2022**, *9*, 11. [CrossRef]
65. Hénault, F. Quantum physics and the beam splitter mystery. In Proceedings of the Nature of Light: What are Photons? VI, San Diego, CA, USA, 10–13 August 2015; SPIE: Bellingham, WA, USA, 2015; Volume 9570, pp. 199–213.
66. Yang, X.; Kweun, J.M.; Kim, Y.Y. Theory for perfect transmodal Fabry-Perot interferometer. *Sci. Rep.* **2018**, *8*, 69. [CrossRef] [PubMed]
67. Bennett, W.R., Jr. Hole burning effects in a He-Ne optical maser. *Phys. Rev.* **1962**, *126*, 580. [CrossRef]
68. Haken, H.; Sauermann, H. Frequency shifts of laser modes in solid state and gaseous systems. *Z. Für Phys.* **1963**, *176*, 47–62. [CrossRef]
69. Bebawi, J.A.; Kandas, I.; El-Osairy, M.A.; Aly, M.H. A comprehensive study on EDFA characteristics: Temperature impact. *Appl. Sci.* **2018**, *8*, 1640. [CrossRef]
70. Ma, Y.; Stewart, L.; Armstrong, J.; Clarke, I.G.; Baxter, G. Recent progress of wavelength selective switch. *J. Lightw. Technol.* **2020**, *39*, 896–903. [CrossRef]
71. Zhang, Y.; He, Z.; Wu, Z.; Ren, J.; Liu, B.; Lei, T.; Yuan, X. Ultrabroadband 2D-Dispersive Wavelength Selective Switch with 1.57 GHz Hyperfine Spectral Resolution and >1600 Channels. *Laser Photon. Rev.* **2023**, *17*, 2300232. [CrossRef]

applied sciences

Article

Wavelength-Dependent Bragg Grating Sensors Cascade an Interferometer Sensor to Enhance Sensing Capacity and Diversification through the Deep Belief Network

Shegaw Demessie Bogale, Cheng-Kai Yao, Yibeltal Chanie Manie, Zi-Gui Zhong and Peng-Chun Peng *

Department of Electro-Optical Engineering, National Taipei University of Technology, Taipei 10608, Taiwan
* Correspondence: pcpeng@ntut.edu.tw; Tel.: +886-2-2771-2171 (ext. 4671)

Abstract: Fiber-optic sensors, such as fiber Bragg grating (FBG) sensors and fiber-optic interferometers, have excellent sensing capabilities for industrial, chemical, and biomedical engineering applications. This paper used machine learning to enhance the number of fiber-optic sensing placement points and promote the cost-effectiveness and diversity of fiber-optic sensing applications. In this paper, the framework adopted is the FBG cascading an interferometer, and a deep belief network (DBN) is used to demodulate the wavelength of the sampled complex spectrum. As the capacity of the fiber-optic sensor arrangement is optimized, the peak spectra from FBGs undergoing strain or temperature changes may overlap. In addition, overlapping FBG spectra with interferometer spectra results in periodic modulation of the spectral intensity, making the spectral intensity variation more complex as a function of different strains or temperature levels. Therefore, it may not be possible to analyze the sensed results of FBGs with the naked eye, and it would be ideal to use machine learning to demodulate the sensed results of FBGs and the interferometer. Experimental results show that DBN can successfully interpret the wavelengths of individual FBG peaks, and peaks of the interferometer spectrum, from the overlapping spectrum of peak-overlapping FBGs and the interferometer.

Keywords: fiber Bragg grating sensor; interferometer; deep belief network; wavelength detection

1. Introduction

Fiber-optic sensing has been highly favored by researchers in recent years, and a large number of related studies have shown that fiber-optic sensing has excellent performance in various measurement applications due to its small size, sensitivity, and immunity to electromagnetic interference [1–5]. Fiber-optic sensors mainly use the position or power changes of specific wavelengths in the spectrum measured after the probing beam reaches the measurement position of the fiber-optic sensor to interrogate changes in various environmental physical parameters. FBGs based on the Bragg grating architecture are simple-to-manufacture fiber-optic sensors made by transversely exposing the core of single-mode fiber to intense ultraviolet light with a periodic pattern [6], which can be used to sense temperature [7], strain [8], pressure [9], vibration [10], acceleration [11], and more. Usually, the strain sensitivity is about a 1 pm wavelength shift with one microstrain ($\mu\varepsilon$) applied change, and the temperature sensitivity is about a 13 pm wavelength shift with a 1 °C change in environmental temperature for a standard FBG sensor around the 1550 nm wavelength [12]. On the other hand, the production process of a fiber-optic interferometer is more complicated than that of FBG, which usually requires the use of fiber fusion splicing equipment to splice different types of fibers, taper, side polishing, etching, etc. [13–15]. By changing the shape of these optical fibers, the path of light beam transmission is altered to form an optical path difference. This allows changes in different environmental physical parameters to shape different light beam paths, creating different interference states for sensing. Fiber-optic interferometers typically exhibit heightened sensitivity in measurements. Their accuracy for sensing temperature [16] and strain [17] is significantly greater—ranging from tens to hundreds of times higher—than that of standard FBGs.

In addition, standard FBGs are usually only capable of measuring either strain alike or temperature alone, requiring the use of another FBG for temperature compensation in the case of strain alike measurement [9]. Fiber-optic interferometers can overcome this problem by using the Vernier effect to measure temperature, strain, or other dual parameters simultaneously [18,19]. However, the main common disadvantages of fiber-optic interferometers are the high optical loss and the complexity of the optical spectrum, which make it difficult to cascade multiple fiber-optic interferometers to perform sensing tasks at the same time. In contrast, multiple FBGs can be cascaded in series to perform multi-point sensing tasks. Thus, if multiple FBGs and fiber-optic interferometers are cascaded in series, it is possible to meet the requirements of measurement conditions that require multiple measurement points, high sensitivity, and diversity of physical parameter measurement. In the previous instance, multiple FBGs cascaded an interferometer for sensing [20]. To increase the number of FBG sensing points, it is necessary to arrange the wavelengths of the FBGs very close to each other, and the overlapping of the wavelengths of the FBGs may occur during the sensing task, making it difficult to interpret the sensing results. In previous studies, it has been demonstrated that the overlapping of FBGs can be solved using machine learning to accurately interpret the sensed information [7,8]. However, FBGs cascade an interferometer framework for sensing tasks with the wavelength overlap of FBGs, which has not been explored.

In this paper, the main goal is to improve the sensing architecture of FBGs cascade with an interferometer by utilizing machine learning technology, specifically the DBN, to address the issue of FBGs' wavelength overlap within this sensing architecture. The focus is on the power variations of different wavelengths caused by the interferometer spectrum drift, which results in different spectral profiles for the same overlapped FBG wavelengths through power modulation. Therefore, in order to interpret the center wavelength of FBGs in various situations based on the different interference spectra, it is essential to predict the central wavelength of FBGs through machine learning. DBN specializes in feature extraction, which is particularly useful for identifying different spectral shapes due to different interferometric and FBGs' overlap patterns. The experimental results show that DBN can successfully decipher the overlap of different wavelengths of FBGs in different interference spectra. Therefore, machine learning will be beneficial for the sensing points' capacity and sensing diversity in FBGs cascading an interferometer sensing architecture.

2. Experimental Setup

Figure 1 shows the experiment setup of wavelength-dependent Bragg grating sensors cascading an interferometer. In this scheme, two FBGs with the same initial wavelength are used for strain sensing, and the applied strain causes the FBG wavelengths to shift. Therefore, the center wavelengths of the two FBGs can overlap partially or completely. In addition, a tunable delay-line interferometer (TDI) is used to simulate the spectral changes of the fiber-optic interferometer during sensing. The TDI consists of two beam splitters placed at a positive and negative angle of 45 degrees, a reflector placed at a negative angle of 45 degrees, and a prism placed at a positive angle of 180 degrees [21]. The TDI splits a beam into two beams and creates a difference in the optical path, resulting in an interferometric spectrum. Furthermore, the TDI can cause a change in the optical path difference between the two beams by moving the position of the prism [21]. This shows that the optical principle of TDI is the same as that of the fiber-optic interferometer described in the previous section; only the medium of light transmission is different. Therefore, since the TDI has the same spectral properties as the fiber-optic interferometer and can adjust its overall spectral position arbitrarily, it is very easy to simulate the spectral shift caused by the fiber-optic interferometer during sensing.

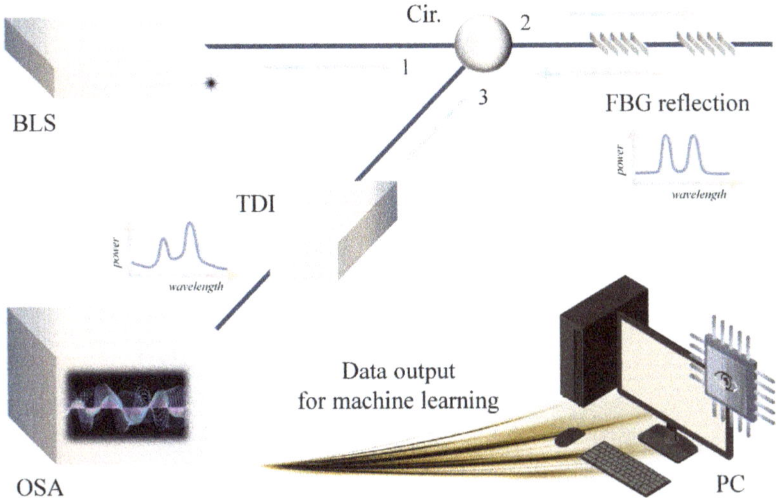

Figure 1. An experimental framework for two FBGs cascading an interferometer for sensing. The sensing data are used for model training and testing of machine learning (BLS: broadband light source; Cir.: circulator; TDI: tunable delay-line interferometer; OSA: optical spectral analyzer; PC: personal computer).

Data acquisition consists of a broadband light source (BLS) emitting a beam that passes through port 1 of the circulator to port 2 of the circulator and senses the strain via the FBG. The reflectance spectra of the strain sensed on the FBG pass through port 2 of the circulator to port 3 of the circulator, where they then pass through a TDI and are then read by an optical spectrum analyzer (OSA). The collected spectra with specific wavelength reflections of FBG overlap interference patterns of the TDI are transmitted to a personal computer (PC) for machine learning model training and for machine learning models to verify the usefulness of the trained models. In addition, the strain application of the FBG is realized by the displacement variation of the linear stage (LS) to change the fixed position of the FBG. The brand models of the main equipment and components (BLS, Cir., FBG, LS, TDI, and OSA) are UNICE. Inc. NA0101 in Taoyuan, Taiwan (wavelength band: 1520 nm to 1570 nm), FOCI Inc. in Hsinchu, Taiwan (wavelength band: 1470 nm to 1610 nm), 3L Technologies Inc. in Miaoli County, Taiwan (standard FBG with 1 pm/$\mu\varepsilon$ sensitivity), Onset Inc. CT02A in Taipei, Taiwan (0.01 mm moving resolution), Kylia Inc. WT-MINT in Paris, France, and Anritsu Inc. MS9740A in Atsugi, Japan (0.03 nm lowest resolution), respectively.

3. Data Collection Setup and Analysis

In the strain sensing experiment, the initial center wavelength of both FBGs was 1544.73 nm. The setup used was to fix the center wavelength of one of the FBGs at 1545.1 nm by applying strain. The other FBG was subjected to the strain from the initial center wavelength position of the FBG, and a total of 12 steps of strain application were performed. Finally, the center wavelength position of the FBG was located at 1545.45 nm. Each step was about 60 $\mu\varepsilon$ of strain application with a 0.06 nm wavelength shift, since the wavelength shift of 1 pm corresponds to about 1 $\mu\varepsilon$ of strain application. The 60 $\mu\varepsilon$ strain was realized by shifting the LS by 20 μm. Since the distance between the fixed points on both sides of the FBG was 33 cm, the microstrain on the FBG was 20 μm divided by 33 cm, which is equal to 60 $\mu\varepsilon$. However, the moving operation of LS was via manual adjustment of the micrometer head, and with the limitations of the actual resolution of the OSA, there may have been an error of 5 $\mu\varepsilon$ to 10 $\mu\varepsilon$ in the application of each strain. From this setup, the spectrum of FBGs can be obtained with no overlap, partial overlap, and almost complete overlap. In order to verify that the overlapped spectra of FBG wavelengths can be successfully interpreted in different interference patterns obtained by fiber-optic interferometry, three different interferometric states of TDI were used to verify the success of the FBG spectra. Figure 2 shows the spectra of the FBG from 0 με to 720 με strain and the unstrained FBG in three different interferometric settings of TDI. The appearance of the interference spectrum is very similar to that of a sinusoidal wave, which shows periodic power modulation as the wavelength changes, and the displacement between the wavelength positions of the troughs is defined as the free spectral range (FSR). Here, the FSR of the three different interference states set by TDI is 2 nm. It can be observed that in the cases of Figure 2a–c, the wavelengths of the unstressed FBGs fall in the left, center, and right halves of one of the 2 nm wide lobes in the interference pattern. The FBG energy is higher in the case of Figure 2b and lower in the cases of Figure 2a,c because the interference power tunes the reflected power of the FBG. Therefore, it can be found that the power of the other FBG under strain will be modulated according to the wavelength position of the FBG and the corresponding power of the set interference.

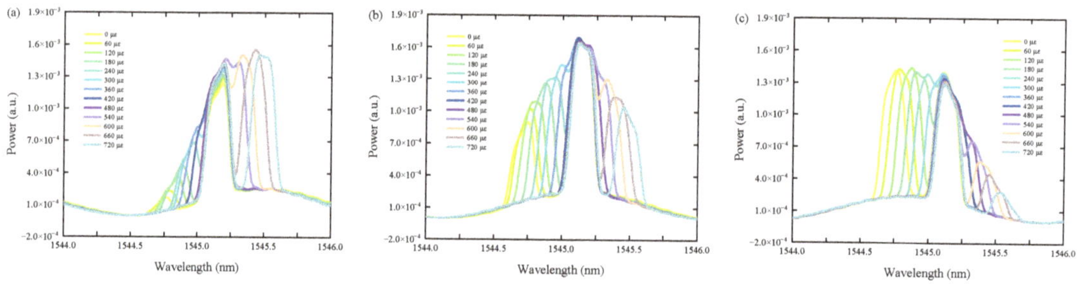

Figure 2. (**a**) The unstrained FBG wavelength falls to the left of the peak of the lobe of the interference pattern. (**b**) The unstrained FBG wavelength falls to the peak of the lobe of the interference pattern. (**c**) The unstrained FBG wavelength falls to the right of the peak of the lobe of the interference pattern.

From the above, it can be seen that even if the FBG wavelengths are at the same position, they will be affected by different interferences and thus show different spectra. Figure 3 shows this concept more clearly. Figure 3a–c show the FBG spectra of the first, eighth, and last steps of the strain in three different interferometric states, respectively. It can be seen that the spectra of the two FBGs, both superimposed and un-superimposed, are very different in the three different interference states. Furthermore, as mentioned before, optical sensing mainly utilizes specific wavelength variations for sensing, and sensing based on interferometric spectroscopy either utilizes the wavelength at a certain peak of the spectrum to obtain sensing information or utilizes the wavelength at a certain dip of the spectrum to obtain sensing information. Hence, in this scheme, the wavelength value at the peak of the interference's lobe on the right side of the FBGs in Figure 3 is assumed to be the main wavelength for determining the sensed information during sensing. The peak wavelengths of the three selected lobe interferences are 1546.88 nm (interference 1), 1547.11 nm (interference 2), and 1547.39 nm (interference 3). Figure 4 clearly shows the fixed sensing wavelengths of the interferometric spectrum for different interference states at each strain step, the theoretical values of the two FBG wavelengths, and the overlapping areas of the two FBG wavelengths under different strain steps.

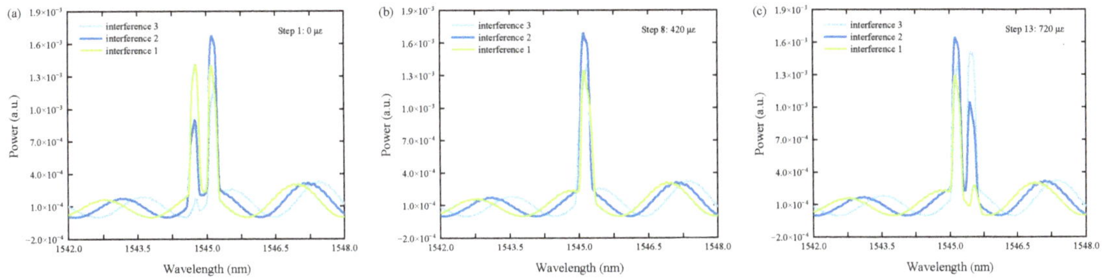

Figure 3. (**a**) The strain on the FBG that senses the strain is 0 με for the three different interference scenarios. (**b**) The strain on the FBG that senses the strain is 420 με for the three different interference scenarios. (**c**) The strain on the FBG that senses the strain is 720 με for the three different interference scenarios.

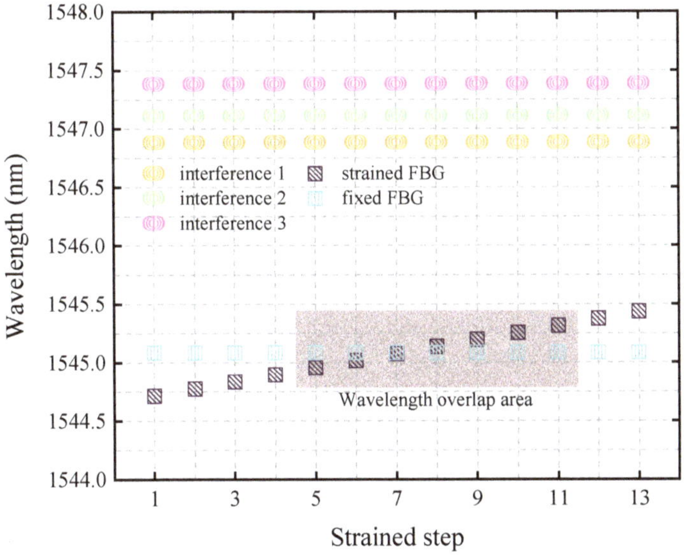

Figure 4. At each strain step, the wavelength information of the two FBG wavelengths and the wavelength information of the interferometric spectrum which is assumed to be used for sensing.

4. Deep Belief Network for Peak Wavelength Prediction

A DBN is a generative graphical model consisting of multiple hidden unit layers. Each pair of connected layers forms a restricted Boltzmann machine (RBM) [22–26], which serves as the core building block of a DBN. An RBM is a two-layer network capable of learning to represent complex data distributions [22,25]. It comprises a visible layer (v) for input data and a hidden layer (h) for learned features. The nodes in these layers are interconnected by undirected edges, with no connections within a layer, simplifying the learning process and rendering the RBM efficient for training. The energy of a configuration, including visible and hidden states in an RBM, is determined by [25]:

$$E(\mathrm{v},\mathrm{h}) = -\sum_i a_i v_i - \sum_j b_j h_j - \sum_{i,j} v_i w_{i,j} h_j \quad (1)$$

where v_i is visible units, h_j is hidden units, a_i is bias for visible units, b_j is bias for hidden units, and $w_{i,j}$ is the weight matrix between visible and hidden units.

The architecture of a DBN is constructed by stacking several RBMs on top of each other, as shown in Figure 5. The visible layer of the first RBM serves as the input layer of the DBN. The hidden layer of the first RBM becomes the visible layer for the next RBM in the stack, and this process continues for each subsequent layer. This hierarchical structure allows the DBN to learn increasingly abstract representations of the input data as it moves up the layers. The training of a DBN typically involves a two-step process: pre-training and fine-tuning. During the pre-training phase, each RBM is trained independently in a greedy, layer-by-layer manner, starting from the bottom-most RBM. This unsupervised pre-training step is crucial in effectively initializing the network's parameters, like the network's weights, in a way that captures the underlying structure of the data. The pre-training acts as a form of feature extraction, allowing the network to learn useful representations of the data [23].

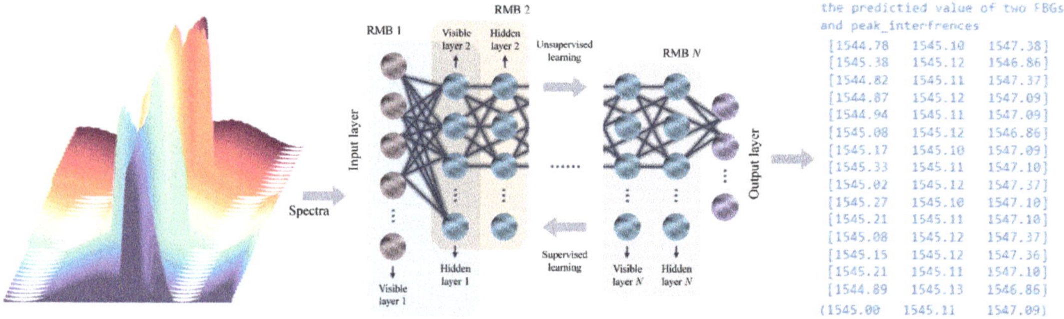

Figure 5. DBN for peak wavelength detection in FBGs and interferometric spectra.

Once all the layers are pre-trained, the entire DBN can be fine-tuned using supervised learning techniques if labeled data are available. This involves adding a final output layer and using backpropagation to adjust the weights across the entire network to minimize the error in predictions [24]. The fine-tuning process ensures that the learned representations are optimized for the specific task at hand, such as classification or regression. The combination of unsupervised pre-training and supervised fine-tuning makes DBNs powerful tools for learning from complex, high-dimensional data, providing a robust framework for capturing intricate patterns and structures in the data. Their ability to learn hierarchical representations of data makes them powerful for handling complex datasets. The hierarchical nature of DBNs means that higher layers capture more abstract features, enabling more accurate and efficient learning and prediction [23,24].

The process of utilizing a DBN for peak wavelength detection in FBGs and interferences involves several well-defined steps, each crucial for building a robust and accurate model. There were 3 different interferometric modes and 13 different FBG strained settings in the experiment, and the total number of interferometric spectral modes and FBG spectral modes of union was 39 different spectra. In order to predict the wavelength accurately in the final DBN model derived from the DBN model training, a total of 20 collections were made for each spectrum, and 780 spectral samples were finally obtained. From these samples, the dataset was split into training (80%) and test sets (20%). Once data were collected, the next phase was data preprocessing, which ensured the raw data were in a suitable form for model training. Thus, the training dataset (i.e., captured reflected spectrum) was preprocessed and normalized between 0 and 1. This involved normalizing the interference patterns to achieve consistency across the dataset. The most critical part of preprocessing is feature extraction, where relevant features indicative of peak wavelengths are extracted from the interference patterns. These features are essential, as they provide the DBN with the necessary information to learn and predict accurately.

The proposed DBN model was implemented using the TensorFlow framework, along with the Keras and Sklearn libraries. The training was conducted on a PC equipped with an Intel Core i7-4790 3.60 GHz CPU and 20.0 GB of RAM. Figure 5 illustrates the architecture and training process of the proposed DBN. The preprocessed reflection spectra of the FBGs with interferences were used as inputs to train the DBN algorithm, with the corresponding peak wavelengths of the FBGs and interferences serving as target values. Each layer of the DBN was pre-trained individually using RBMs. This was done in an unsupervised manner, utilizing contrastive divergence to initialize the network weights. This step is essential in capturing the underlying structure of the data without requiring labeled information. After pre-training, the RBMs were stacked to form the complete DBN. The entire network was then fine-tuned using supervised learning with a labeled dataset of known peak wavelengths. This fine-tuning adjusts the weights and biases across all layers to optimize the network for peak wavelength detection. The mean squared error (MSE), mean absolute error (MAE), and root mean square errors (RMSE) loss function were used for measuring the difference between predicted and actual wavelengths, and the optimizer Adam was used to minimize this loss, The MSE, MAE, and RMSE are expressed in the following equations:

$$MSE = \frac{1}{n}\sum_{i=1}^{n}(y_i - y)^2 \tag{2}$$

$$MAE = \frac{1}{n}\sum_{i=1}^{n}|y_i - y| \tag{3}$$

$$RMSE = \sqrt{\frac{\sum_{i=1}^{n}(y_i - y)^2}{n}} \tag{4}$$

where n is the number of predicted values, y_i is the actual value, and y is the predicted value, respectively. During the training of the DBN, various parameters such as the number of epochs, batch sizes, hidden layers, hidden units, and optimizer and activation functions were adjusted to achieve optimal values and to prevent overfitting. Tuning these parameters resulted in different training times, and MSE, MAE, and RMSE values.

5. Results and Discussions

The different parameters, such as epochs, hidden layers, batch sizes, and optimizer and activation functions, were adjusted until an optimal or well-trained model was obtained. The optimal parameters for the well-trained DBN model are crucial in achieving high accuracy in peak wavelength detection. Thus, the optimal parameters used for training the proposed DBN included 3 hidden layers, 32 batch sizes, 150 epochs, the Adam optimizer, and ReLU activation functions in the hidden layers that introduced non-linearity, improving the network's ability to model complex relationships in the data. Figure 6a shows the DBN's training performance evaluated through its training and validation losses, as well as training and validation accuracy using these optimal parameters. The training loss and validation loss curves provide insights into the model's performance during training and on unseen data, respectively. The training loss quickly declined and stabilized near zero at epoch 150 on the bottom x-axis, indicating effective learning of training data patterns. The validation loss also decreased significantly early on and remains stable, showing good generalization to unseen data without overfitting. The training accuracy reached nearly 1.0 (100%) by epoch 150 at the top x-axis, and the validation accuracy also approached 1.0 (100%), though with some variability. In general, the DBN demonstrated strong performance, effectively learning and generalizing from the data, with high accuracy and low loss metrics.

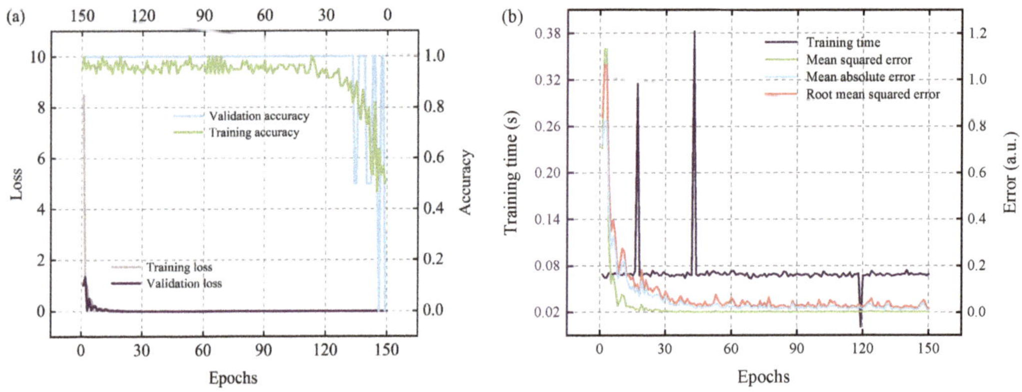

Figure 6. (**a**) Training performance of the DBN model in terms of loss and accuracy. (**b**) Performance of DBN in terms of prediction error and compilation time.

Figure 6b shows that the performance of the DBN for peak wavelength detection of FBG strain sensors was evaluated in terms of prediction error, measured as MSE, MAE, and RMSE, and training time, measured in seconds per epoch. As shown in the figure, the training process showed stabilization of training time after initial fluctuations, indicating computational efficiency. Key error metrics, including MSE, MAE, and RMSE, at epoch 0 were relatively high. However, as training progressed, all significantly decreased, remaining close to zero at 150 epochs, which signifies that the model effectively minimized prediction errors and indicates the DBN's ability to accurately learn and predict peak wavelengths, demonstrating its reliability and effectiveness for this application. Furthermore, to determine the best optimizer, the model was trained and tested with different optimizers, and their performance was compared based on MSE, MAE, and RMSE. According to Figure 7, Adam achieved lower MSE, MAE, and RMSE values of 0.0012, 0.015, and 0.0187 compared to other optimizers. This was due to its computational efficiency and noise minimization. As a result, the proposed DBN model was trained and tested using the Adam optimizer.

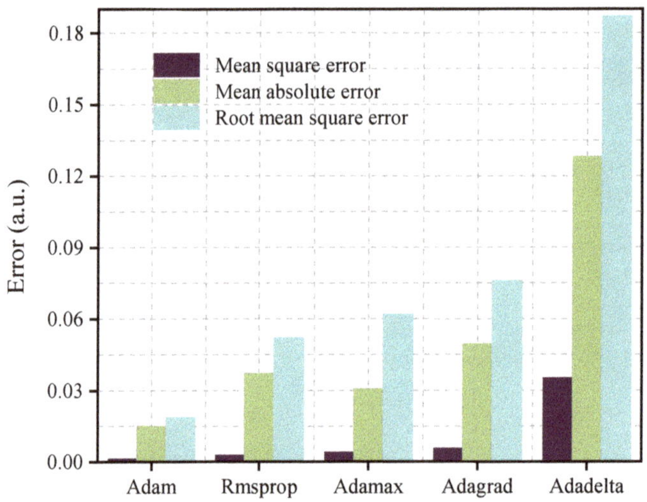

Figure 7. Performance comparison with different optimizers.

Figure 8 shows the predicted peak wavelengths of the strained FBGs, fixed FBGs, and three different interferences. In Figure 8a, the x-axis represents the peak wavelength of the strained FBGs in reality, while the y-axis represents the peak wavelength of the strained FBGs predicted by the DBN model. The yellow symbols on the graph show the ideal values, indicating where the predicted values should be if they exactly match the ideal values. Along this line, 15 randomly selected prediction data points are positioned to demonstrate the accuracy of the DBN model. The closer these points are to yellow symbols, the more accurate the prediction becomes, suggesting that the DBN model is very effective in accurately predicting the peak wavelength of the strained FBG in the case of overlap with the FBG and interferences. The x-axis of Figure 8b is the marked peak wavelength of fixed FBG and three different interferences, which are 1545.1 nm, 1546.88 nm (interference 1), 1547.11 nm (interference 2), and 1547.39 nm (interference 3), respectively. The y-axis corresponds to the predicted wavelength peak of the DBN model. From the results, it can be seen that the DBN model can also accurately predict the marked peak wavelengths of fixed FBG and three different interferences.

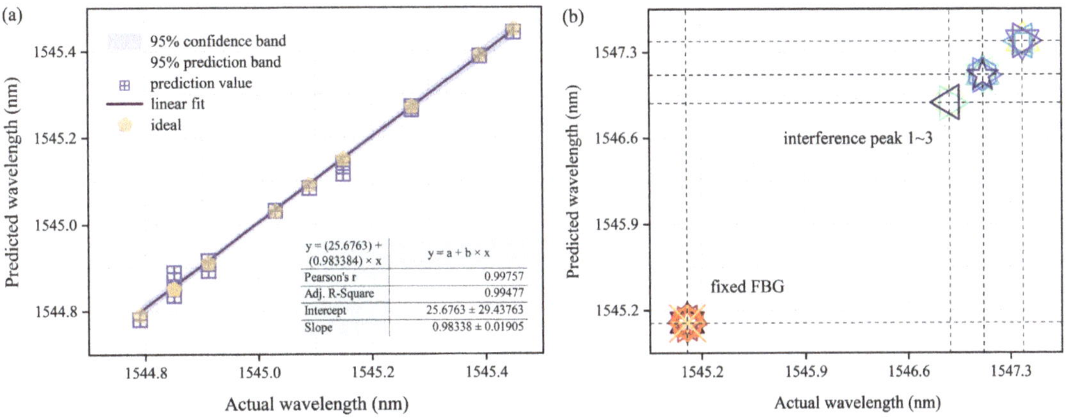

Figure 8. (**a**) Peak wavelength prediction of strained FBG after applying the DBN model. (**b**) Peak wavelength prediction of fixed FBG and three interferences after applying the DBN model.

6. Conclusions

This research centers on the application of DBN to predict sensing outcomes in a flexible sensing design that combines FBG sensors with a simulated fiber-optic interferometer sensor. It is known that the coexistence of FBG sensors and fiber-optic interferometer sensors in the sensing architecture can perform diverse sensing tasks due to the high sensitivity of fiber-optic interferometers and the advantage that multiple FBGs can be arranged in an array. However, in practice, to maximize the number of FBG sensing points, the problem of FBG spectral overlap will inevitably occur. The overlap of the overlapping FBG spectrum and the spectrum sensed by the fiber interferometer will inevitably cause the entire spectrum power to be modulated, making it challenging to accurately extract sensing information. In this case, the use of machine learning, specifically the DBN model, offers an optimal solution for accurately interpreting the sensed information. The DBN model can effectively predict all sensed information for various levels of overlapping FBG spectra under different spectrums sensed by the interferometer, ultimately enhancing the cost-effectiveness, flexibility, and accuracy of the sensing system.

Author Contributions: Conceptualization, S.D.B., C.-K.Y., Y.C.M. and P.-C.P.; methodology, S.D.B., C.-K.Y. and Y.C.M.; software, S.D.B. and Y.C.M.; data curation, C.-K.Y. and Z.-G.Z.; validation, S.D.B., C.-K.Y. and Y.C.M.; formal analysis: S.D.B., C.-K.Y., Y.C.M. and P.-C.P.; investigation: S.D.B., C.-K.Y., Y.C.M., Z.-G.Z. and P.-C.P.; visualization: C.-K.Y.; writing—original draft preparation: C.-K.Y. and S.D.B.; writing—review and editing: S.D.B., C.-K.Y., Y.C.M. and P.-C.P. All authors have read and agreed to the published version of the manuscript.

Funding: This work was supported by the National Science and Technology Council, Taiwan, under Grant NSTC 112-2221-E-027-076-MY2.

Institutional Review Board Statement: Not applicable.

Informed Consent Statement: Not applicable.

Data Availability Statement: The data presented in this study are available in this article.

Conflicts of Interest: The authors declare no conflicts of interest.

References

1. Pendão, C.; Silva, I. Optical Fiber Sensors and Sensing Networks: Overview of the Main Principles and Applications. *Sensors* **2022**, *22*, 7554. [CrossRef] [PubMed]
2. Marques, C.; Leal-Júnior, A.; Kumar, S. Multifunctional Integration of Optical Fibers and Nanomaterials for Aircraft Systems. *Materials* **2023**, *16*, 1433. [CrossRef] [PubMed]
3. Guzmán-Sepúlveda, J.R.; Guzmán-Cabrera, R.; Castillo-Guzmán, A.A. Optical Sensing Using Fiber-Optic Multimode Interference Devices: A Review of Nonconventional Sensing Schemes. *Sensors* **2021**, *21*, 1862. [CrossRef] [PubMed]
4. Taha, B.A.; Ali, N.; Sapiee, N.M.; Fadhel, M.M.; Mat Yeh, R.M.; Bachok, N.N.; Mashhadany, Y.A.; Arsad, N. Comprehensive Review Tapered Optical Fiber Configurations for Sensing Application: Trend and Challenges. *Biosensors* **2021**, *11*, 253. [CrossRef]
5. Li, M.; Singh, R.; Wang, Y.; Marques, C.; Zhang, B.; Kumar, S. Advances in Novel Nanomaterial-based Optical Fiber Biosensors—A Review. *Biosensors* **2022**, *12*, 843. [CrossRef] [PubMed]
6. Kashyap, R. *Fiber Bragg Gratings*; Academic Press: Cambridge, MA, USA, 2009.
7. Arockiyadoss, M.A.; Dehnaw, A.M.; Manie, Y.C.; Hayle, S.T.; Yao, C.K.; Peng, C.H.; Kumar, P.; Peng, P.C. Self-Healing Fiber Bragg Grating Sensor System Using Free-Space Optics Link and Machine Learning for Enhancing Temperature Measurement. *Electronics* **2024**, *13*, 1276. [CrossRef]
8. Dejband, E.; Tan, T.H.; Yao, C.K.; Chang, E.M.; Peng, P.C. Enhancing Multichannel Fiber Optic Sensing Systems with IFFT-DNN for Remote Water Level Monitoring. *Sensors* **2024**, *24*, 4903. [CrossRef] [PubMed]
9. Burhanuddin, W.A.F.W.; Ahmad, H.; Alias, M.A.; Sa'ad, M.S.M.; Sun, S.; Ismail, M.F. Multi-Parameter Fiber Bragg Grating (FBG)-based Sensor Fabricated using 3D Printing Technology for Precise Measurement of Vertical Earth and Pore Pressure. *IEEE Sens. J.* **2024**, *24*, 25793–25801. [CrossRef]
10. Jati, M.P.; Luthfi, M.I.; Yao, C.K.; Dehnaw, A.M.; Manie, Y.C.; Peng, P.C. An Extremely Close Vibration Frequency Signal Recognition Using Deep Neural Networks. *Appl. Sci.* **2024**, *14*, 2855. [CrossRef]
11. Teng, Y.; Zhang, B.; Qiu, Z.; Hu, X.; He, Z. High-Sensitivity Dual-FBG Acceleration Sensor for Low Frequency Vibration Measurement. *J. Opt.* **2024**, *53*, 2361–2373. [CrossRef]
12. Du, J.; He, Z. Sensitivity Enhanced Strain and Temperature Measurements based on FBG and Frequency Chirp Magnification. *Opt. Express* **2013**, *21*, 27111–27118. [CrossRef] [PubMed]
13. Hernández-Ambato, V.; Barrera, D.; Nazemosadat, E.; Moreno, H.; Sales, S. Effect of Taper Waist Diameter on the Sensitivity of Multicore fiber Coupler-based Temperature and Strain Sensors. *J. Light. Technol.* **2024**, 1–8. [CrossRef]
14. Liang, R.; Zhu, H.; Zhu, X.; Guo, M.; Zhang, Q.; Su, Z.; Chen, J.; Hai, Z.; Zheng, Y.; Xue, C. FPI-FBG Cascaded High-temperature Pressure Sensor up to 700 °C Based on Vernier Effect Utilizing Femtosecond Laser. *IEEE Sens. J.* **2024**. [CrossRef]
15. Liu, Q.; Wang, D.; Li, X.; Gao, H.; Yu, D. Simultaneous Measurement of Temperature and Liquid Refractive Index Based on Fiber Open Fabry-Pérot Cavity and Bragg Grating. *Optoelectron. Lett.* **2024**, *20*, 477–482. [CrossRef]
16. Wei, T.; Zhang, Y.; Zhang, L.; Tian, Z.; Nie, Z.; Yao, Y.; Wang, M. Research on Ultra-High Sensitivity Fiber-Optic Cascaded Fabry-Perot Resonator Optical Sensor Based on the Comb-Spectrum Vernier Effect. *J. Light. Technol.* **2024**, 1–7. [CrossRef]
17. Suo, L.; Peng, Y.P.; Yao, C.K.; Ren, S.; Lu, X.; Chen, N.K. High Sensitivity Strain Sensors Using Four-Core Fibers through a Corner-Core Excitation. *Micromachines* **2022**, *13*, 431. [CrossRef]
18. Lu, C.; Su, J.; Dong, X.; Sun, T.; Grattan, K.T. Simultaneous Measurement of Strain and Temperature with a Few-Mode Fiber-Based Sensor. *J. Light. Technol.* **2018**, *36*, 2796–2802. [CrossRef]
19. Lei, X.; Dong, X.; Lu, C.; Sun, T.; Grattan, K.T. Underwater Pressure and Temperature Sensor Based on a Special Dual-Mode Optical Fiber. *IEEE Access* **2020**, *8*, 146463–146471. [CrossRef]
20. Nascimento, M.; Inácio, P.; Paixão, T.; Camacho, E.; Novais, S.; Santos, T.G.; Fernandes, F.M.B.; Pinto, J.L. Embedded fiber sensors to monitor temperature and strain of polymeric parts fabricated by additive manufacturing and reinforced with NiTi wires. *Sensors* **2020**, *20*, 1122. [CrossRef]

21. Yao, C.K.; Dehnaw, A.M.; Peng, P.C. A Multi-Format, Multi-Wavelength Erbium-Doped Fiber Ring Laser Using a Tunable Delay Line Interferometer. *Appl. Sci.* **2024**, *14*, 6933. [CrossRef]
22. Hinton, G.E.; Osindero, S.; Teh, Y.W. A Fast Learning Algorithm for Deep Belief Nets. *Neural Comput.* **2006**, *18*, 1527–1554. [CrossRef]
23. Kale, A.P.; Wahul, R.M.; Patange, A.D.; Soman, R.; Ostachowicz, W. Development of Deep Belief Network for Tool Faults Recognition. *Sensors* **2023**, *23*, 1872. [CrossRef] [PubMed]
24. Larochelle, H.; Bengio, Y.; Louradour, J.; Lamblin, P. Exploring Strategies for Training Deep Neural Networks. *J. Mach. Learn. Res.* **2009**, *10*, 1–40.
25. Li, C.; Wang, Y.; Zhang, X.; Gao, H.; Yang, Y.; Wang, J. Deep Belief Network for Spectral–Spatial Classification of Hyperspectral Remote Sensor Data. *Sensors* **2019**, *19*, 204. [CrossRef]
26. Akbaş, A.; Buyrukoğlu, S. Deep Belief Network Based Wireless Sensor Network Connectivity Analysis. *Balk. J. Electr. Comput. Eng.* **2023**, *11*, 262–266. [CrossRef]

applied sciences

MDPI

Article

A Hybrid GAN-Inception Deep Learning Approach for Enhanced Coordinate-Based Acoustic Emission Source Localization

Xuhui Huang, Ming Han and Yiming Deng *

Department of Electrical and Computer Engineering, Michigan State University, East Lansing, MI 48824, USA; huangxu6@msu.edu (X.H.); mhan@egr.msu.edu (M.H.)
* Correspondence: dengyimi@egr.msu.edu

Abstract: In this paper, we propose a novel approach to coordinate-based acoustic emission (AE) source localization to address the challenges of limited and imbalanced datasets from fiber-optic AE sensors used for structural health monitoring (SHM). We have developed a hybrid deep learning model combining four generative adversarial network (GAN) variants for data augmentation with an adapted inception neural network for regression-based prediction. The experimental setup features a single fiber-optic AE sensor based on a tightly coiled fiber-optic Fabry-Perot interferometer formed by two identical fiber Bragg gratings. AE signals were generated using the Hsu-Nielsen pencil lead break test on a grid-marked thin aluminum plate with 35 distinct locations, simulating real-world structural monitoring conditions in bounded isotropic plate-like structures. It is demonstrated that the single-sensor configuration can achieve precise localization, avoiding the need for a multiple sensor array. The GAN-based signal augmentation expanded the dataset from 900 to 4500 samples, with the Wasserstein distance between the original and synthetic datasets decreasing by 83% after 2000 training epochs, demonstrating the high fidelity of the synthetic data. Among the GAN variants, the standard GAN architecture proved the most effective, outperforming other variants in this specific application. The hybrid model exhibits superior performance compared to non-augmented deep learning approaches, with the median error distribution comparisons revealing a significant 50% reduction in prediction errors, accompanied by substantially improved consistency across various AE source locations. Overall, this developed hybrid approach offers a promising solution for enhancing AE-based SHM in complex infrastructures, improving damage detection accuracy and reliability for more efficient predictive maintenance strategies.

Keywords: acoustic emission; structural health monitoring; generative adversarial network; acoustic emission source localization; data augmentation; fiber optics sensor; t-SNE visualization

1. Introduction

Acoustic emission (AE) is a well-established and powerful non-destructive testing method for structural health monitoring (SHM), offering unique capabilities for real-time damage detection and localization [1]. AE techniques analyze transient elastic waves generated by rapid energy release within materials, providing insights into structural integrity in a variety of structures ranging from wind turbine blades to aerospace components [2,3]. The application of AE to source localization has changed damage detection and monitoring strategies, offering high sensitivity and the ability to detect defects in real-time [4]. Current research aims to develop robust, adaptive, and efficient SHM systems capable of reliably assessing the health of increasingly complex structures in demanding operational environments, focusing on enhancing data quality, improving localization accuracy, and developing more sophisticated predictive models [5–7].

Existing AE data analysis methods employ signal processing techniques such as time-domain, frequency-domain, and time-frequency analysis to extract key features from signals [8]. Conventional approaches such as triangulation and beamforming, while widely used for source localization, face limitations in practical applications, particularly in the complex real-world environment. These techniques rely on the time difference of arrival (TDOA) of signals at multiple sensors and require precise knowledge of wave velocities [9]. Despite their effectiveness in certain scenarios, implementation can be challenging in complex real-world environments, particularly in larger structures or noisy conditions. In response to these challenges, single-sensor AE localization techniques have gained attention, offering more cost-effectiveness and simpler deployment. Ebrahimkhanlou [10,11] pioneered probabilistic frameworks for single-sensor AE source localization in thin metallic plates, relying on the temporal order of signal arrivals rather than absolute time measurements.

In recent years, deep learning approaches have revolutionized AE data analysis, offering powerful tools for SHM and damage detection. These methods have excelled in source localization, signal classification, and characterization of acoustic events [12–15]. Machine learning methods have been widely applied to structural health monitoring (SHM), transforming traditional approaches to damage detection, localization, and prognosis [16,17]. These techniques have demonstrated significant potential in addressing SHM challenges, from vibration-based monitoring to predictive maintenance [18]. Additionally, advanced methods such as the Electro-Mechanical Impedance technique have shown promise in SHM applications, offering new avenues for damage detection and structural assessment [19,20]. Despite these promising advancements, some challenges persist in deep learning for AE data analysis. A fundamental challenge is the limited availability of high-quality data, significantly impacting the development and performance of damage detection and localization algorithms. This scarcity of data stems from the rare and unpredictable nature of AE events, particularly in critical structural failures.

To address data limitations, generative adversarial networks (GANs) have emerged as a powerful tool in data augmentation, offering significant potential for enhancing time series signal analysis. Introduced by [21], GANs comprise two neural networks—a generator and a discriminator—that compete in a zero-sum game, iteratively training each other to produce high-fidelity synthetic data [22]. In the context of AE signal augmentation, GANs present a promising solution to the persistent challenge of limited and imbalanced datasets. By generating synthetic AE signals that closely mimic real data characteristics, GANs can significantly expand training datasets, potentially improving the performance and robustness of damage detection and localization algorithms [23].

2. Related Work

Recent advancements in deep learning have significantly impacted AE source localization techniques. Single-sensor AE source localization, once challenging, has become feasible thanks to deep learning models [24,25], showing particular promise in plate-like structures [22]. The application of these techniques extends beyond conventional materials, demonstrating efficacy in analyzing AE signals from nanoindentation experiments and environmental monitoring applications. However, current approaches face several limitations. A notable limitation is the ability to handle only a restricted number of zones or regions [26]. For example, in ref. [12], a plate was divided into just 13 zones for localization, likely insufficient for complex structures. While authors in [10] achieved better localization with 64 regions, its reliance on simulated edge-reflected waves and strong reflections limits broader application. The complexity of AE signals, which are highly susceptible to noise and environmental factors, further complicates data collection and interpretation [27]. To address these data limitations, researchers have begun exploring innovative approaches to enhance both the quality and quantity of available AE data. Deep learning techniques show promise in detecting acoustic wave reflections and improving signal interpretation [28]. Novel data augmentation techniques and physics-informed approaches are being developed to maximize the utility of limited datasets [29,30]. Improved AE source localization methods, including time-distance domain transformations and wavelet-based noise reduction techniques, aim to extract more information from available signals [31]. Integration of AE with other testing methods and advanced signal processing techniques, such as modified empirical wavelet transforms, is being explored to enhance damage detection capabilities [32].

Among these innovative approaches, Generative Adversarial Networks (GANs) have shown particular promise. GAN-based augmentation typically involves an encoder-decoder structure in the generator, where an input AE signal is encoded into a lower-dimensional manifold, combined with random noise, and then decoded to produce a new, related AE signal. The application of GANs in related fields has demonstrated their versatility and efficacy. In machine fault diagnosis, GAN-based data augmentation has yielded remarkable improvements in classification accuracy, particularly for fault types with sparse samples [33]. In medical imaging, GANs have been employed to generate synthetic radiology images, addressing issues of data scarcity and patient privacy [34]. Within SHM, innovative GAN architectures such as FTGAN have been developed to address imbalanced datasets in bearing fault diagnosis, incorporating both time and frequency domain information [35]. While GANs offer several advantages, including the ability to generate diverse, high-quality samples, learn complex data distributions, and improve model robustness, they also face challenges such as training instability, mode collapse (i.e., where the generator produces limited varieties of samples), and the potential to perpetuate biases present in the original dataset [36]. As the field progresses, addressing these limitations and adapting GAN architectures to the specific nuances of AE signals will be key to fully leveraging their potential. The fundamental GAN process involves the generator creating artificial samples from random noise while the discriminator struggles to differentiate between real and generated data, with both networks refining their performance through this adversarial process.

Building upon these advancements and addressing the identified challenges, we propose a novel hybrid approach for AE source localization, combining a custom-designed GAN for data augmentation with an adapted Inception network for regression-based localization. Our methodology offers several key advancements: (a) We developed and compared four GAN variants (GAN, DCGAN, WGAN, and TSAGAN), customized for AE signal augmentation by incorporating gradient penalty to mitigate mode collapse. This approach generates diverse, high-fidelity synthetic signals, expanding the dataset from 900 to 4500 samples and balancing representation across 35 distinct source locations. (b) We adapted the Inception neural network for AE source localization prediction, modifying the final fully connected layers to predict continuous (x, y) coordinates. This regression architecture preserves multi-scale feature extraction capabilities, which concurrently processes AE signals at three temporal resolutions: 1 μs, 5 μs, and 10 μs simultaneously, enhancing the model's ability to capture complex AE wave propagation characteristics in bounded isotropic plate structures. (c) Our experimental setup employs a novel single fiber-optic AE sensor based on a tightly coiled fiber Fabry-Perot interferometer (FPI) formed by two fiber Bragg gratings (FBGs). This sensor, with a cavity length of 40 cm and center wavelength of 1557.1 nm, offers enhanced sensitivity and electromagnetic interference immunity, enabling precise localization without the need for a multiple sensor array. (d) We implemented a rigorous validation framework comprising t-distributed stochastic neighbor embedding (t-SNE) visualizations and Wasserstein distance analysis to quantitatively assess the quality and diversity of GAN-generated data. The Wasserstein distance between original and synthetic datasets decreased from 19.8 to 3.2 after 2000 training epochs, representing an 83.5% reduction. This numerically confirms the generation of high-fidelity synthetic AE signals and allows for the comparative analysis of different GAN variants' performance. The proposed approach improves AE source localization accuracy by 50% compared to traditional methods, reducing median error from six to three inches. It enables more precise damage detection and localization in plate-like structures, potentially enhancing structural health monitoring practices in various applications.

3. Experiment Setup

The schematic design of a novel fiber-optic coil-based AE sensing and monitoring system is shown in Figure 1. The setup features a grid-marked aluminum plate with a fiber-optic AE sensor, developed in [37], comprising a fiber coil with two identical FBGs that form a low-finesse FPI. The sensor has an 8 mm outer diameter and a 6 mm inner diameter, and is flexibly mounted on the sample surface. The reflection features fine fringes with a free-spectral range of 2 pm within the reflection bandwidth determined by the FBGs. Ultrasound waves impinging on the coiled fiber create refractive index variations, modifying the FPI cavity length and causing spectral fringe shifts. The sensor was interrogated using a modified phase-generated carrier method that allows for the extraction of the AE signal with good linearity and high sensitivity, regardless of the laser wavelength with respect to the sensor fringes [38]. As illustrated in Figure 1, a laser source generates light that passes through a phase modulator driven by a $sin(wt)$ signal. The modulated light travels through a circulator and optical fiber to the FBG-FPI sensor. Reflected light returns through the circulator to a photodetector (PD). The output of the PD is processed through two parallel paths. In the upper path, the signal is mixed with $sin(wt + \varphi_1)$, low-pass filtered at 500 kHz, band-pass filtered from 50–500 kHz, and amplified. The lower path involves mixing with $sin(2wt + \varphi_2)$, followed by similar filtering and amplification steps. Both paths include additional low pass filtering at 25 kHz. The processed signals are then combined to produce the final output $\theta(t)$, which represents the AE signal. This system facilitates high-quality AE signal detection with an omnidirectional response and high ultrasound sensitivity while adapting to environmental perturbations. The AE signals obtained from the fiber-optic AE sensor were used to train and test the proposed hybrid deep learning models for source localization. The Hsu-Nielsen pencil lead break (PLB) test, a widely accepted method for AE signal generation [39], was used in this study. We conducted the test in a sequential manner across 35 distinct locations on the 1/10-inch-thick aluminum plate. The plate was partitioned into 35 distinct locations, as shown in Figure 2. Each location underwent ten PLB tests using a 2H mechanical pencil with a 0.5 mm diameter lead.

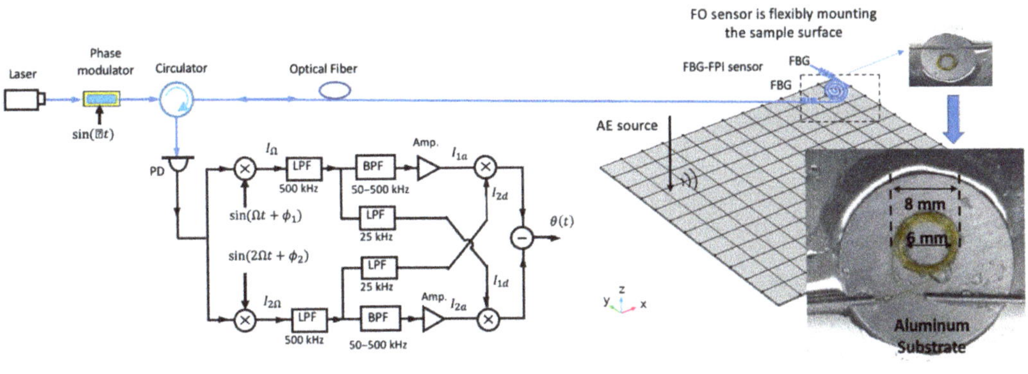

Figure 1. Schematic of the fiber-optic coil-based acoustic emission sensing system. Inset: Close-up image of the sensor, showing the flexible mounting and dimensions (8 mm outer, 6 mm inner diameter).

Figure 2a presents a photograph of the experimental setup, illustrating the aluminum plate divided into a grid for AE testing. This plate is marked with blue numbers indicating specific test points. The numbering is reorganized for simplicity in Figure 2b. The fiber-optic sensor, glued onto an aluminum carrier disk, is affixed to the plate with a liquid couplant to facilitate ultrasonic coupling. Figure 2b provides a schematic representation of the plate, detailing the grid layout and test points. The 35 distinct locations are marked by red dots, corresponding to the PLB test points. This layout ensures the reliability and reproducibility of the collected AE data, crucial for training and testing deep learning models for source localization. The clear separation of test points and consistent methodology across the grid contribute to the robustness of the experimental design.

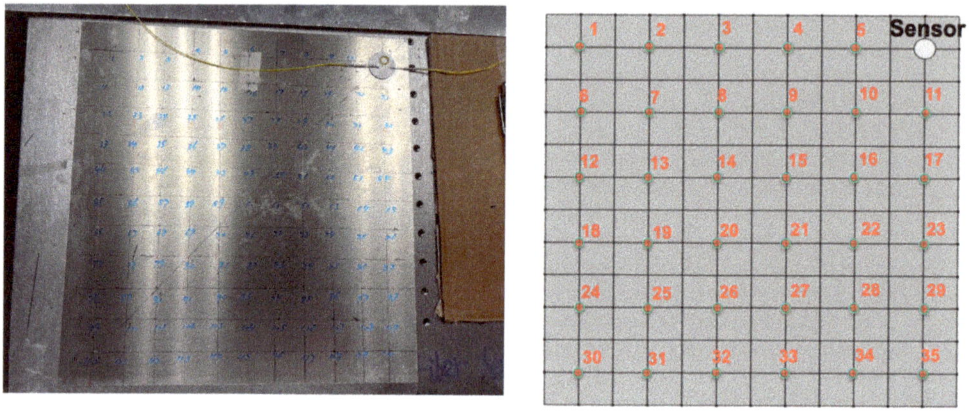

(a) (b)

Figure 2. (**a**) Aluminum plate with the grid and fiber-optic sensor for AE testing (**b**) Schematic representation of the aluminum plate detailing the grid layout and test points.

Figure 3 illustrates the time series augmentation technique employed in this study to balance and expand the dataset, thereby enhancing deep learning model performance for acoustic emission source localization. The image depicts data distribution across 35 different locations, representing distinct labels in our acoustic emission source localization task. The figure highlights the variability in the original dataset, where some locations (e.g., Location 2) have more raw data than others (e.g., Location 1). This variability in the original dataset is caused by the different number of repetitive tests conducted at each location. Due to practical constraints and the nature of the testing process, the number of successful and usable tests varied across locations, resulting in an uneven distribution of raw data, as seen in Figure 3.

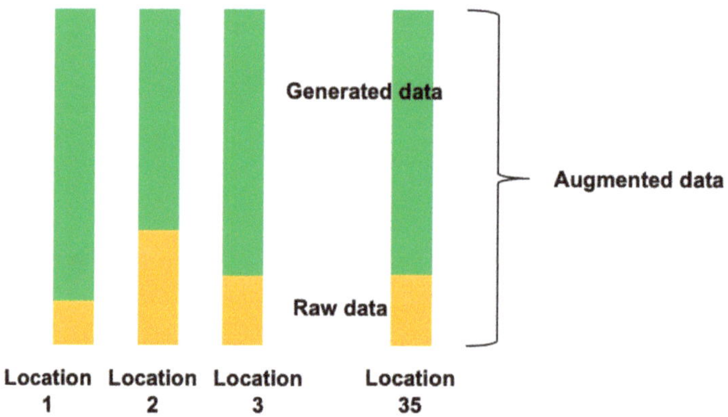

Figure 3. Time series augmentation showing the original data (orange) and generated data (green) to ensure each label has a balanced and sufficient number of samples for improved deep learning model performance.

The proposed augmentation strategy is designed to address these discrepancies by generating high-fidelity synthetic data, with a particular focus on underrepresented locations. By supplementing the original data with these generated samples, we ensure a sufficient and balanced number of time series for each location. The labeling of the augmented dataset is preserved during this process. Each generated AE signal inherits the label (corresponding to the source location) of the original signal it was based on. This approach ensures that the augmented data maintain the same spatial distribution of AE sources as the original dataset, while increasing the volume and diversity of our training data. This data augmentation approach simultaneously tackles two critical challenges in deep learning: it mitigates the risk of overfitting on some particular labels (locations) with abundant data while addressing potential underfitting issues in locations with limited samples. The resulting balanced dataset, a product of this advanced augmentation technique, is anticipated to significantly enhance the model's capacity to accurately localize acoustic emission sources across all 35 locations. This improvement is expected to manifest in enhanced overall performance and superior generalization capabilities, ultimately leading to more robust and reliable acoustic emission source localization in real-world applications.

4. Methodology

This section outlines the overall framework of the innovative hybrid method. Our study employs a novel approach that combines advanced data augmentation techniques with an adapted Inception architecture to enhance the accuracy and robustness of AE source localization in complex structures. This methodology integrates the custom-designed GAN for data augmentation with an Inception network specifically adapted for regression tasks.

The overall process flow is shown in Figure 4a. It commences with the collection of AE signals, each labeled with coordinates in the form of (d_i, θ_i) for 35 distinct positions. These signals undergo a training/test split. The training data are then augmented using four different GAN architectures (GAN, DCGAN, WGAN, and TSAGAN) to address dataset imbalance and scarcity issues. This multi-GAN augmentation approach is crucial as it generates synthetic AE signals that closely emulate the characteristics of real data, effectively expanding the dataset and improving the model's ability to generalize across various AE source locations. Each GAN variant offers unique strengths in data synthesis, allowing for a comprehensive augmentation strategy. The augmented training data from each GAN feed into an Inception network specifically adapted for regression tasks, forming the core of our hybrid approach. This adaptation of the Inception network, originally designed for image classification, enables effective processing of AE signals across multiple scales, capturing both local and global features crucial for accurate localization. This trained network is then utilized to predict AE source locations from the test data. Figure 4b specifies the architecture of the Inception network used for regression. It initiates with an input layer, followed by an Inception Module that processes data through multiple parallel pathways. The Inception Module is particularly adept at AE signal processing as it can simultaneously extract features at different scales, which is essential given the complex nature of AE waveforms. The outputs are concatenated and passed through batch normalization. A ReLU activation function is then applied, followed by global average pooling to reduce spatial dimensions. global average pooling is employed instead of traditional fully connected layers to minimize the number of parameters, mitigate overfitting, and maintain spatial information. Finally, a dense layer produces the output, predicting the AE source location as continuous coordinates.

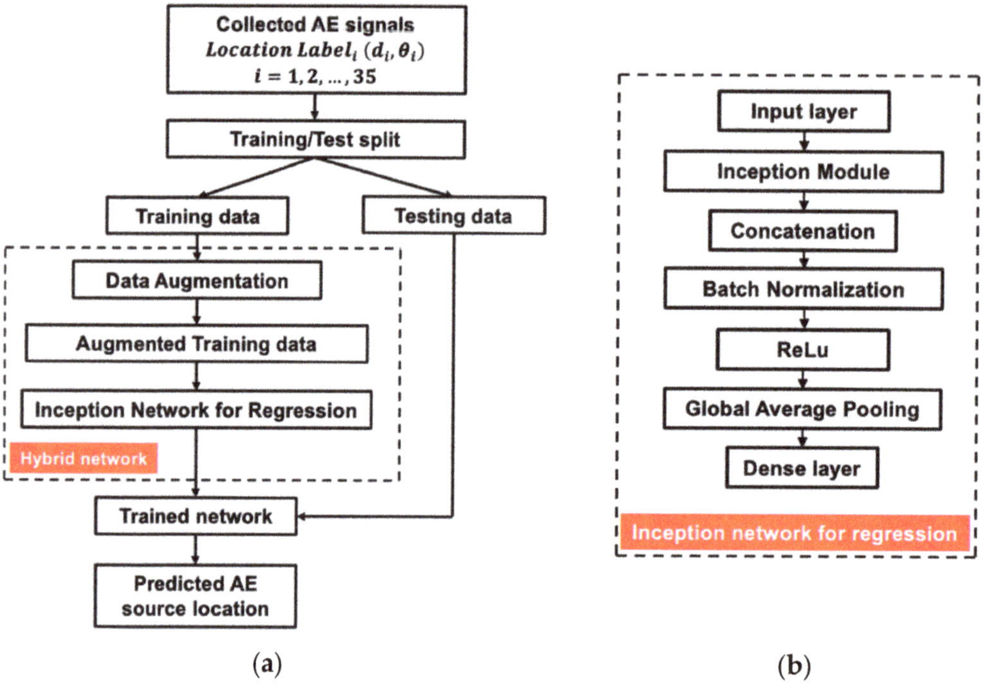

Figure 4. (**a**) Workflow of the hybrid network for AE source localization (**b**) Architecture of the Inception network for regression.

The four GAN architectures—GAN, DCGAN, TSAGAN, and WGAN—share a common foundation but differ in key aspects of their design and training approach, which is shown in Table 1. The original GAN uses fully connected layers in both the generator and discriminator, with a structure of (128, 512, 1024) neurons for the generator and (1024, 512, 64) for the discriminator. It employs LeakyReLU activations throughout, with batch normalization (momentum 0.8) in both networks. The DCGAN introduces convolutional layers, specifically using 1D transposed convolutions in the generator and 1D convolutions in the discriminator, which are particularly effective for capturing spatial or temporal patterns in the data. It typically uses ReLU activations in the generator and LeakyReLU in the discriminator, with a *tanh* activation in the final generator layer. The TSAGAN, tailored for time series data, reverts to a fully connected architecture similar to the original GAN but is optimized for sequential data. The key innovation of the WGAN lies not in its architecture, which is similar to the DCGAN with convolutional layers, but in its use of the Wasserstein loss function and weight clipping in the discriminator (now called a critic). This change allows for more stable training and potentially better-quality results. The WGAN also typically includes dropout in the discriminator, a feature that is not present in the other architectures. All four models use the Adam optimizer with a learning rate of 0.0002, a beta of 0.001, and a batch size of 64, and are trained for 2000 epochs. They all incorporate model collapse monitoring, but the WGAN stands out with its multiple critic updates per generator update. These architectural differences make each variant suitable for different types of data and training scenarios, with the DCGAN and WGAN often performing well on complex, high-dimensional data, while the GAN and TSAGAN can be effective for simpler or specific time-series datasets.

Table 1. Specifications for the GAN Variants.

Feature	GAN	DCGAN	TSAGAN	WGAN
Generator Architecture	Fully connected layers (128, 512, 1024)	1D transposed convolutions, Dense layers (128, 512, 1024)	Fully connected layers (128, 512, 1024)	Fully connected layers (128, 512, 1024)
Input	Noise vector			
Output layer	Dense layer with LeakyReLU	Dense layer with tanh	Dense layer with LeakyReLU	Dense layer with LeakyReLU
Discriminator Architecture	Fully connected layers (1024, 512, 64)	1D convolutions, Dense layers	Fully connected layers (1024, 512, 64)	1D convolutions, Dense layers
Loss Function	Binary cross-entropy	Binary cross-entropy	Binary cross-entropy	Wasserstein loss
Optimizer	Adam (learning rate = 0.0002, beta = 0.001)			
Batch Size	64	64	64	64
Epochs	2000	2000	2000	2000
Special Features	Model collapse monitoring	Model collapse monitoring	Model collapse monitoring	Multiple critic updates, Model collapse monitoring

The GAN architecture for data augmentation, detailed in Figure 5, involves a generator and a discriminator, both of which are sequential models. The generator G transforms a noise vector z into synthetic AE data $G(z)$ that mimics the real data distribution. The discriminator D evaluates both x_{real} and generated data $G(z)$, outputting a probability of the input being real. The training process alternates between updating the discriminator and the generator. The key equations governing this process are presented as follows.

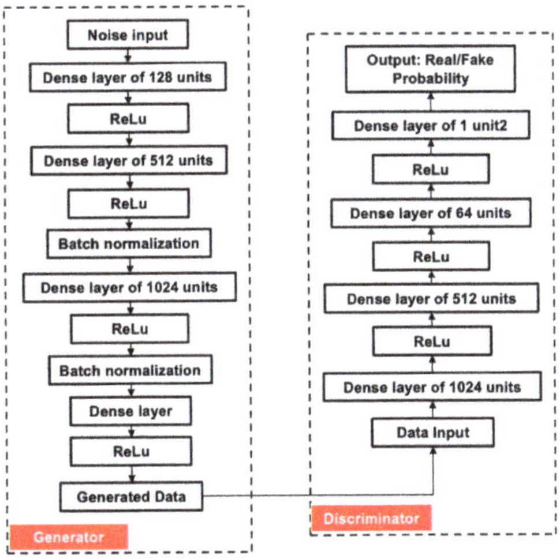

Figure 5. Architecture of the generator and discriminator networks in the GAN for AE signal augmentation.

The first step is to compute the discriminator real loss:

$$\mathcal{L}_{\mathcal{D}_{\text{ra}}} = -\log(D(x_{\text{real}})) \tag{1}$$

Then, we derive the discriminator fake loss as:

$$\mathcal{L}_{\mathcal{D}_{\text{fk}}} = -\log(1 - D(G(z))) \tag{2}$$

By combining the two terms, the overall discriminator loss is:

$$\mathcal{L}_{\mathcal{D}} = 0.5 \times \left(\mathcal{L}_{\mathcal{D}_{\text{ra}}} + \mathcal{L}_{\mathcal{D}_{\text{fk}}}\right) \tag{3}$$

And the overall generator loss is derived from the input noise vector:

$$\mathcal{L}_{\mathcal{G}} = -\log(D(G(z))) \tag{4}$$

The discriminator aims to maximize the probability of correctly classifying real and fake samples, while the generator aims to minimize the probability of the discriminator correctly classifying generated samples as fake. During training, we updated the discriminator by maximizing:

$$E[\log(D(x_{\text{real}}))] + E[\log(1 - D(G(z)))] \tag{5}$$

and then updated the generator by minimizing:

$$E[\log(1 - D(G(z)))] \tag{6}$$

To prevent mode collapse, we implemented a model edge collapse threshold τ. If $|\mathcal{L}_{\mathcal{D}} - \mathcal{L}_{\mathcal{G}}| > \tau$ for a certain number of consecutive iterations, we reinitialized the models. Hyperparameter tuning was conducted using grid search to optimize the learning rates α_D and α_G for the discriminator and generator respectively, as well as the batch size B and a number of epochs N. The optimization problem can be formulated as:

$$\text{minimize } \mathcal{L}_{total}(\alpha_D, \alpha_G, B, N) = \mathcal{L}_{\mathcal{D}} + \mathcal{L}_{\mathcal{G}} \tag{7}$$

This is example 1 of an equation:

$$\text{minimize } \mathcal{L}_{total}(\alpha_D, \alpha_G, B, N) = \mathcal{L}_{\mathcal{D}} + \mathcal{L}_{\mathcal{G}} \tag{8}$$

subject to:

$$\begin{gathered} 0 < \alpha_D, \alpha_G \leq 0.01 \\ 16 \leq B \leq 256 \\ 1 \leq N \leq 2000 \end{gathered}$$

This detailed mathematical approach ensures effective augmentation of the dataset and robust AE source localization, addressing specific challenges that traditional methods struggle with, such as limited data availability and the complex nature of AE signals in real-world applications.

5. Evaluating the GAN Performance for Data Augmentation

To visualize the distribution of synthetic (augmented) and original datasets in a lower-dimensional space, we employed t-Distributed Stochastic Neighbor Embedding (t-SNE), a dimensionality reduction technique. t-SNE maps high-dimensional data to two dimensions, effectively preserving local relationships between points and revealing structure at multiple scales. Figure 6 presents t-SNE visualizations for the four GAN variants (GAN, WGAN, DCGAN, TSAGAN) at their final training epoch (2000), as well as the GAN at epoch 1 and noise-based augmentation for comparison. Original data points are depicted in blue and synthetic data points in red, providing a clear visual distinction.

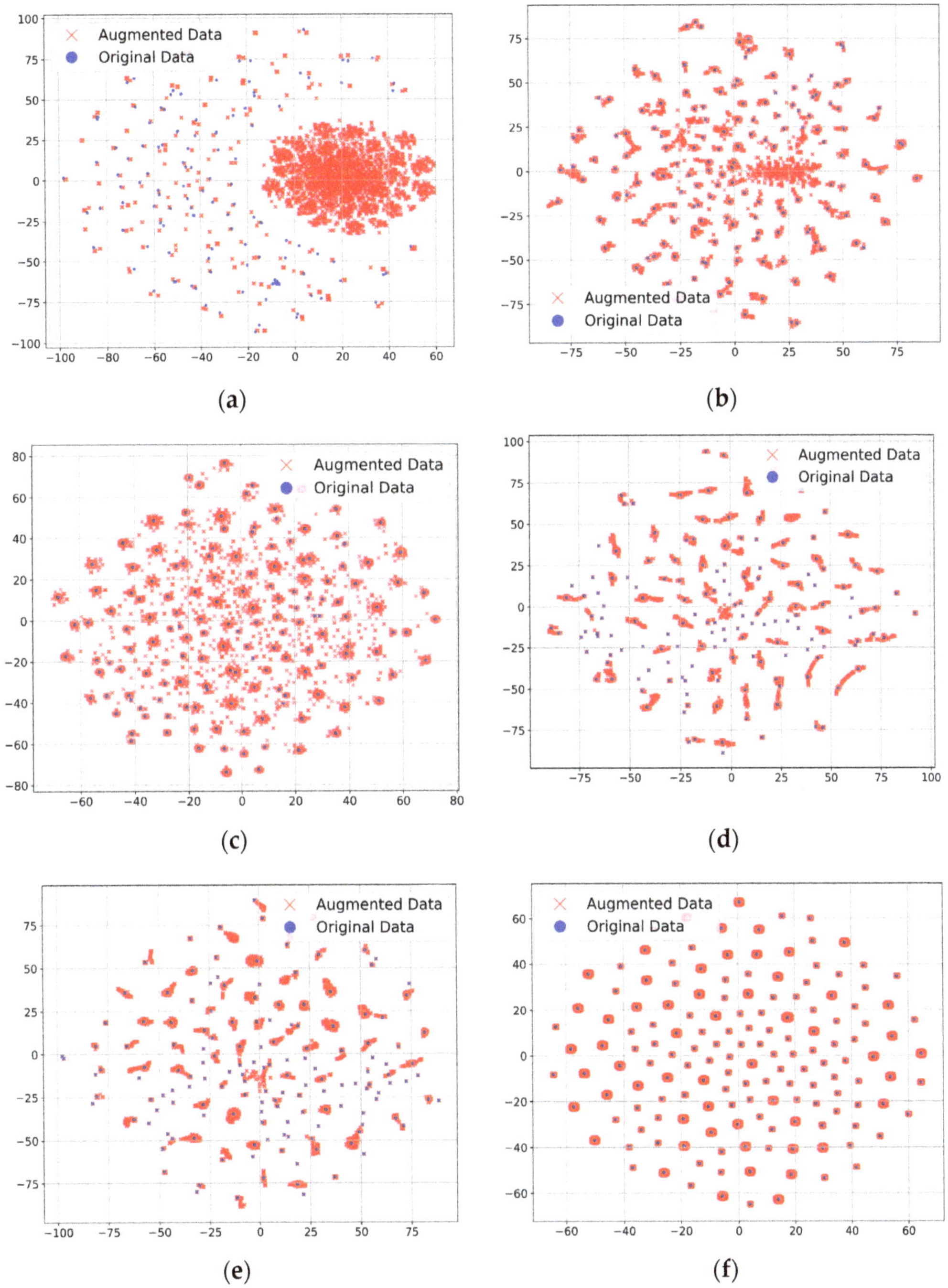

Figure 6. The t-SNE visualization of synthetic and original datasets (**a**) The training epoch of 1 for GAN (**b**) The training epoch of 2000 for GAN (**c**) The training epoch of 2000 for WGAN (**d**) The training epoch of 2000 for DCGAN (**e**) The training epoch of 2000 for TSAGAN (**f**) Augmentation via addition of noise.

In the early stages of GAN training, the augmented data points are sparsely distributed and show little overlap with the original data, indicating poor alignment and high divergence. By epoch 2000, the GAN-augmented data points are well-integrated with the original data, demonstrating the GAN's capability to generate synthetic data that closely resembles the original dataset. The WGAN shows a unique pattern where augmented data forms distinct clusters that encompass the original data points, suggesting it captures the overall distribution well but may over-segment the data space. The DCGAN (Figure 6d) and TSAGAN (Figure 6e) both show good integration of augmented and original data, with the TSAGAN appearing to have a slightly more uniform distribution. In the last image, noise-based augmentation produces a distinct pattern where augmented data form concentric circles around original data points, indicating a simple additive noise approach that does not capture the underlying data distribution as effectively as GAN-based methods. These visualizations highlight the effectiveness of GAN-based augmentation techniques in generating high-quality synthetic data that closely mimics the original dataset. The increasing overlap and similarity in distribution between original and augmented data points across different GAN architectures demonstrate their capability to produce diverse yet representative samples, providing a robust foundation for deep learning models and ensuring balanced representation across all labels.

The Wasserstein distance between the original and synthetic datasets across different epochs for the four GAN variants are illustrated in Figure 7, serving as a metric to quantify the similarity between the two distributions. Lower values indicate higher similarity. The WGAN shows the most rapid convergence, achieving the lowest Wasserstein distance of about 2.5 by epoch 100 and maintaining this level throughout training. The DCGAN and TSAGAN demonstrate similar convergence patterns, starting with high distances but steadily decreasing to around three by epoch 2000. The standard GAN, interestingly, shows the most volatile behavior, with an initial decrease followed by a spike at epoch 500, before eventually converging to a distance similar to the DCGAN and TSAGAN by epoch 2000. This comparison reveals that while all GAN variants eventually achieve similar levels of data similarity, they differ significantly in their convergence paths. The WGAN's rapid and stable convergence suggests it may be the most efficient in generating high-quality synthetic data, despite its underperformance in the final localization task. The standard GAN's volatility indicates a need for careful monitoring during training, although it ultimately achieves competitive results. Overall, the results confirm that with adequate training, the GAN-based augmentation technique significantly enhances the dataset, providing a balanced and high-quality training set for deep learning models.

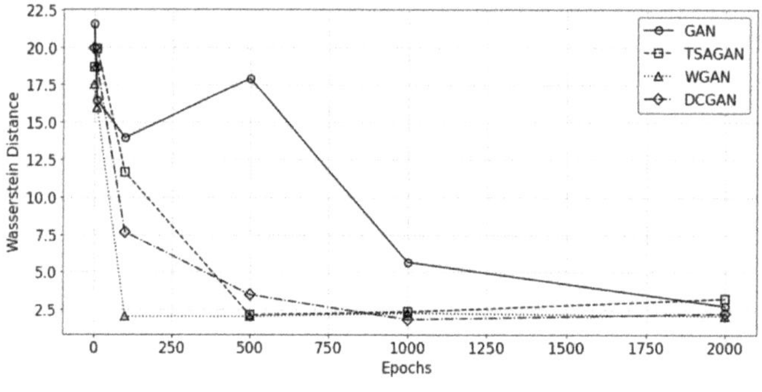

Figure 7. The comparison of Wasserstein distance convergence across epochs for the four GAN variants (GAN, TSAGAN, WGAN, and DCGAN).

6. Results and Discussion

Our novel hybrid approach, combining GAN-based data augmentation with an Inception network for AE source localization, demonstrates significant improvements in accuracy and reliability compared to traditional methods. This section presents a detailed analysis of our results, highlighting the benefits of our approach through comparative studies and error distribution analysis.

6.1. The Comparative Analysis of Localization Performance

Figure 8 presents a comparative analysis of AE source localization performance between two approaches: (a) a hybrid deep learning model incorporating GAN-based data augmentation and an Inception network, and (b) a model using only the Inception network without GAN-based augmentation. In Figure 8a, the hybrid deep learning approach demonstrates remarkable accuracy. The predicted locations (stars) closely align with the actual source locations (squares) across the grid, with minimal deviation. This high accuracy is consistent across the entire plate, suggesting robust performance regardless of the source's position relative to the sensor. In contrast, Figure 8b shows results without GAN-based augmentation. Here, the discrepancies between actual and predicted locations are significantly more pronounced. Many predicted points (stars) are substantially displaced from their corresponding actual locations (squares). Notably, the errors appear to increase towards the edges and corners of the plate, indicating degraded performance for sources further from the sensor. The marked difference in accuracy between these approaches underscores the significant impact of GAN-based data augmentation. By generating additional synthetic training data, the hybrid model appears to have developed a more robust and generalized representation of AE signal characteristics across the entire plate. This results in more accurate predictions, even for locations that might be underrepresented in the original dataset.

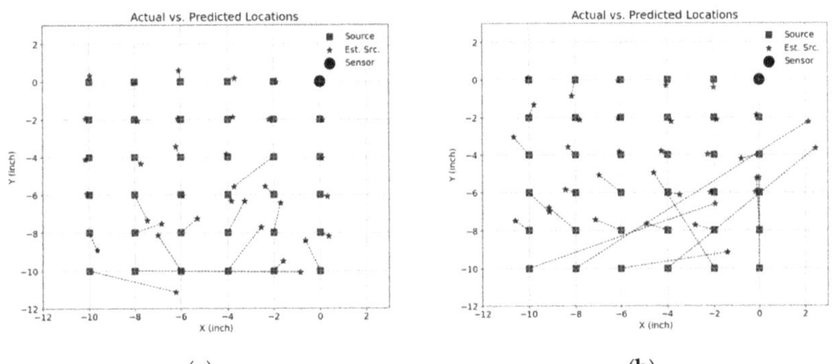

Figure 8. The comparison of acoustic emission (AE) source localization performance. (**a**) Results from the hybrid deep learning model with GAN-based data augmentation and Inception network. (**b**) Results from the Inception network alone without GAN-based augmentation. Square markers represent actual source locations, star markers show predicted locations, and the large circular marker indicates the sensor position. The x and y axes represent dimensions in inches.

6.2. Error Distribution Analysis

To quantify improvements in our hybrid approach, we conducted error distribution analyses across six methods: the original dataset, noise-based augmentation, and four GAN-based techniques (GAN, DCGAN, TSAGAN, WGAN). Each method was evaluated 10 times for model variability. We calculated mean Euclidean distance errors between predicted and actual source locations for each run. Figure 9 presents box plots of these errors, with each box representing the distribution of mean errors across 10 runs per method. This approach provides a statistically sound comparison of augmentation strategies for improving AE source localization accuracy.

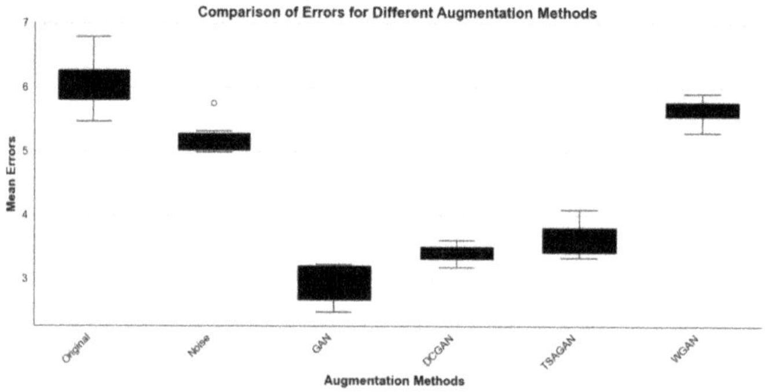

Figure 9. The comparison of errors for the different methods.

The original data show that the highest median error was 6.2 inches with a 1.1-inch interquartile range. Noise-based augmentation marginally improves this to 5.2 inches (median) with a 0.3-inch range, indicating increased consistency but persistent substantial errors. Among the GAN methods, the standard GAN demonstrated superior performance, achieving a median error of 2.9 inches and a 0.7-inch interquartile range. This represents a 53% reduction from the original data and a 44% improvement over noise-based augmentation. The DCGAN and TSAGAN also perform well, with median errors of 3.4 and 3.6 inches respectively, showing 45% and 42% in improvements over the original data. Surprisingly, the WGAN underperforms with a 5.7-inch median error, only marginally better than the original data. This unanticipated performance of the WGAN, which theoretically should provide more stable training, demands further investigations. Its median error of 5.7 inches, only slightly better than noise-based augmentation, may be attributed to specific characteristics of AE signals or potential issues in implementation. We assume that the Wasserstein loss might be less effective in capturing the nuanced features of AE signals compared to the binary cross-entropy loss used in the standard GAN for this particular application.

While our hybrid approach demonstrates significant improvements, it is also important to acknowledge its limitations. This study focused on a specific 24″ × 24″ × 0.1″ aluminum plate configuration, and further research is needed to validate its performance on larger structures and different materials. In conclusion, our innovative hybrid deep learning approach with GAN-based augmentation exhibits superior performance in AE source localization. The standard GAN architecture emerges as the most effective method, reducing median localization error by 53% compared to the original data, and by 44% compared to noise-based augmentation. These quantitative improvements underscore the value of advanced data augmentation techniques in enhancing model accuracy and reliability for structural health monitoring applications.

7. Conclusions

This study presents a comprehensive approach to acoustic emission (AE) source localization, leveraging a novel single fiber-optic coil-based sensing system, advanced signal processing techniques, and data augmentation with GAN. Our hybrid methodology, combining GANs for data augmentation with an adapted Inception network for localization, demonstrated substantial increases in AE source localization accuracy in bounded isotropic plate-like structures. The t-SNE visualizations and Wasserstein distance analysis confirm the high fidelity of the synthetic data generated by the GAN, leading to improved model accuracy and robustness. The comparison of error distributions highlights the superiority of GAN-based augmentation over noise-based augmentation and the original dataset, illustrating an evident reduction in localization errors and improved consistency across various source locations, crucial for detecting fatigue cracks and corrosion in metallic plate structures. The implications of these findings are extensive, potentially transforming structural health monitoring practices across various industries. By addressing the challenges of limited and imbalanced datasets, our approach paves the way for more reliable and efficient damage detection and localization in complex infrastructures. While our study focused on a specific, thin aluminum plate-like structure, future research should extend this methodology to more diverse real-world scenarios, including larger structures of different materials and additive manufacturing. Further exploration of advanced signal processing techniques and alternative deep learning architectures could potentially enhance model performance, particularly in applications such as ship hulls, bridge girders, aircraft wings, and fuselages. In conclusion, the hybrid deep learning architectures we developed here are for more accurate, reliable, and efficient damage detection and localization in plate-like structures, thus contributing to safer and more sustainable infrastructure management. This single-sensor configuration enables precise localization without the need for a multiple sensor array, offering a promising solution for enhancing AE-based SHM in complex infrastructures and improving damage detection accuracy and reliability for more efficient predictive maintenance strategies.

Author Contributions: Conceptualization, X.H. and Y.D.; methodology, X.H.; software, X.H.; validation, X.H.; formal analysis, X.H.; investigation, X.H.; resources, X.H.; data curation, X.H.; writing—original draft preparation, X.H.; writing—review and editing, X.H. and M.H.; visualization, X.H.; supervision, M.H. and Y.D.; project administration, M.H. and Y.D.; funding acquisition, M.H. and Y.D. All authors have read and agreed to the published version of the manuscript.

Funding: This research is funded by the US Office of Naval Research under Award Nos. N00014-20-1-2649, N00014-21-1-2273 and N00014-22-1-2321.

Institutional Review Board Statement: Not applicable.

Informed Consent Statement: Not applicable.

Data Availability Statement: The code used in this study is publicly available at https://github.com/Xuhui94/Hybrid-DL-for-AE-source-localization (accessed on 10 August 2024). The data used in this study has limited access due to privacy/ethical restrictions. Requests for data access can be directed to the corresponding author.

Acknowledgments: The authors would like to express their gratitude for the research support provided by the US Office of Naval Research under Award Nos. N00014-20-1-2649, N00014-21-1-2273, and N00014-22-1-2321, and technical guidance from the Program Manager, Ignacio Perez. The authors are also grateful to Farzia Karim for her assistance in collecting the data which used fiber-optic AE sensors.

Conflicts of Interest: The authors declare no conflicts of interest.

References

1. Wevers, M.; Lambrighs, K. Applications of Acoustic Emission for SHM: A Review. In *Encyclopedia of Structural Health Monitoring*; Wiley: Hoboken, NJ, USA, 2009.
2. Bouzid, O.M.; Tian, G.Y.; Cumanan, K.; Neasham, J. Wireless AE Event and Environmental Monitoring for Wind Turbine Blades at Low Sampling Rates. In *Advances in Acoustic Emission Technology, Proceedings of the the World Conference on Acoustic Emission–2013, New York, NY, USA, 30 September–2 October 2013*; Springer: New York, NY, USA, 2014; pp. 533–546.
3. Holford, K.M.; Pullin, R.; Evans, S.L.; Eaton, M.J.; Hensman, J.; Worden, K. Acoustic emission for monitoring aircraft structures. *Proc. Inst. Mech. Eng. Part G J. Aerosp. Eng.* **2009**, *223*, 525–532. [CrossRef]
4. Bhuiyan, M.Y.; Giurgiutiu, V. Experimental and Computational Analysis of Acoustic Emission Waveforms for SHM Applications. In Proceedings of the 11th International Workshop on Structural Health Monitoring, Stanford, CA, USA, 12–14 September 2017; pp. 12–14.
5. Sen, D.; Nagarajaiah, S. Data-Driven Approach to Structural Health Monitoring Using Statistical Learning Algorithms. In *Mechatronics for Cultural Heritage and Civil Engineering*; Ottaviano, E., Pelliccio, A., Gattulli, V., Eds.; Springer: Cham, Switzerland, 2018; Volume 92, pp. 295–305.
6. Chadha, M.; Yang, Y.; Hu, Z.; Todd, M.D. Evolutionary Sensor Network Design for Structural Health Monitoring of Structures with Time-Evolving Damage. In Proceedings of the 14th International Workshop on Structural Health Monitoring, Stanford, CA, USA, 11–13 September 2023.

7. Malekzadeh, M.; Atia, G.; Catbas, F. Performance-based structural health monitoring through an innovative hybrid data interpretation framework. *J. Civ. Struct. Health Monit.* **2015**, *5*, 287–305. [CrossRef]
8. Dong, S.; Yuan, M.; Wang, Q.; Liang, Z. A Modified Empirical Wavelet Transform for Acoustic Emission Signal Decomposition in Structural Health Monitoring. *Sensors* **2018**, *18*, 1645. [CrossRef] [PubMed]
9. Eaton, M.; Pullin, R.; Holford, K. Towards improved damage location using acoustic emission. *Proc. Inst. Mech. Eng. Part C J. Mech. Eng. Sci.* **2012**, *226*, 2141–2153. [CrossRef]
10. Ebrahimkhanlou, A.; Salamone, S. A probabilistic framework for single-sensor acoustic emission source localization in thin metallic plates. *Smart Mater. Struct.* **2017**, *26*, 095026. [CrossRef]
11. Ebrahimkhanlou, A.; Salamone, S. Acoustic emission source localization in thin metallic plates: A single-sensor approach based on multimodal edge reflections. *Ultrasonics* **2017**, *78*, 134–145. [CrossRef] [PubMed]
12. Ebrahimkhanlou, A.; Salamone, S. Single-Sensor Acoustic Emission Source Localization in Plate-Like Structures Using Deep Learning. *Aerospace* **2018**, *5*, 50. [CrossRef]
13. Mahajan, H.; Banerjee, S. Acoustic emission source localisation for structural health monitoring of rail sections based on a deep learning approach. *Meas. Sci. Technol.* **2023**, *34*, 044010. [CrossRef]
14. Ebrahimkhanlou, A.; Salamone, S. Acoustic Emission Signal Classification Using Feature Analysis and Deep Learning Neural Network. *Fluct. Noise Lett.* **2020**, *20*, 2150030.
15. Daugela, A.; Chang, C.; Peterson, D. Deep learning based characterization of nanoindentation induced acoustic events. *Mater. Sci. Eng.* **2021**, *800*, 140273. [CrossRef]
16. Farrar, C.R.; Worden, K. *Structural Health Monitoring: A Machine Learning Perspective*; Wiley: Hoboken, NJ, USA, 2012.
17. Worden, K.; Manson, G. Machine Learning for Structural Health Monitoring: Challenges and Opportunities. *J. Struct. Monit.* **2007**, *365*, 515–537.
18. Deraemaeker, A.; Preumont, A.; Reynders, E.; De Roeck, G.; Kullaa, J.; Lamsa, V.; Worden, K.; Manson, G.; Barthorpe, R.; Papatheou, E.; et al. Vibration-based structural health monitoring using large sensor networks. *Smart Struct. Syst.* **2010**, *6*, 335–347. [CrossRef]
19. Ai, D.; Zhang, R. Deep learning of electromechanical admittance data augmented by generative adversarial networks for flexural performance evaluation of RC beam structure. *Eng. Struct.* **2023**, *296*, 116891. [CrossRef]
20. Sapidis, G.M.; Kansizoglou, I.; Naoum, M.C.; Papadopoulos, N.A.; Chalioris, C.E. A Deep Learning Approach for Autonomous Compression Damage Identification in Fiber-Reinforced Concrete Using Piezoelectric Lead Zirconate Titanate Transducers. *Sensors* **2024**, *24*, 386. [CrossRef] [PubMed]
21. Goodfellow, I.; Pouget-Abadie, J.; Mirza, M.; Xu, B.; Warde-Farley, D.; Ozair, S.; Courville, A.; Bengio, Y. Generative adversarial networks. *Adv. Neural Inf. Process. Syst.* **2014**, *27*, 8–13. [CrossRef]
22. Ebrahimkhanlou, A.; Dubuc, B.; Salamone, S. A generalizable deep learning framework for localizing and characterizing acoustic emission sources in riveted metallic panels. *Mech. Syst. Signal Process.* **2019**, *130*, 248–272. [CrossRef]
23. Sun, L.; Chen, J.; Xu, Y.; Gong, M.; Yu, K.; Batmanghelich, K. Hierarchical Amortized GAN for 3D High Resolution Medical Image Synthesis. *IEEE J. Biomed. Health Inform.* **2022**, *26*, 3966–3975. [CrossRef]
24. Ebrahimkhanlou, A.; Salamone, S. A Deep Learning Approach for Single-sensor Acoustic Emission Source Localization in Plate-like Structures. *Struct. Health Monit. Int. J.* **2017**, *16*, 293–308.
25. Ebrahimkhanlou, A.; Salamone, S. Single-sensor acoustic emission source localization in plate-like structures: A deep learning approach. In *Health Monitoring of Structural and Biological Systems XII*; SPIE: Bellingham, WA, USA, 2018.
26. Ebrahimkhanlou, A.; Schneider, M.B.; Dubuc, B.; Salamone, S. Single Sensor Localization and Characterization of Acoustic Emission Sources in Metallic Panels: A Deep Learning Approach. *Struct. Health Monit. Int. J.* **2019**, *18*, 248–272.
27. Assarar, M.; Bentahar, M.; El Mahi, A.; El Guerjouma, R. Monitoring of damage mechanisms in sandwich composite materials using acoustic emission. *Int. J. Damage Mech.* **2015**, *24*, 787–804. [CrossRef]
28. Haile, M.A.; Zhu, E.; Hsu, C.; Bradley, N. Deep machine learning for detection of acoustic wave reflections. *Struct. Health Monit.* **2020**, *19*, 1340–1350. [CrossRef]
29. Jung, B.-H.; Kim, Y.-W.; Lee, J.-R. Laser-based structural training algorithm for acoustic emission localization and damage accumulation visualization in a bolt joint structure. *Struct. Health Monit.* **2019**, *18*, 1851–1861. [CrossRef]
30. Jones, M.; Rogers, T.; Cross, E. Constraining Gaussian processes for physics-informed acoustic emission mapping. *Mech. Syst. Signal Process.* **2022**, *188*, 109984. [CrossRef]
31. Gawronski, M.; Grabowski, K.; Russek, P.; Staszewski, W.J.; Uhl, T.; Packo, P. Acoustic emission source localization based on distance domain signal representation. In *Health Monitoring of Structural and Biological Systems 2016*; SPIE: Bellingham, WA, USA, 2016.
32. Ai, L.; Soltangharaei, V.; Bayat, M.; van Tooren, M.; Ziehl, P. Detection of impact on aircraft composite structure using machine learning techniques. *Meas. Sci. Technol.* **2021**, *32*, 084013. [CrossRef]
33. Shao, S.; Wang, P.; Yan, R. Generative adversarial networks for data augmentation in machine fault diagnosis. *Comput. Ind.* **2019**, *106*, 85–93. [CrossRef]
34. Sorin, V.; Barash, Y.; Konen, E.; Klang, E. Creating Artificial Images for Radiology Applications Using Generative Adversarial Networks (GANs)—A Systematic Review. *Acad. Radiol.* **2019**, *27*, 1175–1185. [CrossRef]

35. Wang, H.; Li, P.; Lang, X.; Tao, D.; Ma, J.; Li, X. FTGAN: A Novel GAN-Based Data Augmentation Method Coupled Time–Frequency Domain for Imbalanced Bearing Fault Diagnosis. *IEEE Trans. Instrum. Meas.* **2023**, *72*, 3502614. [CrossRef]
36. Jain, N.; Manikonda, L.; Hernandez, A.O.; Sengupta, S.; Kambhampati, S. Imagining an Engineer: On GAN-Based Data Augmentation Perpetuating Biases. *arXiv* **2018**, arXiv:1811.03751.
37. Liu, G.; Zhu, Y.; Sheng, Q.; Han, M. Polarization-insensitive, omnidirectional fiber-optic ultrasonic sensor with quadrature demodulation. *Opt. Lett.* **2020**, *45*, 4164–4167. [CrossRef]
38. Karim, F.; Zhu, Y.; Han, M. Modified phase-generated carrier demodulation of fiber-optic interferometric ultrasound sensors. *Opt. Express* **2021**, *29*, 25011–25021. [CrossRef] [PubMed]
39. Sause, M.G.R. Investigation of Pencil-Lead Breaks as Acoustic Emission Sources. *J. Acoust. Emiss.* **2011**, *29*, 184–196.

applied sciences

MDPI

Article

Simultaneous Vibration and Temperature Real-Time Monitoring Using Single Fiber Bragg Grating and Free Space Optics

Cheng-Kai Yao, Pradeep Kumar, Bing-Xian Liu, Amare Mulatie Dehnaw and Peng-Chun Peng *

Department of Electro-Optical Engineering, National Taipei University of Technology, Taipei 10608, Taiwan; t109658093@ntut.org.tw (C.-K.Y.); t111999408@ntut.org.tw (P.K.); t111658015@ntut.org.tw (B.-X.L.); mulatieamare7@gmail.com (A.M.D.)

* Correspondence: pcpeng@ntut.edu.tw; Tel.: +886-2-2771-2171 (ext. 4671)

Abstract: Simultaneous monitoring of temperature and vibration of electromechanical systems, public buildings, and volcanic terrain is an important indicator of preventing accidents. This work showcases the simultaneous temperature and vibration measurement using a single-fiber Bragg grating (FBG) sensor. The concurrent interrogation of vibration and temperature by the FBG sensing system can be integrated with free space optics (FSO), which saves on the costs associated with fiber optic cables and overcomes terrain barriers. Furthermore, the real-time measured vibration frequencies and different temperature levels can be recognized by a deep neural network (DNN) sequential model. In addition to the FSO support, the sensing system can be arranged in multiple sensing locations to form a wide range of measurements, offering a highly cost-effective solution.

Keywords: fiber Bragg grating; vibration measurement; temperature measurement; free space optics; deep neural network

1. Introduction

Fiber optic sensing, renowned for its small size, sensitivity, precision, rapid response, and immunity to electromagnetic interference, holds significant promise for a wide range of manufacturing [1,2], biomedical [3,4], and transportation sensing applications [5,6]. The performance of fiber optic sensing technology has witnessed continual enhancements. Despite being typically employed to measure a single physical parameter, fiber optic sensors also possess the capability to concurrently measure multiple parameters if required [7–9]. This feature yields cost savings by obviating the necessity for additional sensors. However, swift identification of the measured parameter's value is imperative when assessing diverse physical parameters.

Theoretically, as long as the optical transceiver is functioning properly and the data sampling points and ranges are appropriate, fiber optic sensors are capable of accurately measuring the frequency of vibrations [10]. Furthermore, fiber optic sensors can be operated over a wide range of temperatures due to the nature of the glass, making fiber optic sensors a very suitable tool for simultaneous monitoring of vibration and temperature. Simultaneous monitoring of vibration and temperature in electromechanical systems, especially within large public facilities, represents a highly effective strategy for ensuring their proper functionality and preventing potential accidents. This study mainly focuses on the simultaneous measurement of temperature and vibration using a solitary FBG sensor. In fiber optic sensing measurements, fiber optic interferometer sensors are the most sensitive choice [11], and they can measure both physical parameters simultaneously [7,8]. Nevertheless, using multiple fiber optic interferometer sensors in tandem is difficult due to complex spectral and optical loss problems. In addition, distributed fiber optic sensing can measure vibration and temperature simultaneously [9]. However, the system equipment is more complicated, and the cost is certainly higher, so it is not cost-effective when the measurement is not over a large area (fiber length of 1 km or more). Standard FBG sensors are capable of measuring a wide range of parameters accurately and can cascade multiple FBG sensors for simultaneous measurements [2]. Furthermore, FBG sensing systems can be integrated with FSO [12,13], allowing the fiber to be easily routed over geographic obstructions and thus eliminating cable routing and maintenance costs. The top half of Figure 1 shows the potential application of an FSO-based FBG sensing system for simultaneous measurement of vibration and temperature. The light emitted from the central office can be passed through the FSO instead of a fiber optic so that the sensed signal is transmitted directly over the air and over obstructions to the target location to be measured. Simultaneous vibration and temperature monitoring by FBGs can play a key role in the proper functioning and safety of tunnels and green energy-related applications.

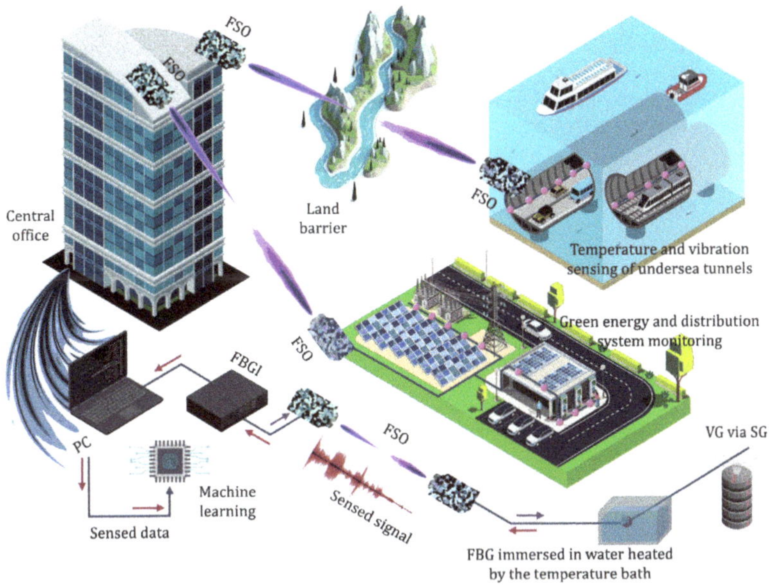

Figure 1. The application vision (upper part) and the experimental setup (lower part) of the FSO-based FBG sensing system for simultaneous monitoring of vibration and temperature. (FBGI: fiber Bragg grating interrogator; PC: personal computer; VG: vibration generator; SG: signal generator).

In a previous study, FBG vibration sensing in a link containing an FSO was demonstrated by a single-frequency laser [14]. However, the previous framework relied on optical power measurement and was incapable of conducting simultaneous multi-site measurements and simultaneous measurements of vibration and temperature. The present work on FBG vibration sensing with FSO link architecture is based on the FBG decoder using wavelength mode signaling, which is capable of measuring vibration and temperature simultaneously and allows simultaneous measurement of multiple points. Moreover, the acquired signals can be learned by machine learning to quickly determine the temperature and vibration frequency, obviating the need for a fast Fourier transform to analyze the vibration frequency. This enhancement renders it more effective for timely vibration and temperature monitoring. While applying fast Fourier transforms may offer greater efficiency compared to training machine learning models, they are limited in their capability to detect variations in temperature and identify subtle frequency differences [10].

2. Measuring Setup, Principle, and Results

The main equipment used in the experiments was an FBG interrogator (Citpo Technologies Co., Taipei, Taiwan) with a wavelength scanning range of 1528~1568 nm, an FSO device (Thorlabs Inc. F810FC-1550 in Newton, NJ, USA), and a vibration generator (Pasco SF-9324, Pasco, Roseville, CA, USA). Other equipment, such as the personal computer, standard FBG, temperature bath, and signal generator, are commonly available on the market. The lower part of Figure 1 shows the experimental setup. First, the FBG interrogator sends the probing beam from the optical fiber via the collimator to the air path, i.e., the 2 m FSO, and then from the collimator to the FBG sensor for sensing. The optical fiber on the side of the FBG is tied to a heavy object and immersed in a temperature bath. The signal generator-driven vibration generator generates different frequencies of vertical vibrations to pull the optical fiber so that the FBG generates a dynamic strain over time. Consequently, this setup allows the FBG to simultaneously measure different vibrations and temperatures. The measured vibration and temperature signals are returned to the FBG interrogator for data acquisition. Subsequently, the sampled data is then used to train a machine learning model to recognize the FBG's signal-sensing information. The computer for machine learning is equipped with a CPU of Intel(R) Core (TM) i7-9700k speed @ 3.60 Hz, installed RAM of 16 GB, and a graphics card of NVIDIA GeForce RTX 2080Ti.

FBG is a periodic grating structure; in temperature measurement, the thermal expansion of the fiber material due to the change of ambient temperature changes the spacing of the grating period and the refractive index of the core material, which results in the Bragg wavelength of the FBG changing with the temperature. In terms of vibration measurement, it relies on the applied dynamic mechanical stress to cause fiber deformation, which dynamically changes the grating period and refractive index, resulting in the FBG Bragg wavelength changing with vibration. That is to say, the FBG sensing temperature and vibration mechanism are different due to the fiber deformation mechanism being different and can be superimposed, so the simultaneous measurement of vibration and temperature will be feasible. Generally speaking, a single FBG cannot measure the strain and temperature change at the same time because the wavelength shift of the FBG when subjected to strain (strain sensitivity is 1 pm/με) will overlap with the wavelength shift of the FBG when subjected to ambient temperature change (temperature sensitivity is 10~13 pm/°C) [15], and it is not possible to recognize the exact amount of the strain and temperature change. However, the difference is that the vibration is a continuous periodic strain response; a vibration signal corresponding to the FBG wavelength change and a temperature change corresponding to the FBG wavelength change will be able to be recognized.

The data collection process consists of two segments: one involving the FSO link, employing an FBG interrogator with a 3 Hz sweeping laser (equivalent to three sampling points per second), and the other without the FSO link, utilizing an FBG interrogator with a 200 Hz sweeping laser (equivalent to 200 sampling points per second). As per the Nyquist sampling theorem [16], the maximum sampling frequencies are restricted to 1.5 Hz and 100 Hz, respectively. In the FSO connection scenario, the FSO collimator may cause wavelength disturbance of the high-speed sweep laser. Consequently, when employing a 200 Hz FBG interrogator, this faster sweeping speed and stability issues with the 200 Hz FBG interrogator result in pronounced wavelength fluctuation, impacting the accurate interrogation of vibration signals. Therefore, to measure higher vibration frequencies in FSO links, in addition to a high-speed swept laser, it must also be stable and prevent wavelength fluctuations. Due to site constraints in the experimental environment, the FSO system has a length of only 2 m, whereas, in normal operation, the theoretical length of the FSO can extend to several kilometers [17,18]. The length discrepancy necessitates a higher laser power to compensate for transmission path loss. Additionally, environmental turbulence in the FSO path will add noise to the FBG spectrum [19], which will cause errors in the read wavelength and produce vibration signals with noise. This issue can be effectively addressed through the application of denoising techniques [20].

Figure 2 shows the results of simultaneous vibration and temperature measurements without the participation of the FSO link. Figure 2a–d display the vibration frequencies of 20 Hz, 40 Hz, and 80 Hz, and the temperatures of 30 °C, 40 °C, 50 °C, and 60 °C are measured for different vibration frequencies. The findings indicate that the vibration signals at various temperatures can be differentiated. The prominently displayed red line represents the vibration signal smoothed using the Savitzky–Golay filter [21], which theoretically will be very close to the wavelength of the FBG under the actual ambient temperature measurement. Since the temperature measurement is without the vibration signal, the corresponding wavelength would be exactly in the middle of the wavelengths corresponding to the peaks and valleys of the hypothetical vibration signal since the vibration is a harmonic motion. It should be clarified that the FBG sensor used for the measurements is used to measure both vibration and temperature, so it is not possible to use the same FBG to measure only the temperature but not the vibration. Also, there may be small temperature differences each time of temperature bath adjustment, and small changes in the fiber optic tension and the water level in the sink may cause the wavelength of the FBGs to fluctuate. Therefore, the FBG wavelength may not fall perfectly in the center of the amplitude level of the vibration signal if the temperature data are measured at different times, so the results of temperature measurement alone are not shown in Figure 2. In addition, one of the main ideas of the scheme is to use the vibration signal directly to recognize the frequency of the vibration and temperature value by machine learning, so consideration of the vibration signal only indicates that the temperature and vibration can be differentiated.

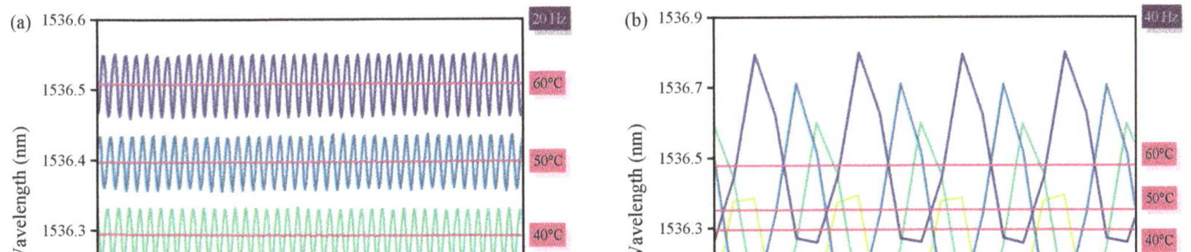

Figure 2. Simultaneous measurements of vibration and temperature at 30 °C, 40 °C, 50 °C, and 60 °C without FSO participation: (**a**) 20 Hz vibration frequency, (**b**) 40 Hz vibration frequency, (**c**) 60 Hz vibration frequency, and (**d**) 80 Hz vibration frequency.

Figure 2 shows that the measured vibration frequency is not affected by the temperature change. The temperature measurements are generally less affected by the frequency of vibration, with wavelengths of 1536.2 nm, 1536.3 nm, 1536.4 nm, and 1536.5 nm for 30 °C, 40 °C, 50 °C, and 60 °C, respectively. Some of the deviations in the temperature measurements are due to the small differences in fiber tension and water level of the bath, as mentioned earlier. Also, the water temperature may not be the same every time the temperature bath is adjusted to the same temperature. In addition, since the sampling point of the FBG interrogator is fixed, the sampling time transients may not necessarily coincide with the peaks and valleys of the mechanical waves generated by the vibration generator, so the waveforms of the data measured at different vibration frequencies may not be perfect sinusoids. This may result in the peaks and valleys of the vibration signal corresponding to the wavelength of the position of the distortion, resulting in the curve of the smoothed vibration signal and the actual temperature measurement of the wavelength of the deviation. Hence, using machine learning to read the vibration signal directly to derive the vibration and temperature information can overcome this problem. Incidentally, the signal amplitude generated by the signal generator is the same, and the length between the FBG and the vibration generator is 30 cm of fiber. Since the resonance length is different for each vibration frequency, the amplitude of the vibration signal will be different.

In practical applications, the operation of the electrical system and changes in the ambient temperature can lead to continuous changes in temperature. For this purpose, the vibration measurements under dynamic temperature were collected as shown in Figure 3, and the frequencies of vibration measurements in Figure 3a–d are 20 Hz, 40 Hz, 60 Hz, and 80 Hz, respectively. Temperature conditions are 31 to 36 °C, 41 to 46 °C, and 51 to 56 °C. Similarly, by smoothing the signal, the wavelength curve closest to the temperature information can be obtained, as shown by the white line. The wavelength response is 1536.2 nm to 1536.26 nm for water heating from 31 to 36 °C, 1536.3 nm to 1536.36 nm for 41 to 46 °C, and 1536.4 nm to 1536.47 nm for 51 to 56 °C. The initial water temperature being 1 °C higher than the condition indicated in Figure 2 has led to a slightly higher corresponding wavelength compared to the measurement result depicted in Figure 2. Additionally, since the water heating time cannot be controlled consistently each time, the sampling time of the signals will be different. The important point is that the level of the vibration signal rises as the temperature rises, indicating that simultaneous measurements of vibration and temperature can still be made under dynamic temperature conditions. The vibration signal and the white smoothed lines indicate minor fluctuations, which are attributed to both the noise from the FBG interrogator itself and the fluctuations in water temperature.

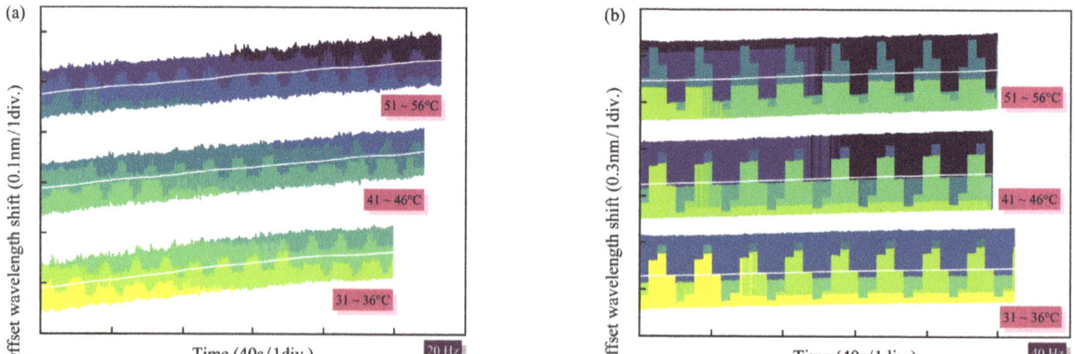

Figure 3. *Cont.*

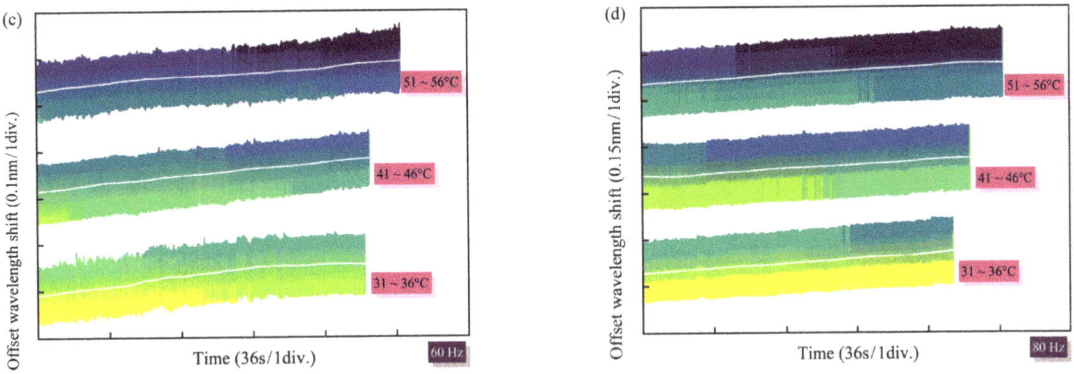

Figure 3. Simultaneous measurements of vibration and dynamic temperature from 31 to 36 °C, 41 to 46 °C, and 51 to 56 °C without FSO participation: (**a**) 20 Hz vibration frequency, (**b**) 40 Hz vibration frequency, (**c**) 60 Hz vibration frequency, and (**d**) 80 Hz vibration frequency. Since the vibration signals are dense, different frequencies can be seen by drawing lines with gradient colors.

Figure 4 shows the simultaneous measurement of vibration and temperature with the participation of FSO. The measured vibration frequencies in Figure 4a–d are 0.1 Hz, 0.5 Hz, 1 Hz, and 1.5 Hz and correspond to four temperatures, namely, 30 °C, 40 °C, 50 °C, and 60 °C. From the results, it can be seen that different vibration frequencies are not affected by temperature changes, and the smoothing result of the vibration signals, i.e., the bright red line, presents temperature-related information that is very close to the measurement results in Figure 2. Besides, the amplitudes of the 0.1 Hz, 0.5 Hz, 1 Hz, and 1.5 Hz vibration signals do not differ much because the vibration frequency is low and cannot cause resonance. The important point is that the FBG can also measure the vibration and temperature signals simultaneously and accurately through the FSO transmission.

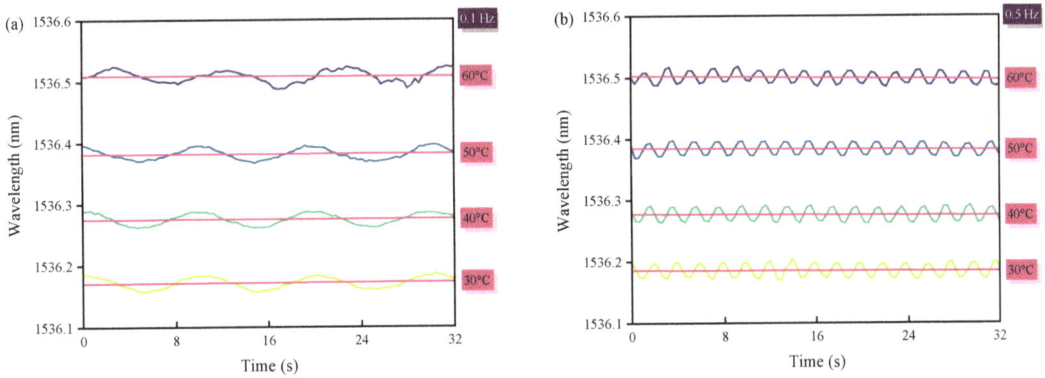

Figure 4. *Cont.*

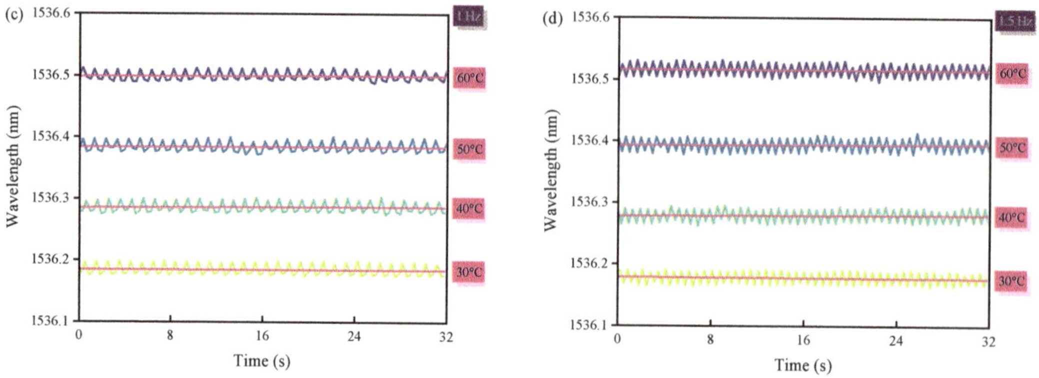

Figure 4. Simultaneous measurements of vibration and temperature at 30 °C, 40 °C, 50 °C, and 60 °C with FSO participation: (**a**) 20 Hz vibration frequency, (**b**) 40 Hz vibration frequency, (**c**) 60 Hz vibration frequency, and (**d**) 80 Hz vibration frequency.

Figure 5 also shows the case with FSO, the temperature conditions are 31 to 36 °C, 41 to 46 °C, and 51 to 56 °C (consistent with the measurement settings in Figure 3), and the vibration frequencies of Figure 5a–d are 0.1 Hz, 0.5 Hz, 1 Hz, and 1.5 Hz, respectively. The smoothed bright red line reveals the wavelength response ranges from 1536.2 nm to 1536.26 nm for water temperatures between 31 °C and 36 °C, from 1536.3 nm to 1536.36 nm for temperatures between 41 °C and 46 °C, and from 1536.4 nm to 1536.47 nm for temperatures between 51 °C and 56 °C. These findings are in line with the outcomes presented in Figure 3. Therefore, it is shown that in a dynamic temperature environment, both vibration and temperature can be measured simultaneously by the FSO transmission.

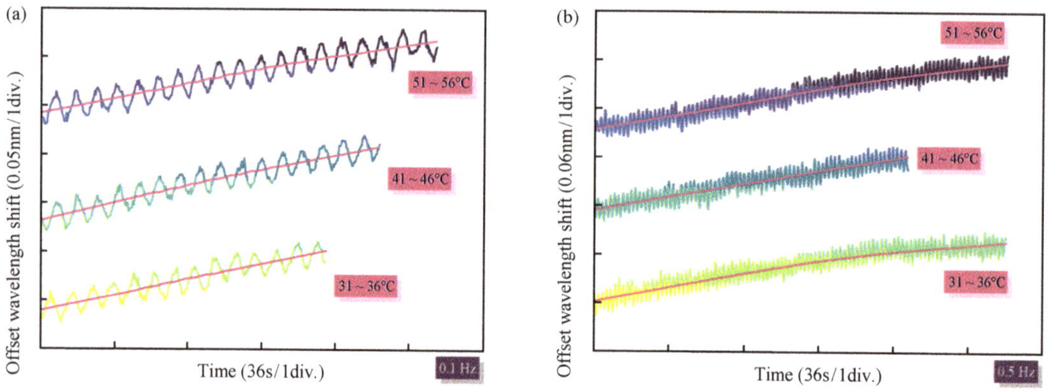

Figure 5. *Cont.*

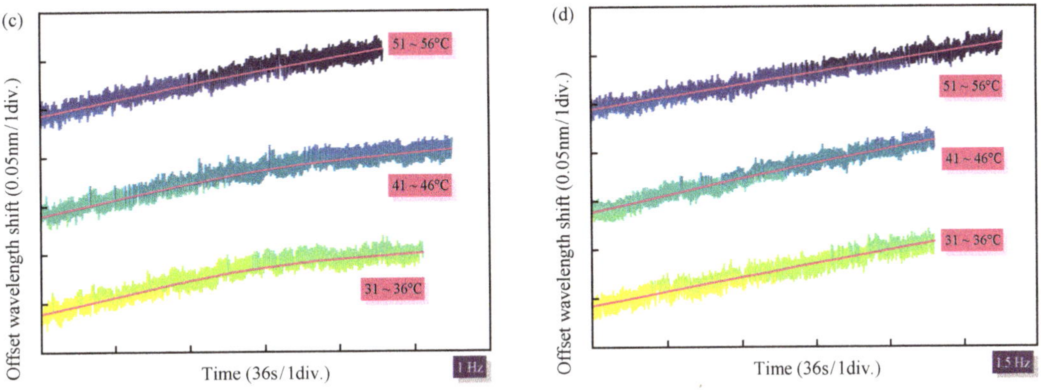

Figure 5. Simultaneous measurements of vibration and dynamic temperature from 31 to 36 °C, 41 to 46 °C, and 51 to 56 °C with FSO participation: (**a**) 20 Hz vibration frequency, (**b**) 40 Hz vibration frequency, (**c**) 60 Hz vibration frequency, and (**d**) 80 Hz vibration frequency.

3. Machine Learns to Recognize Signals

Machine learning is a robust tool for understanding signal characteristics and effectively identifying signals detected across various sensing tasks, such as vibration signals for fault detection [10,22–24], through trained machine learning models. Notably, exploring vibration signals in varied temperature environments, particularly the concurrent measurement of vibration and temperature through fiber optic sensing and the subsequent signal recognition using machine learning, remains an area yet to be thoroughly investigated. This report employs a DNN sequential model [25] to discern the vibration frequency and temperature characteristics of the vibration signals. The main code of the DNN sequential model is defined as follows:

```
model = Sequential([Input([X.shape(1)]), Dense(500, activation='relu'),
BatchNormalization(), Dropout(0.2), Dense(500, activation='relu'),
BatchNormalization(), Dropout(0.2), Dense(128, activation='relu'),
BatchNormalization(), Dropout(0.2), Dense(y_one_hot.shape(1),
activation='softmax')]).
```

The sequential model comprises stacked layers arranged linearly. The model commences with an input layer designed to receive vibration frequency signals. The input layer's shape is specified by Input[X.shape(1)], where X.shape(1) denotes the number of features in the input data. In a sequential model, it is not strictly necessary to define the input layer explicitly, but doing so can improve clarity even though Keras can determine the input shape from the first layer. DNN is made up of several hidden layers, each containing neurons that process the weighted sum of inputs from the layer before it using activation functions. Common activation functions include Rectified Linear Unit (ReLU) and sigmoid. These layers are instrumental in discerning intricate patterns and features from the input data. The first hidden layer, Dense (500, activation = 'relu'), is a fully connected (dense) layer housing 500 neurons. The utilization of ReLU as the activation function introduces non-linearity into the model. ReLU outputs the input directly if it is positive; otherwise, it yields zero. The BatchNormalization() layer normalizes the output of the preceding layer, thereby stabilizing and expediting the training process by reducing internal covariate shifts. This normalization technique adjusts the activations of the prior layer within each batch, leading to a potential acceleration in convergence. Additionally, the Dropout(0.2) layer randomly deactivates 20% of the input units during training, mitigating overfitting by preventing the model from heavily relying on any single feature. Subsequently, the second dense layer, Dense(500, activation = 'relu'), the subsequent BatchNormalization() layer, and the following Dropout(0.2) layer serve similar functions to their respective initial layers.

The third dense layer consists of 128 neurons and employs ReLU activation. Its primary function is to act as a bridge to the output layer. Following this, the third batch normalization, achieved through BatchNormalization(), standardizes the output of the third dense layer. Additionally, the third dropout layer, implemented using Dropout(0.2), serves to mitigate overfitting. The model learns to identify relevant features from the vibration signals as the data moves through the hidden layer. This could include the ability to recognize unique frequency patterns or anomalies that correspond to vibrations that take place at different temperatures. The final layer, Dense(y_one_hot.shape(1), activation='softmax'), functions as the output layer, containing many neurons equivalent to the classes present in the target variable, denoted by 'y_one_hot.shape(1)'. This layer employs the softmax activation function, which is well-suited for addressing multi-class classification tasks by generating class probabilities that sum to 1.

Overall, this DNN model has multiple dense layers with dropout and batch normalization layers in between to increase training stability and decrease overfitting. The model is designed for a classification task, as indicated by the softmax output layer. Every component is essential to achieving both efficient learning from the input data and maintaining generalization to new data. The model is trained using labeled data, where each input signal is associated with a known class. During training, the model fine-tunes its weights through backpropagation to minimize the variance between its predictions and the actual labels. The accuracy of the model can then be verified by evaluating its performance on unseen data (20% of the total data). Based on the learned features, the model can predict the signal's class when new vibration signals are fed in.

Figure 6 shows the training and validation results of using the DNN sequential model to recognize different vibration frequencies under various ambient temperatures. The training and validation results are based on the data results presented in Figures 2 and 4. Figure 6a–d shows the training accuracy, training loss, validation accuracy, and validation loss, respectively. It can be found that the results of 0.1 Hz, 0.5 Hz, 1 Hz, 1.5 Hz, and 20 Hz after 2000 epochs are significantly better than the results of 40 Hz, 60 Hz, and 80 Hz after 2000 epochs at four temperatures, namely, 30 °C, 40 °C, 50 °C, and 60 °C. The validation accuracy of the 0.1 Hz, 0.5 Hz, 1 Hz, 1.5 Hz, and 20 Hz frequencies at different temperatures is 100%, while the validation accuracy of the 40 Hz, 60 Hz, and 80 Hz frequencies at different temperatures is about 92.2%, 87.6%, and 81.7%. This difference is because the 0.1 Hz, 0.5 Hz, 1 Hz, 1.5 Hz, and 20 Hz signals are closer to a sinusoidal waveform, and the 0.1 Hz, 0.5 Hz, 1 Hz, and 1.5 Hz signals have relatively fewer sampling points. Regular and consistent signals tend to be easier for the model to learn and recognize because the regular pattern provides clear features that can be recognized by the model. On the other hand, more complex and irregular signals may introduce a large amount of variability, making it more challenging for the model to accurately recognize and classify them. Furthermore, there are 32 scenarios within eight vibration frequencies and four different ambient temperatures, and the validation accuracy is 89.1%. Figure 7a,b shows the results for 0.1 Hz and 80 Hz vibration frequencies, respectively. Following 1000 epochs, the validation accuracy can reach 95%, respectively. In this scheme, the smallest vibration frequency of 0.1 Hz and the largest vibration frequency of 80 Hz can be recognized under dynamic temperature, which implies that other vibration frequencies can also be recognized successfully. In particular, the validation accuracy of the 80 Hz case is 95.2%, which is higher than the 81.7% validation accuracy result at static temperature. This is due to the fact that the sampling time at dynamic temperatures is longer, and the data points are sufficiently large so that the model can be better trained to have better accuracy.

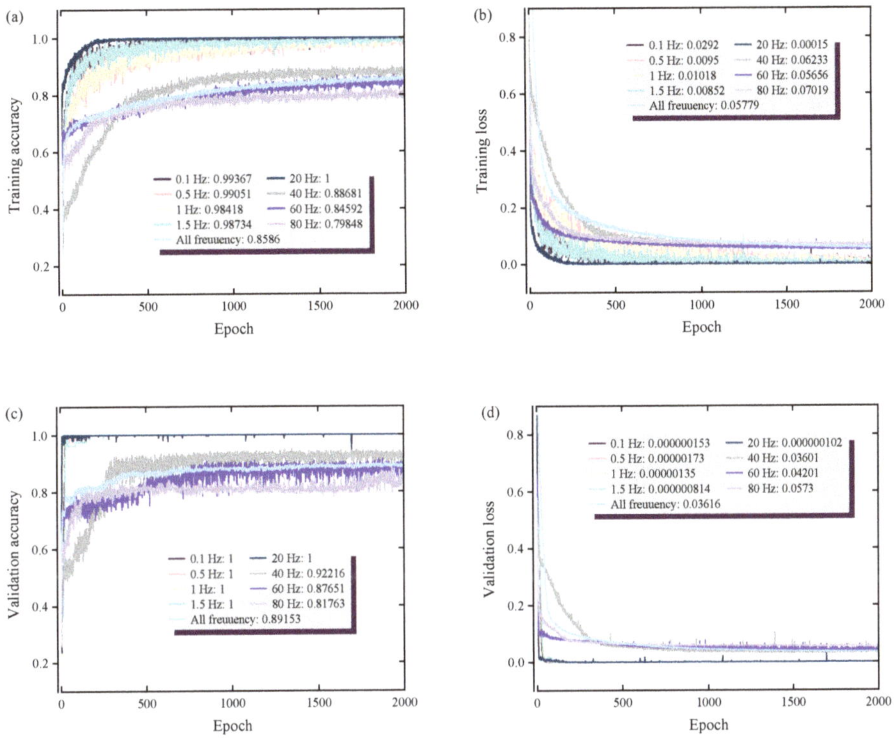

Figure 6. Results of using the DNN model to recognize the vibration frequency and temperature from vibration signals: (**a**) training accuracy, (**b**) training loss, (**c**) validation accuracy, and (**d**) validation loss.

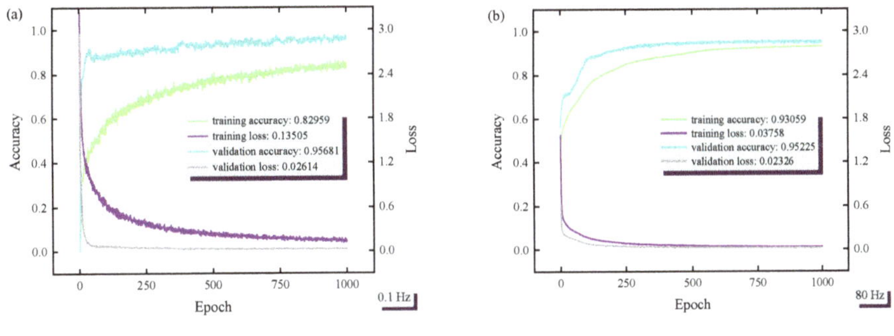

Figure 7. DNN model recognizing vibration measurements at dynamic temperature: (**a**) 0.1 Hz vibration frequency, (**b**) 80 Hz vibration frequency.

4. Conclusions

In this work, a sensing scheme for the simultaneous measurement of vibration and temperature by a single FBG sensor routed through an FSO link is demonstrated for the first time. The vibration measurement signals under different ambient temperatures can be deciphered by a DNN sequential model to obtain the vibration frequency and temperature information. The DNN sequential model recognizes 32 scenarios at various ambient temperatures, i.e., 30 °C, 40 °C, 50 °C, and 60 °C, and at different vibration frequencies, i.e., 0.1 Hz, 0.5 Hz, 1 Hz, 1.5 Hz, 20 Hz, 40 Hz, 60 Hz, and 80 Hz, with an accuracy of up to 89%. In addition, at dynamic temperatures, i.e., 31 to 36 °C, 41 to 46 °C, and 51 to 56 °C, the accuracy of recognition reaches 95% for vibration frequencies of 0.1 Hz and 80 Hz. This solution integrates FSO technology and a single FBG sensor to simultaneously measure vibration and temperature, with the machine learning to recognize the signal data, which is expected to save the cost of optical cable laying and maintenance, the cost of the number of FBG sensors, and the cost of human identification of the signals, and promote the development of related surveillance applications.

Author Contributions: Conceptualization, C.-K.Y., P.K. and P.-C.P.; methodology, C.-K.Y., P.K. and B.-X.L.; software, P.K.; data curation, B.-X.L.; model validation, B.-X.L., C.-K.Y. and P.K.; formal analysis, C.-K.Y., A.M.D.; investigation, C.-K.Y., B.-X.L., P.K., A.M.D. and P.-C.P.; visualization: C.-K.Y.; writing, C.-K.Y.; supervision, P.-C.P. All authors have read and agreed to the published version of the manuscript.

Funding: This work was supported by the National Science and Technology Council, Taiwan, under Grant NSTC 112-2221-E-027-076-MY2.

Institutional Review Board Statement: Not applicable.

Informed Consent Statement: Not applicable.

Data Availability Statement: The data presented in this study are available in this article.

Conflicts of Interest: The authors declare no conflicts of interest.

References

1. Presti, D.L.; Leitao, C.; Noccaro, A.; Tavares, C.; Massaroni, C.; Caponero, M.A.; Antunes, P.; Formica, D.; Schena, E. The Effect of Infill Pattern and Density on the Response of 3-D-Printed Sensors Based on FBG Technology. *IEEE Sens. J.* **2022**, *22*, 19357–19365. [CrossRef]
2. Feng, Y.; Peng, J.; Wu, T.; Xing, F.; Sun, T. A Real-time In-situ Wafer Temperature Measurement System Based on Fiber Bragg Grating Array. *IEEE Sens. J.* **2024**, *24*, 17733–17741. [CrossRef]
3. Presti, D.L.; Massaroni, C.; Leitao, C.S.J.; Domingues, M.D.F.; Sypabekova, M.; Barrera, D.; Floris, I.; Massari, L.; Oddo, C.M.; Sales, S.; et al. Fiber Bragg Gratings for Medical Applications and Future Challenges: A Review. *IEEE Access* **2020**, *8*, 156863–156888. [CrossRef]
4. Roriz, P.; Silva, S.; Frazão, O.; Novais, S. Optical Fiber Temperature Sensors and Their Biomedical Applications. *Sensors* **2020**, *20*, 2113. [CrossRef]
5. Marques, C.; Leal-Júnior, A.; Kumar, S. Multifunctional Integration of Optical Fibers and Nanomaterials for Aircraft Systems. *Materials* **2023**, *16*, 1433. [CrossRef]
6. Xu, H.; Wang, W.; Li, F.; Du, Y.; Tu, H.; Guo, C. Railway Slope Monitoring Based on Dual-Parameter FBG Sensor. *Photonic Sens.* **2024**, *15*, 250121. [CrossRef]
7. Li, Y.; Yu, C.; Lu, P. An Optical Fiber Sensor for Axial Strain, Curvature, and Temperature Measurement based on Single-Core Six-Hole Optical Fiber. *Sensors* **2022**, *22*, 1666. [CrossRef]
8. Lu, C.; Su, J.; Dong, X.; Sun, T.; Grattan, K.T.V. Simultaneous Measurement of Strain and Temperature with a Few-Mode Fiber-Based Sensor. *J. Light. Technol.* **2018**, *36*, 2796–2802. [CrossRef]
9. Zhou, Z.; Tian, L.; Han, Y.; Yang, X.; Liao, F.; Zhang, D.; Sha, Y.; Feng, X.; Zhu, J.; Zheng, X.; et al. Distributed Vibration and Temperature Simultaneous Sensing Using One Optical Fiber. *Opt. Commun.* **2021**, *487*, 126801. [CrossRef]
10. Jati, M.P.; Luthfi, M.I.; Yao, C.-K.; Dehnaw, A.M.; Manie, Y.C.; Peng, P.-C. An Extremely Close Vibration Frequency Signal Recognition Using Deep Neural Networks. *Appl. Sci.* **2024**, *14*, 2855. [CrossRef]
11. Zhu, K.; Zhang, L.; Yao, Y.; Li, X.; Li, J.; Liu, Y.; Wang, M. Two-Dimensional Displacement Sensor with High Resolution and Simple Structure. *IEEE Sens. J.* **2024**, *24*, 34145–34150. [CrossRef]
12. Yu, Y.L.; Liaw, S.K.; Chou, H.H.; Le-Minh, H.; Ghassemlooy, Z. A Hybrid Optical Fiber and FSO System for Bidirectional Communications Used in Bridges. *IEEE Photonics J.* **2015**, *7*, 1–9. [CrossRef]
13. Kanwal, F.; Atieh, A.; Ghafoor, S.; Haq, A.U.; Qureshi, K.K.; Aziz, I.; Mirza, J. Remote Monitoring of Sleep Disorder Using FBG Sensors and FSO Transmission System Enabled Smart Vest. *Eng. Res. Express* **2024**, *6*, 025337. [CrossRef]
14. Yao, C.-K.; Manie, Y.C.; Peng, P.-C. Free Space Optical Communication for Vibration Sensing of Multiple-Channel Fiber Bragg Grating Array. In Proceedings of the Photonic Networks and Devices, Washington, DC, USA, 24–28 July 2023.
15. Du, J.; He, Z. Sensitivity Enhanced Strain and Temperature Measurements based on FBG and Frequency Chirp Magnification. *Opt. Express* **2013**, *21*, 27111–27118. [CrossRef]
16. Vaidyanathan, P.P. Generalizations of the Sampling Theorem: Seven Decades After Nyquist. *IEEE Trans. Circuits Syst. I Fundam. Theory Appl.* **2001**, *48*, 1094–1109. [CrossRef]

17. Hayle, S.T.; Lu, H.-H.; Lin, H.-M.; Wang, C.-P.; Li, C.-Y.; Wu, T.-M.; Lin, C.-H.; Chen, W.-X.; Jin, J.-L.; Xu, Y.-Z. Two-way 5G NR FSO-HCF-UWOC Converged Systems with R/G/B 3-wavelength and SLM-based Beam-tracking scheme. *Sci. Rep.* **2024**, *14*, 22252. [CrossRef]
18. Fernandes, M.A.; Fernandes, G.M.; Brandão, B.T.; Freitas, M.M.; Kaai, N.; Tomeeva, A.; Wielen, B.V.D.; Reid, J.; Raiteri, D.; Monteiro, P.P.; et al. 4 Tbps+ FSO Field Trial Over 1.8 km with Turbulence Mitigation and FEC Optimization. *J. Light. Technol.* **2024**, *42*, 4060–4067. [CrossRef]
19. Bogale, S.D.; Yao, C.-K.; Manie, Y.C.; Dehnaw, A.M.; Tefera, M.A.; Li, W.-L.; Zhong, Z.-G.; Peng, P.-C. Demodulating Optical Wireless Communication of FBG Sensing with Turbulence-Caused Noise by Stacked Denoising Autoencoders and the Deep Belief Network. *Electronics* **2024**, *13*, 4127. [CrossRef]
20. Nguyen, C.D.; Ahmad, Z.; Kim, J.M. Gearbox Fault Identification Framework Based on Novel Localized Adaptive Denoising Technique, Wavelet-based Vibration Imaging, and Deep Convolutional Neural Network. *Appl. Sci.* **2021**, *11*, 7575. [CrossRef]
21. Schafer, R.W. What is a Savitzky-Golay Filter? [lecture notes]. *IEEE Signal Process. Mag.* **2011**, *28*, 111–117. [CrossRef]
22. Lin, W.J.; Lo, S.H.; Young, H.T.; Hung, C.L. Evaluation of Deep Learning Neural Networks for Surface Roughness Prediction Using Vibration Signal Analysis. *Appl. Sci.* **2019**, *9*, 1462. [CrossRef]
23. Toh, G.; Park, J. Review of Vibration-based Structural Health Monitoring Using Deep Learning. *Appl. Sci.* **2020**, *10*, 1680. [CrossRef]
24. Bagri, I.; Tahiry, K.; Hraiba, A.; Touil, A.; Mousrij, A. Vibration Signal Analysis for Intelligent Rotating Machinery Diagnosis and Prognosis: A Comprehensive Systematic Literature Review. *Vibration* **2024**, *7*, 1013–1062. [CrossRef]
25. Qaddoura, R.M.; Al-Zoubi, A.; Faris, H.; Almomani, I. A Multi-layer Classification Approach for Intrusion Detection in Iot Networks Based on Deep Learning. *Sensors* **2021**, *21*, 2987. [CrossRef]

applied sciences

MDPI

Article

Side-Illuminating Optical Fiber for High-Power-Density-Mediated Intraluminal Photoacoustic Imaging

Nidhi Singh [1,2,*], Carlos-Felipe Roa [1,2], Emmanuel Chérin [2], Lothar Lilge [1,3] and Christine E. M. Demore [1,2]

[1] Department of Medical Biophysics, University of Toronto, Toronto, ON M5S 1A1, Canada; cf.roa@mail.utoronto.ca (C.-F.R.); lothar.lilge@uhn.ca (L.L.); c.demore@utoronto.ca (C.E.M.D.)
[2] Sunnybrook Research Institute, Toronto, ON M4N 3M5, Canada; emmanuel.cherin@sri.utoronto.ca
[3] Princess Margret Cancer Centre, University Health Network, Toronto, ON M5G 2C4, Canada
* Correspondence: nidhi.singh@mail.utoronto.ca

Abstract: Intraluminal photoacoustic (PA) imaging has the potential for providing physiological and functional information in wide-ranging clinical applications. Along with endoluminal ultrasound transducers, these applications require compact light delivery devices which can deliver high-energy ns-pulsed laser to the target region. In this work, we describe the design, method of fabrication and characterization of a new compact, side-fire optical fiber that can deliver high-energy laser pulses for PA imaging. Side-fire illuminators were fabricated using UV laser ablation to create windows on the side of a 1.5 mm diameter single core, multi-mode optical fiber with a reflective silver coating and a beveled end. Devices with 10 mm, 20 mm, and 30 mm window lengths were fabricated and their beam profiles characterized. Elongated side-fire fibers with −6 dB beam size up to 30.79 mm × 5.5 mm were developed. A side-fire to total output ratio of up to 0.69 and a side fire efficiency of up to 40%, relative to a standard front-fire fiber, were achieved. We evaluated the effects of high-energy ns-pulsed light propagation on the fiber by coupling the fiber to 18 mJ or 100 MW/cm^2 (at 750 nm) beam from a Q-switched laser. The PA imaging with the fiber was demonstrated by detecting India ink targets embedded in chicken breast tissue over the full length of a 20 mm illumination window and over a 100° angle and by visualizing in vivo the rat ear vasculature.

Keywords: intraluminal; endovascular light delivery; high-energy laser; photoacoustic imaging; side-fire fibers; laser micromachining

1. Introduction

Photoacoustic (PA) imaging combines optical contrast from pulsed laser illumination of tissue with the imaging field of view of ultrasound. The PA signal arises from wavelength-specific absorption of typically near-infrared wavelength, nanosecond (ns) laser pulses by chromophores in the tissue, including hemoglobin, lipids, melanin, and exogenous contrast agents. With multi-wavelength illumination, PA imaging is capable of differentiating chromophores with distinct absorption spectra, so it can provide functional information such as total hemoglobin concentration, oxygen saturation [1–3], and molecular imaging [4–7], mapped to the anatomical structure shown by conventional B-mode ultrasound. The pressure generated by the absorption of light is given by

$$p_0(r) = \Gamma \mu_a(r) \Phi(r)$$

where the Grüneisen parameter (Γ) relates to the efficiency of converting absorbed optical energy into pressure, $\mu_a(r)$ is the optical absorption coefficient of the chromophore, and $\Phi(r)$ is the local fluence (energy per unit area). As the light travels through the tissue, it is scattered and absorbed, resulting in a gradual reduction in the fluence with distance from the light source; hence, the PA signal is limited by the fluence reaching it. This can degrade the PA imaging performance when large-volume imaging is required or if the imaging target is situated deep in the anatomy [8–10].

PA imaging systems for non-invasive imaging often use optical fiber bundles flanking the ultrasound transducer to illuminate tissue over the ultrasound imaging plane with high peak power laser pulses, typically in MWs. However, the fiber bundle adds to the bulk of the ultrasound probe [1,11]. The PA field of view can match the ultrasound imaging field if the fiber bundles deliver sufficient fluence at depth while being distributed along the length of the US probe. However, for applications that have limited access, such as image-guided intervention, intraluminal, or endovascular imaging, both the ultrasound probe and the light delivery device must be compact. Single optical fibers, with forward illumination from the end or side-fire with a prism, mirror, or bevel at the tip have been developed to investigate these applications [9,12–27]. Small-diameter fibers (e.g., <5 mm) can also enable through-transmission illumination aligned with the ultrasound image plane, for example, intraluminal light delivery separated from the US probe [28–30].

While compact side fire devices have yielded good PA imaging depth [28,29,31], the imaging field of view is restricted by the small spot size and weakly diverging beam, requiring the optical fiber to be mechanically swept to illuminate the full region of interest. This may significantly increase the imaging time and make it difficult to use in a clinical setting. Further, the laser pulse energy may have to be restricted so that the fluence at the small aperture is below safety limits to avoid localized tissue heating or damage. Therefore, a compact light delivery device with an extended illumination aperture is needed.

Radial fiber diffusers deliver light circumferentially along a section of the fiber [17,27,32]. Such designs can have an extended aperture and provide large illumination volumes. However, the cylindrical beam is not an efficient use of the laser energy for PA imaging using a finite ultrasound imaging plane. Furthermore, diffusers based on light scattering centers in the fiber core, such as those used for photodynamic therapy applications [32,33] are not able to withstand high peak powers. Therefore, we propose a compact side-fire fiber with an elongated illumination aperture that can carry high peak power (MW) ns pulses. Elongated side-fire illuminators have been developed by creating multiple notches along the fiber to reflect the light [34], by depositing materials with different refractive indices on the core [35], by laser-ablating micro-holes the generate multiple side-fire beams [36,37], or by adding a conical mirror to a cylindrical diffuser [38]. However, high power laser pulse transmission has not been demonstrated with the first three, while the diameter of the latter [38] is 8 mm. The challenge of generating uniform side-fire illumination from a compact (e.g., <3 mm diameter) device capable of carrying high pulse power remains.

In this paper, we present an elongated side-fire fiber comprising rectangular apertures, which were laser micromachined on one side of an optical fiber. By ablating into the core, light from the low order propagation modes can be redirected out of the machined window. UV laser ablation was used to create a roughened window on the side of a 1.5 mm diameter optical fibers. The fibers were coated with silver to redirect light out of the ablated region. Fibers with windows up to 30 mm long were developed and characterized. We measured the beam profile and the side-firing efficiency using a low-power CW laser (660 nm). The fabricated fibers were coupled to a 100 MW/cm^2 beam from a Q-switched PA imaging laser (680–970 nm), and the effect of repeated transmission of the laser pulses on the emission characteristics of the window was evaluated. PA imaging with a fabricated fiber was demonstrated with ink tubes in a tissue phantom and then by scanning a rat ear in vivo.

2. Materials and Methods

2.1. Side-Fire Fiber Design

The forward light propagation in multi-mode optical fibers can be altered by modifying the refractive index or the physical structure of the fiber core or its surface through methods such as chemical etching, mechanical abrasion, UV laser ablation, and femtosecond laser writing [17,32,39,40]. These modifications cause scattering of optical energy and redirection of some light propagation modes, allowing photons to escape the core and be emitted from the fiber's side, such as in circumferential diffusers. In our work, we used UV laser ablation to create corrugated windows along the side of single core 1.5 mm diameter multi-mode optical fibers (FT1500UMT, Thorlabs, Inc., Newton, NJ, USA), which have a damage threshold of 1 GW cm^{-2}. We made side-emitting fibers with 10 mm (n = 2), 20 mm (n = 2), and 30 mm (n = 2) long windows. Three successive ablations were performed to enhance the side emission by creating deeper structures in the fiber, enabling more transmission modes to interact with the window. The ablated window was positioned 5 mm from the fiber's distal end, leaving an intact segment for quantifying emission properties, including the side-to-front fire ratio, as described below. The goal of these modifications is to disrupt the total internal reflection at the modified core surface, causing a constant redistribution of the modes traveling in the core.

Reflective silver coatings were used to further redirect modes that do not satisfy the optical fibers' NA and would not be emitted at the machined window. Silver films are highly reflective across the visible and infrared spectrum down to 400 nm and minimally absorbed in the visible range [41]. While optical properties depend on film thickness, stability in transmittance and reflectance is achieved once a minimum thickness of 12–18 nm is reached [42].

2.2. Side-Fire Fiber Fabrication

The optical fiber's buffer coating was stripped using a fiber stripping tool, keeping the cladding intact, and the fiber was secured on a glass slide. To address the challenge of focusing the ablation laser on the fiber's curved surface, one side was flattened using a dicing saw (DAD3240, DISCO Corporation, Tokyo, Japan). A 200 µm thick, 500-grit diamond blade was used to make 150-µm-deep cuts perpendicular to the fiber axis, spaced 100 µm apart, creating overlapping passes that flattened the surface by removing the cladding and exposing the core (Figure 1a,b). Additionally, a 45° bevel was made at the distal end with the same blade to reflect light back into the fiber core and to reduce forward emissions.

The fiber was then secured on a glass slide and mounted on a four-axis servomotors-operated stage. The diced surface was patterned using an ArF excimer laser (wavelength = 193 nm; IX-255, IPG Photonics, Oxford, MA, USA), emitting 10 ns pulses at a 100 Hz repetition rate. We focused a 90 µm square laser beam spot onto the fiber, setting the ablation energy at 6.42 J/cm^2. The laser focus was set to the mean depth of the ablated surface, which was found by adjusting the focus of the inspection camera of the laser micromachining system at three positions along the window. As the four-axis stage moved the fiber along the x-axis in 3 µm steps, each pulse overlapping the previous one, continuous ablation lines were created along the fiber's length. We patterned seven parallel lines across the fiber's width (y-axis) to achieve a 0.8 mm window width (Figure 1c). This process was repeated twice to ablate deeper into the fiber (Figure 1d). For each successive ablation, the width of the cut (along the y-axis) was reduced by 180 µm, giving the ablated region a u-shape as shown in Figure 1e.

After laser ablation, we cleaned the fibers with isopropyl alcohol, masked the window area with masking tape, and wrapped the proximal end of the fiber in aluminum foil, leaving only the distal end exposed for silver coating. A 3000 Å thick layer of silver was deposited using a magnetron sputtering system (Orion3, AJA International, Scituate, MA, USA). Fibers, between 12 cm and 19 cm long to fit within the sputtering system, were placed horizontally on the carrier plate, and sputtering was performed twice to coat the entire fiber circumference. After sputtering, the tape covering the window was removed, and the distal end of the fiber, including the ablated window, was encapsulated in a clear fused quartz tubing (Technical Glass Products Inc., Painesville, OH, USA) with 2.2 mm (inner diameter) × 3 mm (outer diameter) dimensions. The final outer diameter of the enclosed fiber was 3 mm.

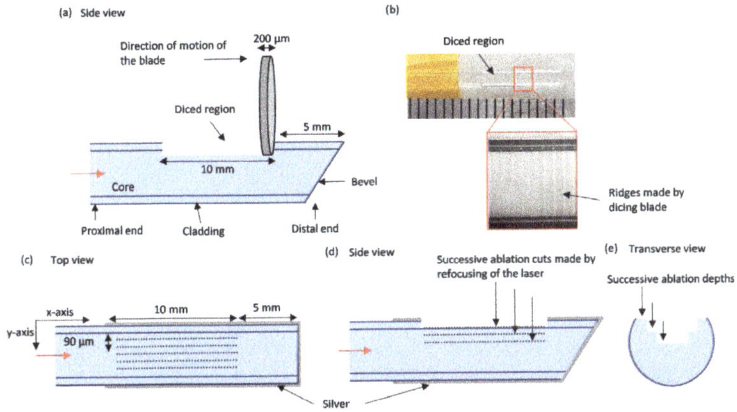

Figure 1. The micromachined side-fire optical fiber. (**a**) Side-view sketch showing the flattened region and beveled tip machined with a dicing saw. (**b**) A fiber before silver coating showing the surface ridges made by the dicing blade. (**c**) The top view sketch illustrates the laser ablation pattern along the length of the fiber (dotted lines). (**d**) A side view sketch illustrates the refocusing of the micromachining laser focus on ablating successive layers deeper into the fiber. (**e**) Transverse view sketch illustrating the successive ablations.

2.3. Fiber Emission Characterization

Fiber characterization involved measuring the beam profile in the longitudinal and azimuth directions. Additionally, the percentage of light emitted from the side of the fiber was measured relative to the total light output from the fiber to estimate the proportion diverted through the side-fire window. The percentage of side-emitted light was also compared to that of a naïve optical fiber of the same model to estimate the combined light lost in back-reflections and front leakage.

To measure the longitudinal profile, we scanned an isotropic detection fiber (IP85, Medlight, Ecublens , Switzerland, tip diameter 850 µm) over a 30 mm × 30 mm area in the x-longitudinal and y-transverse directions with a step size of 500 µm in both directions, recording the relative irradiance emitted by the area side-fire fiber regardless of the emission angle. The detection fiber, which was positioned at a distance of 5 mm from the side-fire fiber, was connected to a photodiode (DET10, Thorlabs, Newton, NJ, USA) to record the output voltage with an oscilloscope and was positioned at a distance of 5 mm from the side-fire fiber. The side-fire fiber was connected to a 660 nm laser source (S1FC660, Thorlabs, Newton, NJ, USA) via a coupling assembly consisting of a collimator (F810SMA-780, Thorlabs, Newton, NJ, USA) and a focusing lens (LA1560-B, Thorlabs, Newton, NJ, USA). This laser wavelength is close to the Q-switched PA imaging laser wavelengths, and the wavelength difference does not significantly change the refractive index, hence the scattering at the window can be expected to be similar for all the PA wavelengths. The Numerical Aperture (NA) of the coupling assembly was 0.10 and the fiber was spatially overfilled by placing it behind the lens focus. The coupling assembly was mounted on a computer-controlled 2D motor stage for automated data collection by synchronizing the oscilloscope output with a 2D motor stage. The "front" beam profile corresponds to measurements taken with the light-emitting window facing the isotropic detection fiber, while the "back" profile corresponds to measurements when the detection fiber faces the silver-coated region. Figure 2a illustrates this setup.

For azimuthal profile measurements, the side-fire fiber was fixed vertically on a mount, while the detection fiber was mounted on a rotation stage. The middle of the machined window was visually aligned to face towards the detection fiber. The distance between the two fibers was set at 1 mm, and data points were acquired at 5° intervals while manually rotating the detection fiber around the stationary side-fire fiber. Azimuth measurements were taken at the proximal, mid, and distal regions of the side-fire window, as shown in Figure 2b. The beam spread in the azimuth direction was calculated by the difference between the maximum and minimum angle at 50% of the peak intensity.

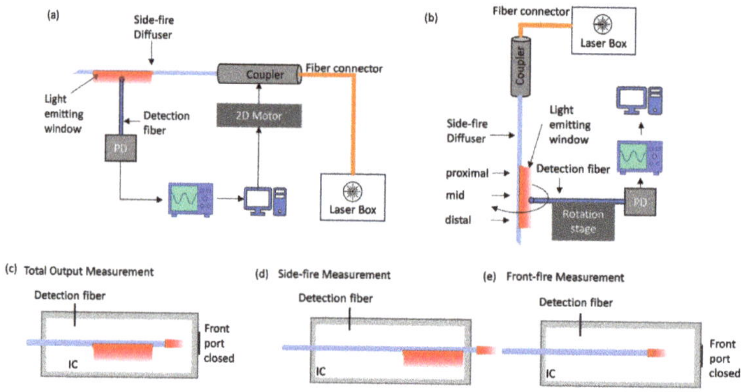

Figure 2. Experiment setup for fiber characterization. (**a**) Two-dimensional beam profile measurement is shown here for the "front" profile. (**b**) Azimuthal profile measurement. Schematic diagram of the experiment setup showing the position of the optical fiber inside an integrating sphere for measuring: (**c**) Total output I_t (**d**) Side fire I_s. (**e**) Front fire I_f from a non-modified front firing optical fiber.

The side-fire ratio is the ratio of the light emitted from the side of the fiber, I_s, relative to the total light exiting fiber I_t, and it evaluates the proportion of the total emitted light that is coming from the machined window. This is critical to assess how much light is lost in forward propagation. The side-fire efficiency is the side fire emission I_s relative to the light emitted from a multi-mode straight-cut front-firing fiber of the same model (I_f). This measurement evaluates the proportion of light diverted to the side-fire window compared to the total light in the fiber. This is useful for assessing the amount of light lost in forward propagation and back reflections from the bevel. These measurements were performed using an integrating cylinder (IC) manufactured from ultra-high-density polyurethane (Gigahertz-Optik, Munich, Germany). A fraction of the light inside the IC was recorded via a detector fiber (FT600UMT, Thorlabs, Newton, NJ, USA) inserted via a small port hole. The measurement configurations for I_t, I_s, I_f are illustrated in Figure 2c–e. For side-fire measurement, the fiber extends through the IC such that its distal end passes through the front hole, otherwise the front hole is sealed with reflective tape. An average of 10 readings was acquired for each prototype characterized and each measurement configuration.

2.4. Coupling PA Laser into Fiber

A Q-switched Nd:YAG OPO tunable laser (LAZR, FUJIFILM VisualSonics, Inc., Toronto, ON, Canada) was used as the PA laser source. The laser system produces 40 mJ pulses with a duration of 5–10 ns in the near-infrared range (680–970 nm) at 20 Hz repetition rate. For multispectral photoacoustic (PA) imaging using the Vevo 3100 system (FUJIFILM VisualSonics, Inc., Toronto, ON, Canada), up to 550 frames can be collected during one acquisition, with 4 pulses required for each frame. Thus, a total of 2200 pulses can be fired during one PA acquisition. To demonstrate the fiber's mechanical resilience, it was tested by coupling more pulses than needed for one PA acquisition; thus, a total of 6000 shots at 18 mJ pulse energy (λ = 750 nm) were transmitted into the fiber. A custom coupling assembly was developed as described in our previous studies [28] and had an NA of 0.10, as used in the characterization setup. The light from the coupling assembly was directed into a 1-meter-long patch cable of the same specification as the fabricated side-fire fibers. The fabricated fibers were then butt coupled to this long fiber connected to the PA laser with an index matching gel (G608N3, Thorlabs, Inc., Newton, NJ, USA); we measured an energy loss of 10–16% due to the butt connection. The patch fiber was spatially overfilled by placing the coupling end slightly beyond the focal point of the focusing lens. The per-pulse energy emitted from the window region was measured using a power/energy meter (Maestro; Gentec, OR, USA; 5 cm diameter) before and after PA laser pulse transmission as part of standard measurement prior to imaging.

2.5. Phantom and In Vivo PA Imaging Using the Side-Fire Fiber

PA imaging of a phantom using the side-fire fiber was demonstrated by imaging absorbing targets embedded in chicken breast tissue. The targets were polyurethane tubes (0.015″ ID and 0.033″ OD) filled with India ink (DeSerres, Vancouver, BC, Canada). The fiber with the 20 mm long window was coupled to the tunable laser (LAZR, FUJIFILM VisualSonics, Inc., Toronto, ON, Canada). The energy output at the 750 nm wavelength used for imaging was 3–5 mJ, as measured from the window region. B-mode and PA-mode images were acquired with a Vevo 3100 ultrasound imaging system (FUJIFILM VisualSonics, Inc., Toronto, ON, Canada) and a 20 MHz linear array (MX250; max. image width 23 mm) positioned to image the tube cross-sections. To evaluate the illumination along its length, the optical fiber was positioned parallel to the image plane, i.e., perpendicular to the tubes. The fiber was then aligned perpendicular to the image plane (i.e., parallel to the target tubes) and positioned centrally approximately 10 mm from the tubes to evaluate the tissue illumination by the fiber in the azimuth direction. A motor stage was used to translate the transducer along its elevation direction with a step size of 150 µm and acquire a volumetric image of the tube targets.

For in vivo PA imaging, a rat ear was placed on an anechoic and optically transparent support that held the optical fiber such that the ear was within the large illumination area of a 20 mm window side-fire fiber. Anesthesia of the animal was initiated by 5% isoflurane in oxygen and maintained at 2% throughout the imaging. B-mode and PA-mode imaging was performed with the same laser and ultrasound acquisition setup as for the phantoms described above. The optical fiber was inserted into a channel 5 mm below the top surface of a block of plastisol ($50 \times 50 \times 50$ mm^3) [28,43] to separate visually the strongly reflecting fiber from the thin ear in the images. The transducer was held by a motor stage above the ear, with ultrasound gel coupling between the ear and the plastisol block, and up to 8 mm thick above the ear. A 3D scan of the ear was set up to acquire an image volume of 25 mm (image width) $\times$ 16 mm (image depth) $\times$ 10 mm (scan length) with a step size of 150 µm. Animal experiments were performed under the approved animal use protocol of Sunnybrook Research Institute. Image reconstruction and analysis for both phantom and in vivo imaging were performed using the Vevo LAB software (v5.6.1, FUJIFILM VisualSonics, Inc., Toronto, ON, Canada).

3. Results

3.1. Side-Fire Fiber Fabrication

Three variations of the side-fire fibers were fabricated, sporting 10 mm, 20 mm, and 30 mm long windows on the fiber's side. Figure 3a–c shows microscope images of the laser machined region on a fiber. The ablated surface of the fiber is visibly roughened, and a pit was observed along the central axis of the fiber (Figure 3b). Figure 3c shows the cross-sectional view of the fiber showing a u-shaped valley created as a result of three successive ablations. Figure 3d,e shows the side and top view of a 20 mm long window fiber after coating it with silver, encasing in a 3 mm (OD) glass tube. A red laser was coupled manually to the fiber, and the light exiting over the full length of the window was observed. As shown in Figure 3d, some light can be seen exiting from the silver coated bevel and sideways from the window region. More light was observed visually towards the proximal end of the window compared to the distal end (Figure 3e). The maximum ablation depth into the fiber, estimated with the laser machining system inspection camera, is reported in Table 1.

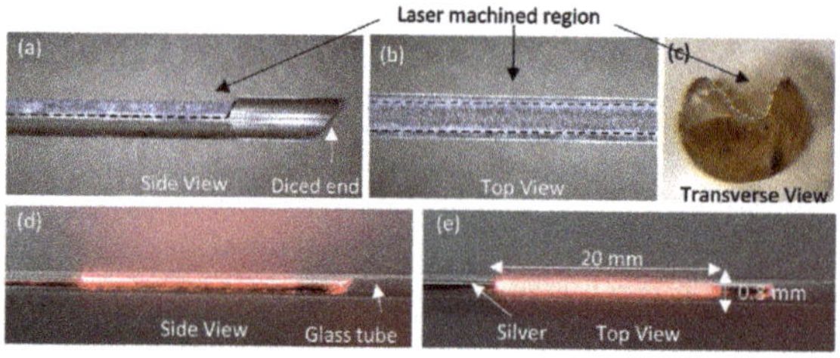

Figure 3. Laser machined side-fire fiber: (**a**) side view showing the machined window demarcated by black dashed lines in the 1.5 mm OD fiber and the 45° diced end, (**b**) top view showing the width of the cut indicated by black dashed lines, and (**c**) transverse view of the window cut along the width showing a valley created by the laser machining. (**d**) Side and (**e**) top view images of the fiber with silver coating, except in the ablated window, placed in a glass tube and with a red laser pointer coupled to it to show the light emitting from the machined window.

Table 1. Summary of beam characteristics for fabricated prototypes.

Prototype#	Est. Ablation Depth (μm)	2D Beam Size (mm)	Azimuth Angle (Degrees)	Side-Fire Ratio (I_s/I_t)	Side-Fire Efficiency (I_s/I_f)
A: 10 mm window	(i) 588 (ii) 500	(i) 12.29 × 6.89 (ii) 11.3 × 7.38	(i) 104 (ii) 112	(i) 0.62 (ii) 0.59	(i) 20% (ii) 18%
B: 20 mm window	(i) 780 (ii) 750	(i) 20.6 × 7.08 (ii) 20.1 × 7.37	(i) 109 (ii) 109	(i) 0.69 (ii) 0.66	(i) 40% (ii) 37%
C: 30 mm window	(i) 623 (ii) 631	(i) 30.07 × 6.0 (ii) 30.79 × 5.5	(i) 81 (ii) 81	(i) 0.38 (ii) 0.44	(i) 29% (ii) 31%

3.2. Side-Fire Fiber Characterization

The beam profiles of fibers with window lengths of 10 mm, 20 mm, and 30 mm are shown in Figure 4, and their characteristics are summarized in Table 1. The distal and proximal ends of the windows, relative to the light travel direction in the core, are indicated by yellow arrows; intensities are normalized to the peak in each dataset. Figure 4b,e, and k show the plots of the normalized intensity along the longitudinal and transverse directions relative to the fiber axis, indicated by the black dotted lines in the 2D beam plots. The maximum intensity in the beam profile was observed near the center of the ablated aperture for the 10 mm window prototypes. The maximum intensity was about 5 mm from the proximal window end for the 20-mm window prototypes, with the signal intensity tapering slowly to about 4 mm from the distal end then dropping quickly (Figure 4e). The beam profile for the 30 mm prototype has peaks within 4 mm of each end of the window and a 30% intensity drop in the middle; the highest intensity was measured near the distal edge. The pair of 10 mm window prototypes had substantially similar beam profiles, as did the 30 mm window pair. The pair of 20 mm window prototypes had differing slopes in the plateau over the machined window, with the beam profile of the second prototype.dropping to 70% of the peak before the steep drop at the distal edge of the window.

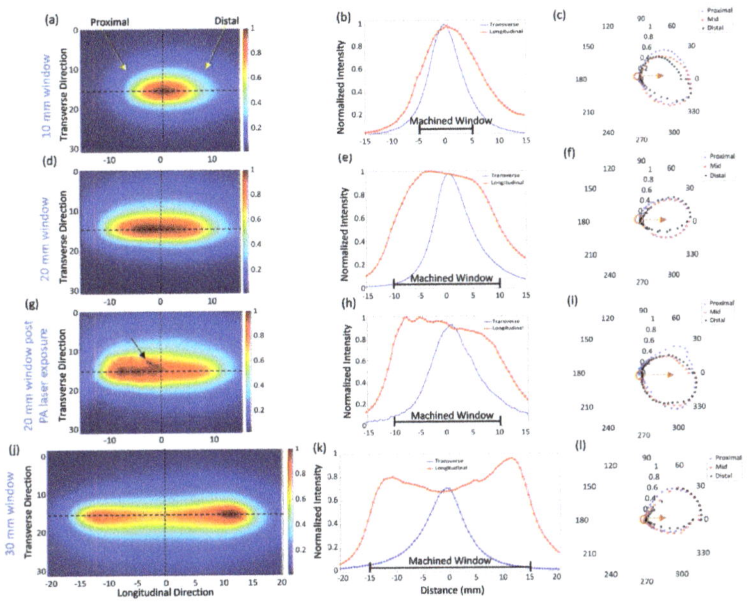

Figure 4. Normalized measured beam profiles for side-fire fibers: (**left**) two-dimensional beam profile, (**mid**) longitudinal and transverse intensity plots along black dotted lines in beam profile, and (**right**) azimuthal beam profile at proximal, central, and distal positions within the ablated aperture of fibers with (**a–c**) a 10 mm window, (**d–f**) a 20 mm window, (**g–i**) a 20 mm window fiber after receiving 6000 PA laser pulses, and (**j–l**) a 30 mm window.

The azimuth beam profile at the distal, middle, and proximal window positions for the three window lengths are shown in Figure 4c,f,l. The intensity values are normalized to the maxima at each longitudinal position. The average azimuth angle span was 108°, 109°, and 82° for the 10 mm, 20 mm, and 30 mm window prototypes, respectively. The prototypes achieved a side fire ratio ranging from 38% for a 30 mm window and up to 69% for a 20 mm window. The side-fire efficiency, compared to a naïve fiber with emission only at the flat distal end, was a maximum of 40% for one of the 20 mm prototypes and as low as 18% for a 10 mm window.

3.3. Beam Profile Before and After PA Laser Transmission

Figure 4g shows the 2D beam profile of the 20 mm long window prototype fiber after launching 6000, 18 mJ (λ = 750 nm) pulses into the fiber. A clear difference was observed in the 2D beam profiles pre- and post-PA laser coupling, as seen in Figure 4d,e and Figure 4g,h, respectively. The beam width (FWHM) after the PA laser coupling fiber expanded by 2.8 mm in the longitudinal direction and 1.63 mm in the transverse direction. After PA laser exposure, the location of the maximum intensity shifted proximally, as can be seen from the 2D beam plot graphs in Figure 4e,h. A small diagonal crack, indicated by an arrow in Figure 4i, appeared on the window after PA laser transmission and was also observed with the naked eye as a bright spot when the 660 nm test laser was coupled to the fiber. The azimuth plots indicated an expansion of the beam divergence by 10° at the proximal position. The per-pulse energy from the window region, as measured by the energy meter, decreased from 8 mJ at the beginning of the session to 5 mJ after transmitting the 6000 pulses.

3.4. Phantom and In Vivo PA Imaging Using the Side-Fire Fiber

Figure 5 shows B-mode and PA-mode images of the tissue phantom with embedded tubes filled with India ink. The tubes are imaged in cross-section and visible in the B-mode as bright features; as expected, both the top and bottom walls of the tube are visible as two echoes for each tube. When the optical fiber is perpendicular to the tubes and parallel to the transducer (Figure 5a), it is faintly visible in B-mode because it is slightly out of plane (Figure 5b). In the corresponding PA-mode image (Figure 5c), all targets above the optical fiber are detected, including the tubes furthest (9 mm) from the fiber. When the optical fiber is aligned parallel to the ink-filled tubes, nd perpendicular to the image plane (Figure 5d), it is visible as a bright feature in the B-mode image (Figure 5e). Tubes spanning 102° are visible in the PA-mode (Figure 5f), which matches the 109° azimuthal beam profile of the fiber (Figure 4f). The PA signals from the tube delineating the top and bottom of the ink can be seen in the inset in Figure 5f; this is expected when imaging contrast-filled tubes due to the N-shaped PA waveform [44]. Since the light source is below the tubes, the bottom surface of the ink yields a stronger signal. The volumetric scan has the fiber aligned approximately parallel to the tubes (Figure 5g), and demonstrates PA signals are detected from these tubes along the 10 mm scan (Figure 5h). The top and bottom of the ink-filled tubes is also detected, appearing as a doublet for tubes that received sufficient energy to generate a strong PA signal at the top surface.

The PA-mode image of the rat ear overlaid on the B-mode image is shown in Figure 6. The PA image is acquired with laser per pulse energy of 3 mJ measured at 750 nm. The optical fiber is visualized in the B-mode as a strong reflector at the bottom of the image; it is tilted due to the deformation of the plastisol block in which the fiber is inserted. The shape of the ear is visible in the B-mode, with anechoic gel and plastisol above and below the ear, respectively. No signal was observed from plastisol as it was devoid of any optical or acoustic scatterers. Due to hemoglobin absorption, blood vessels appear as bright regions in the PA-mode, which are, as expected, confined within the ear. PA signals from the blood vessels were observed over 19 mm apart in the central cross-section of the ear, matching the site of the machined window.

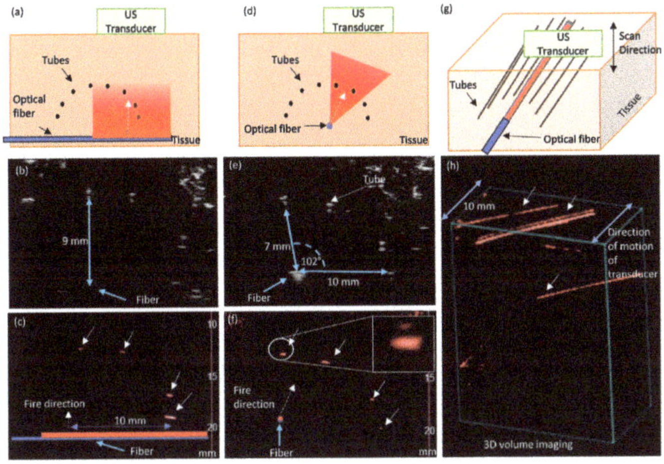

Figure 5. PA imaging of polyurethane tubes filled with India ink embedded in chicken breast tissue. (**a**) Schematic diagram, (**b**) B-mode US image and (**c**) PA-mode image of the tube cross-sections with optical fiber perpendicular to the tubes (parallel to image plane), demonstrating the longitudinal illumination field, and (**d**) Schematic diagram, (**e**) B-mode US image, and (**f**) PA-mode image with the optical fiber parallel to the tubes (perpendicular to the plane) showing the angular illumination field; inset in (**f**) is zoomed-in view of PA signal from top and bottom surfaces of ink in the tube. Multi-plane imaging with 10 mm scan (**g**) schematic and (**h**) volume rendered PA-mode image showing both angular and longitudinal illumination field. Solid white arrows indicate tubes visible in PA-mode images; dotted white arrows indicate approximate orientation of the machined window and direction of the emitted laser beam.

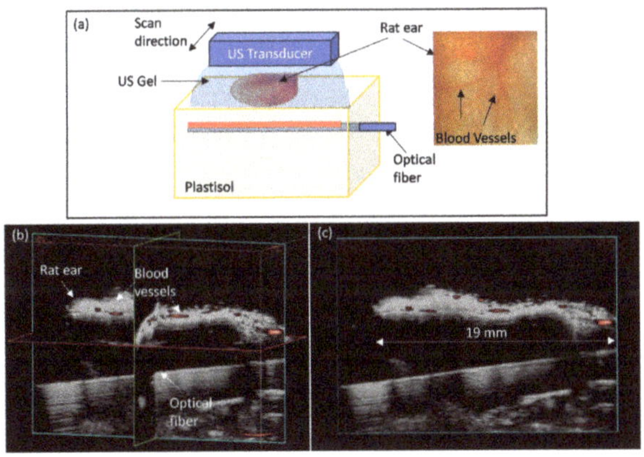

Figure 6. (**a**) Setup for imaging rat ear in vivo, with optical fiber placed 5 mm below the surface of a plastisol block supporting the ear. Ultrasound gel couplant surrounds the ear. (**b**) Volume rendering of 3D B-mode and PA-mode scan of a rat ear. Pseudo-color PA-mode image (red) is overlaid on the grayscale B-mode image, and three perpendicular cross-sections are shown, intersecting near the middle of the ear. (**c**) The transverse view shows the same imaging plane as in (**b**). The image dimensions are 23 mm (lateral) × 16 mm (depth), and the length of the imaged ear is 19 mm.

4. Discussion and Conclusions

Intraluminal or interstitial photoacoustic imaging of tissues and organs requires the delivery of high peak power (MW) over the field of view of the ultrasound probe while minimizing the cross-sectional dimensions of the light delivery device for ease of access. As PA imaging requires the delivery of a large number of sub-micro-second pulses, the light delivery system must be resilient to high transient optical power densities. At the same time, the local fluence near the illumination device must be kept below safety limits. Since off-the-shelf cylindrical diffusers are not suitable due to their low power damage threshold, an elongated illuminator that can carry high peak power and scatter optical energy from the core is needed. Furthermore, the light emission should be directional and confined within the US imaging plane. In order to obtain light out from the side of the fiber, the modes in the core need to be disturbed. While structures or modifications on the fiber surface can redistribute energy from higher modes, alteration of the material close to the fiber center is required to redistribute the energy from lower modes and redirect the majority of the energy out of the fiber.

4.1. Observations and Insights

In this work, single-core, multimode fused silica fibers with high damage threshold (1 GW cm^{-2}) were utilized to develop side-fire elongated illuminator using ArF excimer laser machining. The developed device was compact with a 1.5 mm OD or a 3 mm OD with a glass tube housing, had extended side-fire illumination demonstrated up to 30 mm in length, and was capable of carrying 18 mJ per-pulse energy or 100 MW cm^{-2} power density. As shown in Figure 4, for all the prototypes, the −6 dB beam size in the longitudinal direction matched the machined window length. The fibers with 10 mm long windows showed a central peak in intensity, whereas the longer windows had steep rises in intensity within 5 mm of the window ends and tilted plateaus or dips (>10% drop within 10 mm) over the middle of the window. The peaks are likely due to scattering of the higher modes on interaction with the machined material, whether at the proximal end of the window or at the distal window end after reflection of forward propagating modes at the bevel. The decreasing intensity with distance from the peaks is likely due to the reducing energy in the higher order modes after interaction with the machined surface and therefore decreasing scattering and escaping light as a function of position. As expected, the silver deposition led to negligible light intensity detected opposite the window. The lateral window dimension was the same for all prototypes, corresponding with reasonably consistent transverse beam spread (80–110°), though the effect of widow width was not investigated here. It should be noted that the optical coupling used in the current setup has an NA of 0.10 which is less than the NA of the fiber of 0.39, which leads to mode under-filling. While mode filling was not investigated here, if we overfill the modes, we can expect more light interacting with the machined window region. This may result in more light escaping out from the side and less light propagating unperturbed and improved side-fire efficiency.

The ablation depth in the fiber core is a crucial factor in the beam profile and side-fire efficiency, with deeper ablation corresponding with higher side-fire efficiency irrespective of window length (Table 1). For example, the 30 mm window was expected to allow more light to exit the fiber but had lower efficiency than the 20 mm fibers, corresponding with shallower ablation depths. Hence, greater ablation depths and the consequent disruption closer to the center of the core is a larger determinant of side-fire efficiency than the window size. The u-shaped ablation trench, created by three successive machining layers, was selected to retain as much of the fiber material for structural integrity while ablating into the core in order to scatter the lower modes. The cross-sectional geometry, such as the opening angle of the fabricated prototypes, is expected to affect the emitted azimuthal beam profile but was not investigated in this study. All fibers were structurally intact after machining but were fragile and could easily break during testing. In some early prototypes, the trench was encapsulated in an optically transparent epoxy to strengthen the fiber, but charring was observed after coupling to the PA imaging laser, caused by air bubbles not fully removed during the epoxy casting process. However, once they are encased in a glass tube, they maintain mechanical integrity through all the characterization and imaging experiments.

Compared to our earlier prototype [45], this new design includes a bevel at the distal end of the fiber. This bevel, which redirects the reflected light into the fiber, allowing for further interaction with the machined region, resulted in a 30% increase in the side-fire ratio compared to the 10 mm window prototype with a flat-cut end. When the device has both a beveled tip and a conformal silver coating to reflect light from the fiber surface, up to 40% of the light emitted from a flat ended fiber is redirected outside of the machined fiber. Some of the light is lost in back reflections from the beveled tip that also does not interact with the machined window, such that the side-fire efficiency, relative to a flat-cut fiber, is lower than the ratio of side-to-total light emitted from the tip.

We also observed some light transmission at the bevel end, which increased after 6000 PA laser pulses were transmitted into the fiber. Upon visual inspection, the silver covering the beveled tip was completely eroded after a set of PA laser transmits was used in the test and imaging demonstration of the fibers. This emission from the bevel results in a decrease in the side-fire efficiency of the device. This could be remedied by replacing the silver coating by a 45° rod reflector at the terminating end of the fiber [29].

The optical fiber developed in this study is designed for Acoustic Resolution–Photoacoustic Imaging (AR-PAI) applications, where image resolution is determined by the ultrasound transducer and system and not the light beam dimension [28,46].While light intensity drops by up to 30% along the window (for the 30 mm windows), tissue scattering is expected to significantly homogenize the beam, reducing the effect of beam intensity variation on PA signal uniformity. As with any light source, fluence is highest near the emitting surface and decreases as light travels through tissue due to absorption and scattering, leading to stronger PA signals near the firing window, as observed in Figure 5c. However, apart from this expected PA signal intensity decrease with distance, no significant effect on the spatial registration of PA signal due to window size is anticipated.

4.2. Challenges and Limitations

We used an excimer laser to fabricate the side-fire fiber prototypes because it is a versatile and programmable micromachining tool available in our lab [47]. This allowed us to investigate the side-fire fiber concept. This laser was able to ablate the core materials and generate a roughened surface. However, the cylindrical shape of the optical fiber makes it challenging to focus the machining laser at the surface and causes insufficient and non-uniform ablation at the sides of the fiber. This challenge was overcome by dicing the optical fiber surface prior to laser machining. The dicing saw creates a planar surface with ridges, as shown in Figure 1b, so that the ArF radiation could be focused on the intended region of the fiber.

The focus of the excimer laser to remove the successive layers of glass was set using the inspection camera. This has limited accuracy because of the roughened surface of the ablated window, so the six characterized fibers have variable ablation depths with a mean and standard deviation of 645 ± 95 μm, consistent with an imprecise laser focal position relative to the surface height and the effect of it is reflected in a variation in the fiber side-fire efficiency (Table 1).

The nanosecond PA imaging laser pulses have the potential to effectively continue altering the fiber structurally. The long-duration ns pulse allows for the heat to diffuse in the fiber material, which may result in melting and fracturing [48] in the fibers. As an example, a diagonal crack (black arrow, Figure 4g) near the center of the window was observed following the 6000 pluses of the PA imaging laser. This may have been a result of the elongation of an existing crack or the formation of a new crack as the ns laser pulses interact with the modified fiber structure. The beam profile was also altered, increasing by 12% and 18% in the longitudinal and transverse directions, respectively, and the peak intensity shifted proximally (Figure 4g–i). This can be expected as the proximal region of the window interacts with the highest power density, redirecting a portion of the energy out of the fiber, while more distal regions experience lower power densities. The azimuth emission profile was also affected most at the proximal window position (Figure 4i), increasing the beam spread by 10°. Overall, the light emitted from the window side decreased by 37% following the initial sequence of imaging laser pulses, and a drop in the energy was observed after each phantom and in vivo imaging session. Repeated use of the fibers may change the beam characteristics and reduce side-fire efficiency significantly, which may limit their use to single use only.

4.3. Optimization and Future Directions

While it is possible to fabricate the elongated side-fire fibers with an excimer machining laser, and process optimization can improve some of the limitations discussed above, the method leads to fragile fibers that degrade when transmitting ns PA imaging laser pulses. Micromachining processes like femto-second (fs) laser would be better suited for altering the optical fiber core because of the short pulse duration and their ability to perform subsurface ablations without affecting the fiber surface [49,50]. The pulse duration of fs lasers is shorter than the thermalization time scale, resulting in smaller temperature diffusion length and a more precise ablation [51]. This is expected to reduce fractures in the fiber and improve strength and robustness post fabrication. Another advantage of using fs laser ablation is that patterns of ablated points can be created across the full diameter of the fiber so that light in all the modes in the fiber can be redistributed. This has the potential to significantly improve the side-fire efficiency of the fibers if an appropriate pattern and density of ablated spots can be created.

We demonstrated PA imaging of India ink-filled tubes embedded in chicken breast tissue and vasculature in a rat ear in vivo using the side-fire fiber prototypes. Targets were visible in the PA-mode over an angle of 102° and for the full width of the US transducer field of view. The initial device design considered optical fibers positioned within and parallel to the ultrasound probe imaging plane so as to benefit from the illumination from the full longitudinal length of the side-fire window. Yet, the 3D imaging of the tissue and ink phantom (Figure 5f) demonstrates the feasibility of PA imaging of a volume using these extended illumination side-fire optical fibers aligned perpendicular to the image plane. In this setup, the optical fiber can be stationary, illuminating a constant segment of tissue, and the ultrasound probe is translated to cover its entire length. In the end, PA imaging of the phantom with India ink tubes and rat ear in vivo was performed at 5 mJ and 3 mJ, respectively, which is quite low when compared to the 35 mJ delivered by the commercial PA system used in our lab. Yet, we were able to visualize tubes in the chicken at 10 mm depth and blood vessels in the rat ear, demonstrating a potential for deeper PA imaging with this elongated side-fire fiber design.

4.4. Conclusions

In this work, we introduced a compact (OD = 1.5 mm without casing, 3 mm with casing) side-fire fiber capable of carrying 100 MW/cm^2 peak power density for at least 6000 pulses. Fibers of different illumination lengths were fabricated and characterized. We showed the improvement in side-fire coupling of the light in fibers for an increased illumination window size. The fiber was coupled to the PA imaging laser, and variations in beam profile due to high-energy PA laser pulse transmission were evaluated. The fibers were shown to be structurally stable over multiple imaging sessions. We demonstrated the

PA imaging capability of the fiber by imaging India ink targets embedded in chicken breast and the vasculature in a rat ear. The compact device developed here has the potential to be used for intraluminal optical illumination for PA imaging in applications in the brain, prostate, esophagus, colon, and blood vessels or for minimally invasive PA imaging.

Author Contributions: Conceptualization, N.S., C.-F.R., L.L. and C.E.M.D.; funding acquisition, C.E.M.D.; methodology, N.S., C.-F.R., E.C. and L.L.; writing—original draft, N.S.; writing—review and editing, N.S., C.-F.R., E.C., L.L. and C.E.M.D.; supervision, C.E.M.D.; project administration, C.E.M.D. All authors have read and agreed to the published version of the manuscript.

Funding: This work was supported by the Terry Fox Research Institute New Frontiers Program Project Grant-1075 and Natural Sciences and Engineering Research Council of Canada (NSERC) through the Discover Grant RGPIN- 2017-06834.

Institutional Review Board Statement: The animal study protocol was approved by the Institutional Review Board of Sunnybrook Research Institute (protocol code 21739, approved November 2023, renewed November 2024).

Informed Consent Statement: Not applicable.

Data Availability Statement: The original contributions presented in this study are included in the article. Further inquiries can be directed to the corresponding author.

Acknowledgments: The authors acknowledge Aaron Boyes for helping with the fiber fabrication, Yohannes Soenjaya for helping with animal handling for rat imaging, and Robin Marjoribanks and his team for discussions on light–matter interaction and laser ablation physics.

Conflicts of Interest: The authors declare no conflicts of interest. The funders had no role in the design of the study; in the collection, analyses, or interpretation of data; in the writing of the manuscript; or in the decision to publish the results.

References

1. Needles, A.; Heinmiller, A.; Sun, J.; Theodoropoulos, C.; Bates, D.; Hirson, D.; Yin, M.; Foster, F. Development and initial application of a fully integrated photoacoustic micro-ultrasound system. *IEEE Trans. Ultrason. Ferroelectr. Freq. Control* **2013**, *60*, 888–897 . [CrossRef] [PubMed]
2. Liu, C.; Liang, Y.; Wang, L. Single-shot photoacoustic microscopy of hemoglobin concentration, oxygen saturation, and blood flow in sub-microseconds. *Photoacoustics* **2020**, *17*, 100156. [CrossRef] [PubMed]
3. Lin, X.; Yang, C.; Lv, Y.; Zhang, B.; Kan, J.; Li, H.; Tao, J.; Yang, C.; Li, X.; Liu, Y. Preclinical multi-physiologic monitoring of immediate-early responses to diverse treatment strategies in breast cancer by optoacoustic imaging. *J. Biophotonics* **2024**, *17*, e202300457. [CrossRef]
4. Emelianov, S.Y.; Li, P.C.; O'Donnell, M. Photoacoustics for molecular imaging and therapy. *Phys. Today* **2009**, *62*, 34–39.
5. Mallidi, S.; Luke, G.P.; Emelianov, S. Photoacoustic imaging in cancer detection, diagnosis, and treatment guidance. *Trends Biotechnol.* **2011**, *29*, 213–221. [CrossRef]
6. Luke, G.P.; Yeager, D.; Emelianov, S.Y. Biomedical Applications of Photoacoustic Imaging with Exogenous Contrast Agents. *Ann. Biomed. Eng.* **2012**, *40*, 422–437. [CrossRef]
7. Weber, J.; Beard, P.C.; Bohndiek, S.E. Contrast agents for molecular photoacoustic imaging. *Nat. Methods* **2016**, *13*, 639–650. [CrossRef]
8. Singh, M.K.A.; Parameshwarappa, V.; Hendriksen, E.; Steenbergen, W.; Manohar, S. Photoacoustic-guided focused ultrasound for accurate visualization of brachytherapy seeds with the photoacoustic needle. *J. Biomed. Opt.* **2016**, *21*, 120501. [CrossRef]
9. Xia, W.; Nikitichev, D.I.; Mari, J.M.; West, S.J.; Pratt, R.; David, A.L.; Ourselin, S.; Beard, P.C.; Desjardins, A.E. Performance characteristics of an interventional multispectral photoacoustic imaging system for guiding minimally invasive procedures. *J. Biomed. Opt.* **2015**, *20*, 86005. [CrossRef]
10. Benavides-Lara, J.; Manwar, R.; McGuire, L.S.; Islam, M.T.; Shoo, A.; Charbel, F.T.; Menchaca, M.G.; Siegel, A.P.; Pillers, D.A.M.; Gelovani, J.G.; et al. Transfontanelle photoacoustic imaging of intraventricular brain hemorrhages in live sheep. *Photoacoustics* **2023**, *33*, 100549. [CrossRef]
11. Daoudi, K.; van den Berg, P.J.; Rabot, O.; Kohl, A.; Tisserand, S.; Brands, P.; Steenbergen, W. Handheld probe for portable high frame photoacoustic/ultrasound imaging system. In Proceedings of the Photons Plus Ultrasound: Imaging and Sensing,

San Francisco, CA, USA, 3–5 February 2013; Oraevsky, A.A., Wang, L.V., Eds.; International Society for Optics and Photonics; SPIE: Bellingham, WA, USA, 2013; Volume 8581, p. 85812I. [CrossRef]
12. Jansen, K.; Wu, M.; van der Steen, A.F.; van Soest, G. Photoacoustic imaging of human coronary atherosclerosis in two spectral bands. *Photoacoustics* **2013**, *2*, 12–20. [CrossRef] [PubMed]
13. Wu, M.; FW van der Steen, A.; Regar, E.; van Soest, G. Emerging Technology Update Intravascular Photoacoustic Imaging of Vulnerable Atherosclerotic Plaque. *Interv. Cardiol.* **2016**, *11*, 120–123. [CrossRef] [PubMed]
14. Schneider, M.K.; Wang, J.; Kare, A.; Adkar, S.S.; Salmi, D.; Bell, C.F.; Alsaigh, T.; Wagh, D.; Coller, J.; Mayer, A.; et al. Combined near infrared photoacoustic imaging and ultrasound detects vulnerable atherosclerotic plaque. *Biomaterials* **2023**, *302*, 122314. [CrossRef]
15. He, H.; Englert, L.; Ntziachristos, V. Optoacoustic Endoscopy of the Gastrointestinal Tract. *ACS Photonics* **2023**, *10*, 559–570. [CrossRef]
16. Zhang, K.; Qiu, J.; Yang, F.; Wang, J.; Zhao, X.; Wei, Z.; Ge, N.; Chen, Y.; Sun, S. Photoacoustic endoscopy and EUS: Shaking the future of multimodal endoscopy. *Endosc. Ultrasound* **2022**, *11*, 1. [CrossRef]
17. Bungart, B.; Cao, Y.; Yang-Tran, T.; Gorsky, S.; Lan, L.U.; Roblyer, D.; Koch, M.O.; Cheng, L.; Masterson, T.; Cheng, J.X. Cylindrical illumination with angular coupling for whole-prostate photoacoustic tomography. *Biomed. Opt. Express* **2019**, *10*, 1405–1419. [CrossRef]
18. Lin, L.; Xia, J.; Wong, T.T.W.; Li, L.; Wang, L.V. In vivo deep brain imaging of rats using oral-cavity illuminated photoacoustic computed tomography. *J. Biomed. Opt.* **2015**, *20*, 016019. [CrossRef]
19. Lediju Bell, M.A.; Ostrowski, A.K.; Li, K.; Kazanzides, P.; Boctor, E.M. Localization of Transcranial Targets for Photoacoustic-Guided Endonasal Surgeries. *Photoacoustics* **2015**, *3*, 78–87. [CrossRef]
20. Zhang, P.; Li, L.; Lin, L.; Hu, P.; Shi, J.; He, Y.; Zhu, L.; Zhou, Y.; Wang, L.V. High-resolution deep functional imaging of the whole mouse brain by photoacoustic computed tomography in vivo. *J. Biophotonics* **2018**, *11*, e201700024. [CrossRef]
21. Graham, M.T.; Sharma, A.; Padovano, W.M.; Suresh, V.; Chiu, A.; Thon, S.M.; Tuffaha, S.; Bell, M.A.L. Optical absorption spectra and corresponding in vivo photoacoustic visualization of exposed peripheral nerves. *J. Biomed. Opt.* **2023**, *28*, 097001. [CrossRef]
22. Xia, W.; West, S.J.; Nikitichev, D.I.; Ourselin, S.; Beard, P.C.; Desjardins, A.E. Interventional multispectral photoacoustic imaging with a clinical linear array ultrasound probe for guiding nerve blocks. In Proceedings of the Photons Plus Ultrasound: Imaging and Sensing 2016, San Francisco, CA, USA, 14–17 February 2016; p. 97080C. [CrossRef]
23. Mari, J.M.; Xia, W.; West, S.J.; Desjardins, A.E. Interventional multispectral photoacoustic imaging with a clinical ultrasound probe for discriminating nerves and tendons: An ex vivo pilot study. *JBO* **2015**, *20*, 110503. [CrossRef] [PubMed]
24. Maneas, E.; Xia, W.; Singh, M.K.A.; Sato, N.; Agano, T.; Ourselin, S.; West, S.J.; David, A.L.; Vercauteren, T.; Desjardins, A.E. Human placental vasculature imaging using an LED-based photoacoustic/ultrasound imaging system. In Proceedings of the Photons Plus Ultrasound: Imaging and Sensing, San Francisco, CA, USA, 28 January–1 February 2018; SPIE: Bellingham, WA, USA, 2018; Volume 10494, pp. 69–73. [CrossRef]
25. Shubert, J.; Bell, M.A.L. Photoacoustic imaging of a human vertebra: Implications for guiding spinal fusion surgeries. *Phys. Med. Biol.* **2018**, *63*, 144001. [CrossRef] [PubMed]
26. Wiacek, A.; Wang, K.C.; Wu, H.; Bell, M.A.L. Dual-wavelength photoacoustic imaging for guidance of hysterectomy procedures. In Proceedings of the Advanced Biomedical and Clinical Diagnostic and Surgical Guidance Systems XVIII, San Francisco, CA, USA, 2–4 February 2020; Volume 11229, pp. 73–79. [CrossRef]
27. Li, M.; Lan, B.; Liu, W.; Xia, J.; Yao, J. Internal-illumination photoacoustic computed tomography. *J. Biomed. Opt.* **2018**, *23*, 030506. [CrossRef] [PubMed]
28. Singh, N.; Chérin, E.; Roa, C.F.; Soenjaya, Y.; Wodlinger, B.; Zheng, G.; Wilson, B.C.; Foster, F.S.; Demore, C.E. Adaptation of a Clinical High-Frequency Transrectal Ultrasound System for Prostate Photoacoustic Imaging: Implementation and Pre-clinical Demonstration. *Ultrasound Med. Biol.* **2024**, *50*, 457–466. [CrossRef]
29. Lediju Bell, M.A.; Guo, X.; Song, D.Y.; Boctor, E.M. Transurethral light delivery for prostate photoacoustic imaging. *J. Biomed. Opt.* **2015**, *20*, 36002. [CrossRef]
30. Lediju Bell, M.A.; Kuo, N.P.; Song, D.Y.; Kang, J.U.; Boctor, E.M. In vivo visualization of prostate brachytherapy seeds with photoacoustic imaging. *J. Biomed. Opt.* **2014**, *19*, 126011. [CrossRef]
31. Ai, M.; Youn, J.i.; Salcudean, S.; Rohling, R.; Abolmaesumi, P.; Tang, S. Photoacoustic tomography for imaging the prostate: A transurethral illumination probe design and application. *Biomed. Opt. Express* **2019**, *10*, 2589. [CrossRef]
32. Vesselov, L.; Whittington, W.; Lilge, L. Design and performance of thin cylindrical diffusers created in Ge-doped multimode optical fibers. *Appl. Opt.* **2005**, *44*, 2754–2758. [CrossRef]
33. Kostanski, L.; Pope, M.; Hrymak, A.; Gallant, M.; Whittington, W.; Vesselov, L. Development of Novel Tunable Light Scattering Coating Materials for Fiber Optic Diffusers in Photodynamic Cancer Therapy. *J. Appl. Polym. Sci.* **2009**, *112*, 1516–1523. [CrossRef]
34. Appeldorn, R.H.; Hulme-Lowe, A.G.; Lea, M.C. Illumination Devices and Optical Fibres for Use Therein. US Patent US5432876A, 11 July 1995.

35. Mori, K. Apparatus for Uniform Illumination Employing Light Diffuser. US Patent US4460940A, 17 July 1984.
36. Nguyen, H.; Parvez Arnob, M.M.; Becker, A.T.; Wolfe, J.C.; Hogan, M.K.; Horner, P.J.; Shih, W.C. Fabrication of multipoint side-firing optical fiber by laser micro-ablation. *Opt. Lett.* **2017**, *42*, 1808. [CrossRef]
37. Volkov, V.V.; Loshchenov, V.B.; Konov, V.I.; Kononenko, V.V. Fibreoptic diffuse-light irradiators of biological tissues. *Quantum Electron.* **2010**, *40*, 746–750. [CrossRef]
38. Ai, M.; Tang, S.; Salcudean, T.; Rohling, R.; Abolmaesumi, P. Transurethral illumination probe design for deep photoacoustic imaging of prostate. In Proceedings of the Photons Plus Ultrasound: Imaging and Sensing, San Francisco, CA, USA, 28 January–1 February 2018; p. 11. [CrossRef]
39. Kosoglu, M.A.; Hood, R.L.; Rossmeisl, J.H.; Grant, D.C.; Xu, Y.; Robertson, J.L.; Rylander, M.N.; Rylander, C.G. Fiberoptic microneedles: Novel optical diffusers for interstitial delivery of therapeutic light. *Lasers Surg. Med.* **2011**, *43*, 914–920. [CrossRef] [PubMed]
40. Ströbl, S.; Wäger, F.; Domke, M.; Rühm, A.; Sroka, R. Homogeneously Emitting, Mechanically Stable, and Efficient fs-Laser-Machined Fiber Diffusers for Medical Applications. *Lasers Surg. Med.* **2022**, *54*, 588–599. [CrossRef] [PubMed]
41. Sun, X.; Hong, R.; Hou, H.; Fan, Z.; Shao, J. Optical properties and structures of silver thin films deposited by magnetron sputtering with different thicknesses. *Chin. Opt. Lett.* **2006**, *4*, 366–369.
42. Liu, P.; Shi, Z.; Teng, D.; Liu, F.; Cao, Y.; Lin, Y.; Yang, Z.; Yang, A.; Zheng, Y.; Chen, L. Optical Characteristics of Silver Thin Films from Island to Percolation in the Ultra-Wide Infrared Spectral Range. *Coatings* **2023**, *13*, 1910. [CrossRef]
43. Spirou, G.M.; Oraevsky, A.A.; Vitkin, I.A.; Whelan, W.M. Optical and acoustic properties at 1064 nm of polyvinyl chloride-plastisol for use as a tissue phantom in biomedical optoacoustics. *Phys. Med. Biol.* **2005**, *50*, N141–N153. [CrossRef]
44. Karpiouk, A.; Aglyamov, S.; Mallidi, S.; Scott, W.; Rubin, J.; Emelianov, S. Combined ultrasonic and photoacoustic imaging to age deep vein thrombosis: Preliminary studies. In Proceedings of the IEEE Ultrasonics Symposium, Rotterdam, The Netherlands, 18–21 September 2005; Volume 1, pp. 399–402. . [CrossRef]
45. Singh, N.; Roa, C.F.; Lilge, L.; Demore, C.E. Development of a Side-Fire Optical Fiber Diffuser for High-Energy Light Transport for Intraluminal Photoacoustic Imaging. In Proceedings of the 2024 Photonics North (PN), Vancouver, BC, Canada, 28–30 May 2024; pp. 1–2. [CrossRef]
46. Yao, J.; Wang, L.V. Photoacoustic tomography: Fundamentals, advances and prospects. *Contrast Media Mol. Imaging* **2011**, *6*, 332–345. [CrossRef]
47. Roa, C.F.; Chérin, E.; Singh, N.; Yin, J.; Boyes, A.; Foster, F.S.; Demore, C.E.M. Development of a Small-Footprint 50 MHz Linear Array: Fabrication and Micro-Ultrasound Imaging Demonstration. *Sensors* **2024**, *24*, 1847. [CrossRef]
48. Haglund, R.F. 2. Mechanisms of Laser-Induced Desorption and Ablation. In *Experimental Methods in the Physical Sciences*; Miller, J.C., Haglund, R.F., Eds.; Laser Ablation and Desorption; Academic Press: Cambridge, MA, USA, 1997; Volume 30, pp. 15–138. [CrossRef]
49. Marjoribanks, R.S.; Tang, J.; Dzelzainis, T.; Prickaerts, M.; Lilge, L.; Akens, M.; Veevers, C.; Gharabaghi, N.; King, A.; Hitzler, A.; et al. Plasma persistence, accumulated absorption, and scattering: What physics lets us control the heat left behind in ultrafast-pulse burst-mode laser surgery. In Proceedings of the Frontiers in Ultrafast Optics: Biomedical, Scientific, and Industrial Applications XXIV, San Francisco, CA, USA, 28–30 January 2024; Herman, P.R., Osellame, R., Ben-Yakar, A., Eds.; International Society for Optics and Photonics; SPIE: Bellingham, WA, USA, 2024; Volume 12875, p. 1287504. [CrossRef]
50. Marjoribanks, R.S.; Dille, C.; Schoenly, J.E.; McKinney, L.; Mordovanakis, A.; Kaifosh, P.; Forrester, P.; Qian, Z.; Covarrubias, A.; Feng, Y.; et al. Ablation and thermal effects in treatment of hard and soft materials and biotissues using ultrafast-laser pulse-train bursts. *Photonics Lasers Med.* **2012**, *1*, 155–169. [CrossRef]
51. Rethfeld, B.; Sokolowski-Tinten, K.; von der Linde, D.; Anisimov, S. Timescales in the response of materials to femtosecond laser excitation. *Appl. Phys. A* **2004**, *79*, 767–769. [CrossRef]

applied sciences

MDPI

Article

Novel Fiber Bragg Grating Sensing Structure for Simultaneous Measurement of Inclination and Water Level

Cheng-Kai Yao, Yao-Jen Chung, Yong-Chang Xu, Pradeep Kumar and Peng-Chun Peng *

Department of Electro-Optical Engineering, National Taipei University of Technology, Taipei 10608, Taiwan; t111659004@ntut.org.tw (C.-K.Y.)
* Correspondence: pcpeng@ntut.edu.tw; Tel.: +886-2-2771-2171 (ext. 4671)

Abstract: In the context of significant climate change, monitoring inclination, water levels, and temperatures in public buildings and surrounding environments is sensible. This paper presents a pair of fiber Bragg grating (FBG) subsidence sensor systems designed to simultaneously measure tilt and water levels and explore the system's potential to detect temperature variations. The configuration of the FBG subsidence sensor is intentionally skewed to enhance measurement sensitivity. The system is capable of concurrently detecting a 0.5 cm variation in water level and a 0.424° change in tilt, with tilt measurements spanning from −1.696° to 1.696°. Furthermore, the measurement system can be integrated with free-space optics (FSO), which is anticipated to address the challenges associated with installing fiber optic cables. Consequently, the proposed innovative FBG sensor system can measure multiple parameters using fewer sensors, thereby improving sensing capacity and cost-efficiency.

Keywords: fiber Bragg grating; inclination sensing; water level sensing; free-space optics

1. Introduction

Global extreme climatic events such as hurricanes, floods, and heat waves present significant challenges to both human life and the economy. The monitoring of water levels and the structural integrity of critical public facilities and infrastructure is essential for the prevention and mitigation of such disasters. Fiber optic sensors are particularly advantageous as they are power-free, compact in design, possess long operational lifespans, and are resistant to electromagnetic interference [1]. Additionally, they are highly sensitive and capable of measuring a wide range of physical parameters [2–5]. These characteristics render fiber optic sensors an optimal choice for monitoring efforts toward achieving net-zero carbon emissions.

Recent advancements in fiber optic sensors for liquid level or hydrological-related measurement have garnered significant attention [6–20], attributed to their numerous advantages. However, existing research predominantly addresses the measurement of a singular liquid level parameter. In practical applications, variables such as the liquid's refractive index, temperature fluctuations, or the inclination of the monitoring structure may compromise this focus. Fiber optic interferometer sensors exhibit excellent measurement resolution, wide-band measurability [21–24], and the capacity to assess multiple parameters [25–27] simultaneously. However, due to the inherent complexity of the spectrum, employing multiple fiber optic interferometer sensors for concurrent measurements can often present significant challenges. Distributed fiber optic sensors possess the capability to measure multiple sites and various parameters simultaneously at a single location [28–31]. However, it is important to note that these sensors are typically more expensive and may not provide a cost-effective solution for measurements over small areas.

Appl. Sci. **2025**, *15*, 4819 https://doi.org/10.3390/app15094819

Standard FBG sensors can be configured in arrays to provide measurements akin to those obtained through distributed fiber optic sensing and can be effectively integrated with FSO to address cable routing challenges [32,33]. When FSO technology is integrated with sensors in applications where optical communication over distances of several hundred meters is feasible [34,35]. Commercially available FSO systems can achieve ranges of several kilometers or more, demonstrating their capability to function effectively even under adverse weather conditions [36]. This resilience highlights the utility of FSO technology in a diverse range of applications requiring reliable data transmission across considerable distances.

The liquid level measurement technique founded on the Archimedean buoyancy principle enables FBG sensors to accurately gauge liquid levels through a stretching mechanism [6]. This approach circumvents the influence of refractive index variations that may occur when the sensor comes into contact with the liquid [6]. Furthermore, intensity-based measurement sensors are often susceptible to variations in light source power; in contrast, FBG-based sensors exhibit resilience to such fluctuations [6]. By utilizing the center wavelength of the FBG sensor as a reference point, a single FBG is capable of simultaneously measuring both vibration and temperature [37]. However, simultaneous measurement of strain and temperature is not feasible unless spectral interrogation techniques are implemented in lieu of relying solely on the center wavelength value of the FBG [38].

In Archimedean buoyancy principle-based FBG subsidence sensor applications, two physical quantities, specifically changes in water level and tilt, are measured through the degree of stretching of the FBG indicated by the center wavelength. Consequently, the main limitation of FBG sensor measurements is that additional sensors must be arranged to measure more than two physical quantities simultaneously, owing to the same mechanism for FBG spectrum variation of these quantities. Nonetheless, it is important to note that FBG sensors retain significant potential for multi-site measurement applications within compact measurement environments. Moreover, the investigation of FBG subsidence sensors utilizing conventional FBGs for the simultaneous measurement of multiple parameters has not been conducted.

The primary distinction between the current tilt measurement study and the similar previous study [39] lies in the following point: the prior research utilized only two FBGs to ascertain both the tilt angle and the tilt direction. This study explores a modular system comprising two FBG subsidence sensors designed for the simultaneous measurement of water level and inclination while also investigating the feasibility of concurrently measuring various water temperatures. It is important to clarify that the upper limit of the measurement range and the measurement sensitivity of water level and tilt are not the primary focus of this study. Concentrating on these aspects would shift attention towards the design of sensor packaging rather than the measurement of multiple parameters. Furthermore, the FBG subsidence sensors were skewed in deployment to enhance the sensitivity of tilt measurements. Additionally, the measurement system can be integrated with the FSO, thereby potentially increasing the system's flexibility for practical applications. Consequently, this study contributes to the cost optimization and efficacy of related monitoring and measurement processes.

2. Experimental Setup

The upper section of Figure 1 illustrates the potential applications of utilizing FBG subsidence sensors in conjunction with FSOs to measure water levels and tilt. Such applications are pertinent in various contexts, including dams, offshore regions, and sea islands, among others. The optical coupler is vital in enhancing the multi-sensory pathways, allowing for the strategic distribution of multiple FBG sensing networks across various locations within measurement applications. When the density of FBG sensors increases, challenges may arise due to wavelength overlap or interference, which can adversely impact the accuracy of wavelength interrogation results. To address this issue, implementing machine learning techniques is recommended to effectively resolve the wavelength interrogation challenges [40]. The primary framework of the experiment is illustrated in the lower section of Figure 1. For real applications, the sensing signal transceiver must be strategically positioned at the sensing center. Through FSO, the optical signal can be captured by the FBG sensor located at the measurement point. To facilitate experimental procedures and enhance operational efficiency, the broadband light source (BLS) emits the optical signal via an optical circulator (Cir.) to the 2 m FSO link in the lab instead of the outdoor corresponding scene. This is subsequently connected to the FBG array sensors for measurement purposes. The sensed signal is then routed back to the optical circulator and directed to the optical spectrum analyzer (OSA) for comprehensive signal observation. The FBG sensing module comprises 3 FBGs; two FBG subsidence sensors are used to measure the water level and tilt simultaneously, and the remaining FBG is used to correct air temperature change. The FBG subsidence sensor consists of a conventional FBG that supports a buoy [41]. Both the FBG and the buoy are contained within a hollow cylindrical bottle, which is designed with an opening to facilitate water flow in and out. Consequently, variations in water level, or changes induced by the tilting of the cylindrical bottle, will affect the buoy's positioning. This, in turn, will stretch the FBG, leading to alterations in the center wavelength of the FBG. For the measurement conditions, FBG subsidence sensors are placed in the temperature bath (TB) to control the water level and temperature for customization of measurement conditions. The tilted measurement setup is made by placing a gasket under the leftmost or rightmost side of the rectangular TB. The tilting of the TB tilts the FBG subsidence sensors, but the water surface remains flat. The length of the TB is 32 cm, and the thickness of the gasket is 2.4 mm, so the calculation shows that each additional gasket used will increase the tilt angle by 0.424°. To enhance measurement sensitivity, a gasket with a thickness of 1.6 mm has been positioned beneath the FBG subsidence sensors. Furthermore, the two FBG subsidence sensors have been strategically placed at opposing ends of the TB apparatus. This arrangement ensures that, in instances of TB tilt, there is a pronounced water level difference between the two sensors, thereby optimizing the accuracy of the measurements. Incidentally, the equipment's brand and model are respectively affiliated with UNICE Inc. (Taoyuan, Taiwan) NA0101 for BLS; Thorlabs Inc. (Newton, NJ, USA) F810FC-1550 for FSO; Citpo Technologies Inc. (Taipei, Taiwan) for the FBG subsidence sensor; and Anritsu Inc. (Atsugi, Japan) MS9740A for OSA.

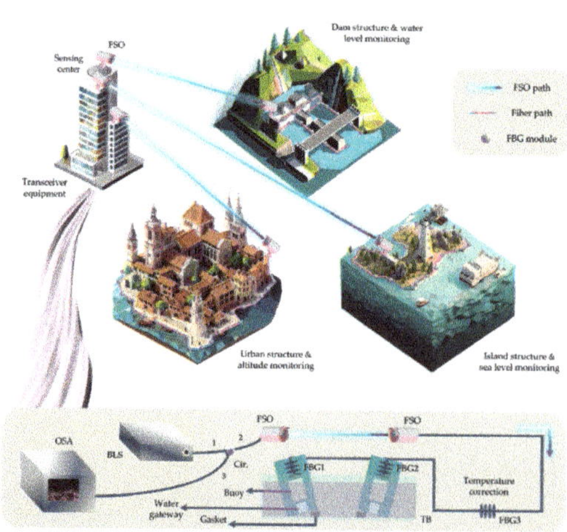

Figure 1. The application vision and experimental setup of using FBG module to integrate with FSO to simultaneously measure inclination angle and water level.

3. Experimental Results

As previously stated, variations in water levels or angular dip result in shifts in the position of the buoys on the FBG subsidence sensors. These shifts subsequently induce strain on the FBGs, leading to alterations in their center wavelengths. Consequently, this mechanism allows for the understanding that, at a consistent water level, variations in the inclination angle of FBG subsidence sensors can alter the buoy's submerged surface area. This, in turn, affects the buoyancy, leading to the FBG's elongation. To enhance measurement resolution, it is advisable to increase the immersion area variation of the buoy whenever the inclination angle varies. As previously indicated, incorporating a gasket at the base of the FBG subsidence sensor can augment its sensitivity. This enhancement is attributed to the buoy's rectangular design; when the FBG subsidence sensor is positioned at an angle, it results in a greater alteration of the buoy's immersed area with each change in tilt, in contrast to a vertically oriented FBG subsidence sensor.

The experimental results are shown in Figure 2, in which the initial wavelength of FBG1 is 1548 nm; the initial wavelength of FBG2 is 1542 nm, so the overall tilting of the TB to the right, such as 0.424°, 0.848°, 1.272°, and 1.696°, will lead to a decrease in the water level at the FBG1, and thus the buoyancy force decreases, so that the FBG wavelength is subjected to long wavelength drift. Conversely, a leftward tilt of the TB will cause the water level at FBG1 to rise and increase the buoyancy, resulting in a reduction of the strain on the FBG and a drift to a shorter wavelength. On the other hand, FBG2 is located on the right side of TB, so when TB is tilted left or right, the water level at FBG2 changes in the opposite direction to that at FBG1, which makes the wavelength drift direction of FBG2 opposite to that of FBG1. Figure 2a–f sequentially presents the tilt measurements at water levels of 1 cm, 1.5 cm, 2 cm, 2.5 cm, 3 cm, and 3.5 cm. It is observed that the drift trends of the FBG wavelengths exhibit consistency across different water levels during the tilting process. A crucial aspect of Figure 2 is that the solid line indicates the FBG wavelength drift observed when the FBG subsidence sensors are installed at an oblique angle. In contrast, the drift represented by the dotted line corresponds to the FBG wavelength changes when the sensors are positioned normally. The measurement results affirm the initial hypothesis, demonstrating that the inclined placement of FBG subsidence sensors enhances measurement sensitivity. Furthermore, the drift in FBG wavelength appears to be a somewhat nonlinear pattern, especially at low water levels. This is because in low water, the buoys are in contact with less water, and the FBG subsidence sensors are placed on both sides of the TB rather than in the center; therefore, when the buoys are tilted left and right at the same water level, the buoys are buoyed differently, thus causing the drift of the FBG wavelength to be not so linear.

The experimental results presented in Figure 2 indicate that the skewed FBG subsidence sensors exhibit greater wavelength variation and enhanced sensitivity. Consequently, subsequent experiments will measure skewed readings at various water levels and temperatures, adhering to the experimental configuration depicted in Figure 1. Figure 3 presents the drift results of the FBG wavelength at 1542 nm, evaluated under varying water temperatures and levels. Given that fluctuations in water temperature also influence the buoyancy of the buoys, it is essential to examine the measurements across different thermal conditions. The water temperatures in Figure 3 correspond to 30 °C, 40 °C, 50 °C, and 60 °C in Figure 3a–d and the water level and inclination settings are consistent with those in Figure 2. The data presented in Figure 3 indicate that, irrespective of water temperature, an increase in water level results in a corresponding reduction in wavelength. This phenomenon can be attributed to the amplification of buoyancy, which consequently decreases the strain experienced by the FBG. In addition, 1542 nm of the FBG2 wavelength is placed on the right side of TB, so TB tilted to the right will cause the water level to rise, i.e., buoyancy will increase so that the FBG wavelength will drift towards the shorter wavelengths, while TB tilted to the left will have the opposite effect. Corresponding to the results in Figure 2, the wavelength drift is linear with respect to the angle of inclination. Furthermore, an increase in water temperature is observed to cause a significant drift of the FBG wavelength towards longer wavelengths. As the temperature of the water rises, the buoyancy typically declines due to a decrease in water density. Consequently, if the buoy volume of discharged water remains constant, the overall buoyancy of the water will diminish.

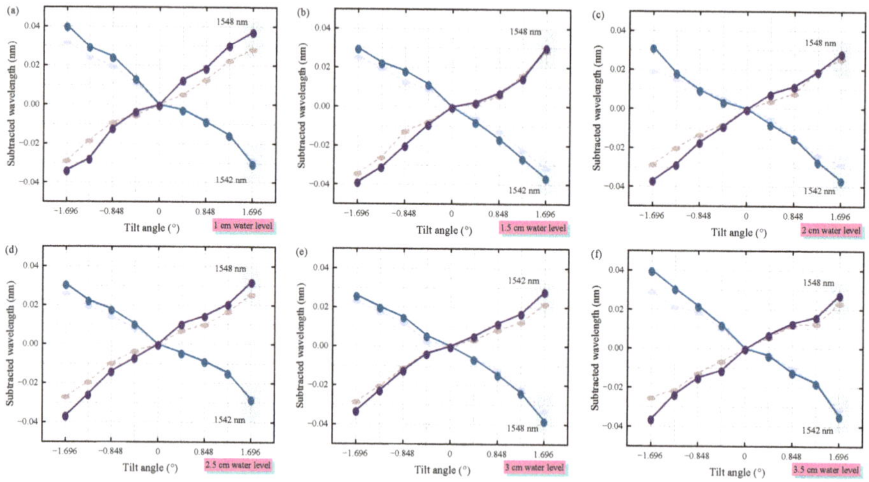

Figure 2. At varying water levels, FBG subsidence sensors have been deployed at both skewed positions (represented by solid line data) and standard positions (denoted by dashed line data) to monitor the wavelength drift resulting from tilting. The measurements correspond to the 30 °C water temperature with the following water levels: (**a**) 1 cm, (**b**) 1.5 cm, (**c**) 2 cm, (**d**) 2.5 cm, (**e**) 3 cm, and (**f**) 3.5 cm.

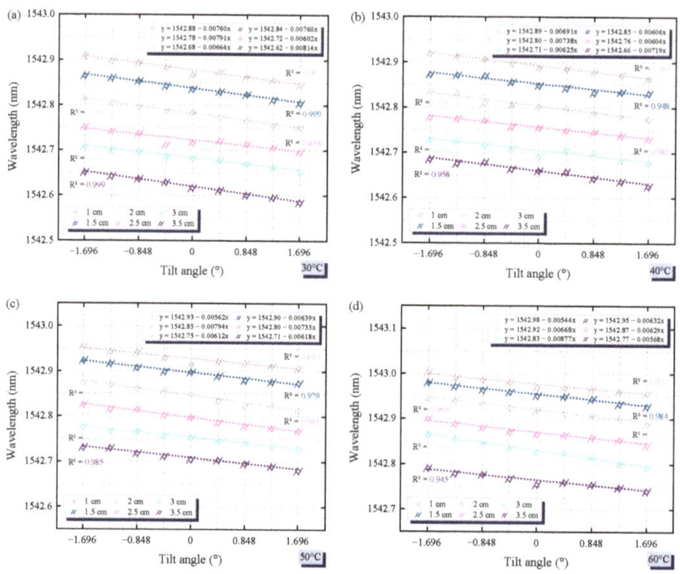

Figure 3. The FBG wavelength shift results of the tilt measurements correspond to 1542 nm at different water levels. The temperature conditions for the water during the measurements were as follows: (**a**) 30 °C, (**b**) 40 °C, (**c**) 50 °C, and (**d**) 60 °C.

As previously noted, the necessity of concurrently measuring water levels and incline variations necessitates the deployment of two FBGs. This requirement arises because the right-leaning condition observed at low water levels may yield wavelength readings for the FBGs that are indistinguishable from those generated under the left-leaning condition at elevated water levels, as illustrated in Figure 3. Consequently, employing two FBGs for measurements can effectively eliminate potential confusion regarding the parameter signals and address this limitation. Figure 4a–d illustrate the wavelength measurements of FBG1, which is based on a reference wavelength of 1548 nm. The measurement setups utilized are identical to those presented in Figure 3a–d. A notable distinction is that FBG1 is positioned on the left side of the TB. Consequently, when the TB is inclined to the left or right, the resulting changes in the water level exhibit an inverse relationship with those observed in FBG2. Therefore, the most significant finding depicted in Figure 4 is that the drift tendency of the FBG wavelength is precisely opposite to that illustrated in Figure 3. By recording the wavelength readings from both FBG1 and FBG2, it is possible to concurrently ascertain the water level and tilt parameters. The findings presented in Figures 3 and 4 indicate that an increase in temperature results in a drift of the FBG towards longer wavelengths. However, while measuring both water level and tilt using two separate FBG sensors is feasible, conducting concurrent measurements of water temperature with the same two FBGs poses significant challenges. This difficulty arises from the fact that the wavelength drift of the FBG, attributed to buoyancy changes stemming from variations in water temperature, is minimal. Furthermore, a more complex consideration is that the salinity of the water influences its density, subsequently affecting buoyancy. It is advisable to incorporate additional sensors to accurately measure water temperature, ensuring that the concurrent measurements of water level and inclination are not compromised by variations in temperature and salinity. Notably, it is feasible to measure either water level and temperature or inclination and temperature simultaneously, provided that salinity remains constant. The feasibility of concurrent measurement of water level, tilt, and temperature is contingent upon a significant temperature differential in the water utilized for the measurement applications, specifically exceeding 30 °C. It is important to note that the data corresponding to the shortest wavelength of subfigure a and the longest wavelength of subfigure d in Figures 3 and 4 do not overlap with other datasets. Thus, under these exceptional conditions, it is possible to accurately obtain the aforementioned measurements simultaneously. It is also worth mentioning that the wavelength of FBG3, utilized for calibration to adjust for wavelength measurement deviations induced by changes in air temperature, is consistently maintained at 1544 nm. Consequently, there is no requirement to compensate for wavelength changes due to fluctuations in air temperature in the context of this experiment.

As previously discussed, the difference in buoyancy of the buoy at the same water level when tilted to the left or right results in a non-linear shift of the FBG wavelength. This phenomenon arises from the tilting of the FBG subsidence sensors; however, it is also primarily attributable to the geometric configuration of both the water container and the buoy utilized in the FBG subsidence sensor. Achieving a consistent FBG wavelength shift for identical tilt angle changes proves to be challenging and impractical, as it necessitates consideration of varying water levels and temperatures. Consequently, it is difficult to design a water container and buoy with an appropriate shape and mass that would ensure a uniform buoyancy change for every equal tilt angle variation, particularly when accounting for the diverse conditions presented by different water levels and temperatures. Consequently, a slight nonlinear wavelength drift is inevitable and may indirectly influence measurement accuracy. However, this challenge is not insurmountable. In comparison to a previous similar investigation [39], the sensitivity of the tilt measurement within this framework still presents opportunities for optimization. These enhancements primarily hinge on the interplay between the design of the buoy and the configuration of the water container. For the existing scheme, if the size of the water vessel, the volume of the buoy, and the length of the TB are increased, the measurement sensitivity can be improved theoretically. Therefore, as measurement sensitivity improves, the resultant increase in the reflected FBG wavelength drift will diminish the likelihood of misinterpreting the measured tilt angle or water level, thereby fostering greater measurement accuracy.

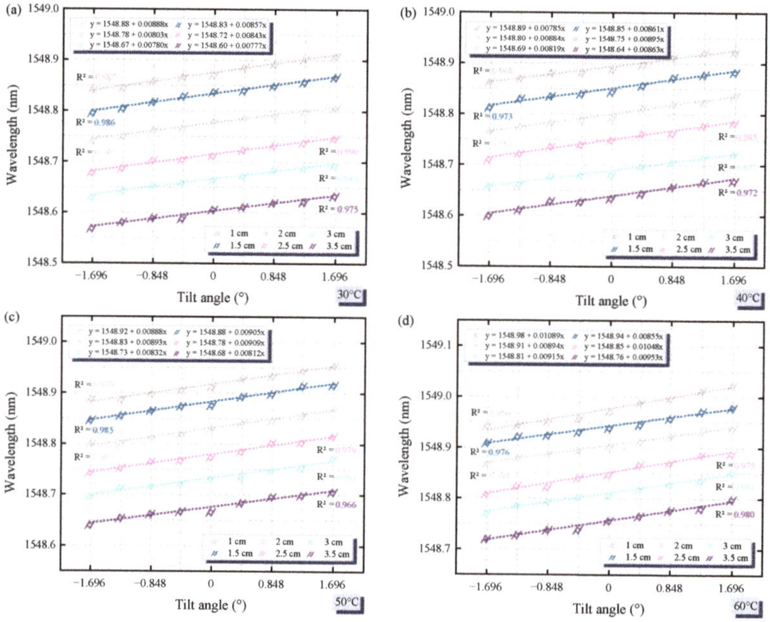

Figure 4. The FBG wavelength shift results of the tilt measurements correspond to 1548 nm at different water levels. The temperature conditions for the water during the measurements were as follows: (**a**) 30 °C, (**b**) 40 °C, (**c**) 50 °C, and (**d**) 60 °C.

4. Conclusions

This paper presents a novel pair of FBG subsidence sensors, which are designed to measure left and right tilt and changes in water levels simultaneously. The sensors are uniquely skewed and configured in opposing directions, facilitating the collection of accurate data on slope variations and water level fluctuations. The sensing system is capable of resolving at least 0.424° of inclination and 0.5 cm of water level change, while also being able to detect variations in water temperature when either water level or inclination measurements are conducted independently. Furthermore, the proposed measurement system is compatible with FSO, streamlining the system by reducing the maintenance and installation costs typically associated with fiber optic cabling. This innovative FBG sensor system enables the measurement of multiple parameters with a reduced number of sensors, significantly enhancing sensing capabilities and maximizing cost efficiency.

Author Contributions: Conceptualization, C.-K.Y.; methodology, C.-K.Y., Y.-J.C. and Y.-C.X.; data curation, Y.-C.X. and C.-K.Y.; model validation, Y.-J.C. and C.-K.Y.; formal analysis, C.-K.Y., Y.-J.C. and P.K.; investigation, C.-K.Y., Y.-J.C., Y.-C.X., P.K. and P.-C.P.; visualization: C.-K.Y.; writing—original draft preparation: C.-K.Y.; writing—review and editing: C.-K.Y.; supervision, P.-C.P. All authors have read and agreed to the published version of the manuscript.

Funding: This work was supported by the National Science and Technology Council, Taiwan, under Grant NSTC 112-2221-E-027-076-MY2.

Institutional Review Board Statement: Not applicable.

Informed Consent Statement: Not applicable.

Data Availability Statement: The data presented in this study are available in this article.

Conflicts of Interest: The authors declare no conflicts of interest.

References

1. Song, J.; Leng, J.; Li, J.; Wei, H.; Li, S.; Wang, F. Application of Acoustic Emission Technique in Landslide Monitoring and Early Warning: A Review. *Appl. Sci.* **2025**, *15*, 1663. [CrossRef]
2. Leal-Junior, A.; Silva, J.; Macedo, L.; Marchesi, A.; Morau, S.; Valentino, J.; Valentim, F.; Costa, M. The Role of Optical Fiber Sensors in the New Generation of Healthcare Devices: A Review. *Sens. Diagn.* **2024**, *3*, 1135–1158. [CrossRef]
3. Cięszczyk, S.; Panas, P.; Skorupski, K.; Kida, M. Fibre Bragg Grating Wavelength Shift Demodulation with Filtering and Enhancement of Spectra by Simple Nonlinear Signal Processing. *Appl. Sci.* **2025**, *15*, 3384. [CrossRef]
4. Leal-Junior, A.; Silveira, M.; Macedo, L.; Frizera, A.; Marques, C. Polarization-Assisted Multiparameter Sensing Using a Single Fiber Bragg Grating. *Opt. Fiber Technol.* **2024**, *84*, 103775. [CrossRef]
5. Mandal, H.N.; Sidhishwari, S. Apodized Chirped Fiber Bragg Grating for Measuring the Uniform and Non-Uniform CharacTeristics of Physical Parameters. *Measurement* **2025**, *240*, 115606. [CrossRef]
6. Ramos, C.C.; Preizal, J.; Hu, X.; Caucheteur, C.; Woyessa, G.; Bang, O.; Rocha, A.M.; Oliveira, R. High Resolution Liquid Level Sensor Based on Archimedes' Law of Buoyancy Using Polymer Optical Fiber Bragg Gratings. *Measurement* **2025**, *252*, 117368. [CrossRef]
7. de Almeida, G.G.; Barreto, R.C.; Seidel, K.F.; Kamikawachi, R.C. A Fiber Bragg Grating Water Level Sensor Based on the Force of Buoyancy. *IEEE Sens. J.* **2019**, *20*, 3608–3613. [CrossRef]
8. Rosolem, J.B.; Dini, D.C.; Penze, R.S.; Floridia, C.; Leonardi, A.A.; Loichate, M.D.; Durelli, A.S. Fiber Optic Bending Sensor for Water Level Monitoring: Development and Field Test: A Review. *IEEE Sens. J.* **2013**, *13*, 4113–4120. [CrossRef]
9. He, R.; Teng, C.; Kumar, S.; Marques, C.; Min, R. Polymer Optical Fiber Liquid Level Sensor: A Review. *IEEE Sens. J.* **2021**, *22*, 1081–1091. [CrossRef]
10. Lee, H.K.; Choo, J.; Kim, J. Multiplexed Passive Optical Fiber Sensor Networks for Water Level Monitoring: A Review. *Sensors* **2020**, *20*, 6813. [CrossRef]
11. Lee, H.K.; Kim, Y.; Park, S.; Kim, J. Passive IoT Optical Fiber Sensor Network for Water Level Monitoring with Signal Processing of Feature Extraction. *Electronics* **2023**, *12*, 1823. [CrossRef]
12. Dejband, E.; Tan, T.H.; Yao, C.K.; Chang, E.M.; Peng, P.C. Enhancing Multichannel Fiber Optic Sensing Systems with IFFT-DNN for Remote Water Level Monitoring. *Sensors* **2024**, *24*, 4903. [CrossRef]
13. Lee, H.K.; Choo, J.; Shin, G.; Kim, J. Long-Reach DWDM-Passive Optical Fiber Sensor Network for Water Level Monitoring of Spent Fuel Pool in Nuclear Power Plant. *Sensors* **2020**, *20*, 4218. [CrossRef]
14. Pereira, K.; Coimbra, W.; Lazaro, R.; Frizera-Neto, A.; Marques, C.; Leal-Junior, A.G. FBG-Based Temperature Sensors for Liquid Identification and Liquid Level Estimation Via Random Forest. *Sensors* **2021**, *21*, 4568. [CrossRef] [PubMed]
15. Schenato, L. A Review of Distributed Fibre Optic Sensors for Geo-Hydrological Applications. *Appl. Sci.* **2017**, *7*, 896. [CrossRef]
16. Shangguan, M.; Yang, Z.; Shangguan, M.; Lin, Z.; Liao, Z.; Guo, Y.; Liu, C. Remote Sensing Oil in Water with an All-Fiber Underwater Single-Photon Raman Lidar. *Appl. Opt.* **2023**, *62*, 5301–5305. [CrossRef]
17. Li, P.; Yan, H.; Zhang, H. Highly Sensitive Liquid Level Sensor Based on an Optical Fiber Michelson Interferometer with Core-Offset Structure. *Optik* **2018**, *171*, 781–785. [CrossRef]
18. Wang, S.; Yang, Y.; Zhang, L.; Mohanty, L.; Jin, R.B.; Wu, S.; Lu, P. High-Precision Fiber Optic Liquid Level Sensor Based on Fast Fourier Amplitude Demodulation in a Specific Range of Spectrum. *Measurement* **2022**, *187*, 110326. [CrossRef]
19. Martins, J.; Diaz, C.A.; Domingues, M.F.; Ferreira, R.A.; Antunes, P.; Andre, P.S. Low-Cost and High-Performance Optical Fiber-Based Sensor for Liquid Level Monitoring. *IEEE Sens. J.* **2019**, *19*, 4882–4888. [CrossRef]
20. Diaz, C.A.; Leal-Junior, A.; Marques, C.; Frizera, A.; Pontes, M.J.; Antunes, P.F.; Andre, P.S.; Ribeiro, M.R. Optical Fiber Sensing for Sub-Millimeter Liquid-Level Monitoring: A Review. *IEEE Sens. J.* **2019**, *19*, 7179–7191. [CrossRef]
21. Lee, C.L.; Yeh, C.Y.; Jiang, Y.X. Effective Liquid-filled Leaky-Guided Fiber Mach-Zehnder Interferometer with a Side-Polished Fiber. *IEEE Sens. J.* **2025**, *25*, 9681–9688. [CrossRef]
22. Zhu, K.; Ren, S.; Li, X.; Liu, Y.; Li, J.; Zhang, L.; Wang, M. Sub-Micron Two-Dimensional Displacement Sensor Based on a Multi-Core Fiber. *Photonics* **2024**, *11*, 1073. [CrossRef]
23. Noori, N.F.; Mansour, T.S. A Review of Recently Optical Hybrid Optical Fiber Interferometers. *J. Opt* **2025**, 1–14. [CrossRef]

24. Wei, Y.; Yang, M.; Shao, Y.; Liu, C.; Ren, P.; Zhang, Z.; Liu, Z. Asymmetric Coated Direction Recognition Fiber Optic Interferometric Curvature Sensor. *J. Light. Technol.* **2025**, 1–6. [CrossRef]
25. Lu, C.; Su, J.; Dong, X.; Sun, T.; Grattan, K.T. Simultaneous Measurement of Strain and Temperature with a Few-Mode Fiber-Based Sensor. *J. Light. Technol.* **2018**, *36*, 2796–2802. [CrossRef]
26. Lei, X.; Dong, X.; Lu, C.; Sun, T.; Grattan, K.T. Underwater Pressure and Temperature Sensor Based on a Special Dual-Mode Optical Fiber. *IEEE Access* **2020**, *8*, 146463–146471. [CrossRef]
27. Duan, S.; Wang, W.; Xiong, L.; Wang, B.; Liu, B.; Lin, W.; Zhang, H.; Liu, H.; Zhang, X. All In-Fiber Fabry–Pérot Interferometer Sensor Towards Refractive Index and Temperature Simultaneous Sensing. *Opt. Laser Technol.* **2025**, *180*, 111551. [CrossRef]
28. Dang, Y.; Zhao, Z.; Wang, X.; Liao, R.; Lu, C. Simultaneous Distributed Vibration and Temperature Sensing Using Multicore Fiber. *IEEE access* **2019**, *7*, 151818–151826. [CrossRef]
29. Lu, L.; Yong, M.; Wang, Q.; Bu, X.; Gao, Q. A Hybrid Distributed Optical Fiber Vibration and Temperature Sensor Based on Optical Rayleigh and Raman Scattering. *Opt. Commun.* **2023**, *529*, 129096. [CrossRef]
30. Zhou, Z.; Tian, L.; Han, Y.; Yang, X.; Liao, F.; Zhang, D.; Sha, Y.; Feng, X.; Zhu, J.; Zheng, X.; et al. Distributed Vibration and Temperature Simultaneous Sensing Using One Optical Fiber. *Opt. Commun.* **2021**, *487*, 126801. [CrossRef]
31. Yao, C.K.; Lin, T.C.; Chen, H.M.; Hsu, W.Y.; Manie, Y.C.; Peng, P.C. Inclination Measurement Adopting Raman Distributed Temperature Sensor. *IEEE Sens. J.* **2023**, *23*, 22543–22555. [CrossRef]
32. Kanwal, F.; Atieh, A.; Ghafoor, S.; Haq, A.U.; Qureshi, K.K.; Aziz, I.; Mirza, J. Remote Monitoring of Sleep Disorder Using FBG Sensors and FSO Transmission System Enabled Smart Vest. *Eng. Res. Express* **2024**, *6*, 025337. [CrossRef]
33. Sadik, S.A. Leveraging Fiber Bragg Grating Sensors for Enhanced Security in Smart Grids. In *5G and Fiber Optics Security Technologies for Smart Grid Cyber Defense*; IGI Global: Hershey, PA, USA, 2024; pp. 259–288.
34. Zhang, Q.; Yu, J.; Long, J.; Wang, C.; Chen, J.; Lu, X. A Hybrid RF/FSO Transmission System Based on a Shared Transmitter. *Sensors* **2025**, *25*, 2021. [CrossRef]
35. Syriopoulos, G.; Kyriazi, E.; Prousalidi, T.; Ntanos, A.; Stathis, A.; Kourelias, P.; Toumasis, P.; Zervos, H.; Giannoulis, G.; Poulopoulos, G.; et al. Co-Existence of SiPh Sensing Link and Real World Traffic over 100 m FSO link for 6G Deployments. *IEEE Photonics Technol. Lett.* **2025**, *37*, 9. [CrossRef]
36. Yao, C.K.; Lin, H.P.; Cheng, C.L.; Chung, M.A.; Lin, Y.S.; Wu, W.B.; Chiang, C.W.; Peng, P.C. Fiber/Free-Space Optics with Open Radio Access Networks Supplements the Coverage of Millimeter-Wave Beamforming for Future 5G and 6G Communication. *Fibers* **2025**, *13*, 39. [CrossRef]
37. Yao, C.K.; Kumar, P.; Liu, B.X.; Dehnaw, A.M.; Peng, P.C. Simultaneous Vibration and Temperature Real-Time Monitoring Using Single Fiber Bragg Grating and Free Space Optics. *Appl. Sci.* **2024**, *14*, 11099. [CrossRef]
38. Choi, B.K.; Kim, J.S.; Ahn, S.; Cho, S.Y.; Kim, M.S.; Yoo, J.H.; Jeon, M.Y. Simultaneous Temperature and Strain Measurement in Fiber Bragg Grating Via Wavelength-Swept Laser and Machine Learning. *IEEE Sens. J.* **2024**, *24*, 27516–27524. [CrossRef]
39. Chao, C.R.; Liang, W.L.; Liang, T.C. Design and Testing of a 2D Optical Fiber Sensor for Building Tilt Monitoring Based on Fiber Bragg Gratings. *Appl. Syst. Innov.* **2017**, *1*, 2. [CrossRef]
40. Bogale, S.D.; Yao, C.K.; Manie, Y.C.; Zhong, Z.G.; Peng, P.C. Wavelength-Dependent Bragg Grating Sensors Cascade an Interferometer Sensor to Enhance Sensing Capacity and Diversification through the Deep Belief Network. *Appl. Sci.* **2024**, *14*, 7333. [CrossRef]
41. Dejband, E.; Manie, Y.C.; Deng, Y.J.; Bitew, M.A.; Tan, T.H.; Peng, P.C. High Accuracy and Cost-Effective Fiber Optic Liquid Level Sensing System Based on Deep Neural Network. *Sensors* **2023**, *23*, 2360. [CrossRef]

MDPI AG
Grosspeteranlage 5
4052 Basel
Switzerland
Tel.: +41 61 683 77 34

Applied Sciences Editorial Office
E-mail: applsci@mdpi.com
www.mdpi.com/journal/applsci

www.ingramcontent.com/pod-product-compliance
Lightning Source LLC
Chambersburg PA
CBHW061617210926
51850CB00009B/2353
9783725859917